Heat and Mass Transfer

Series Editors: D. Mewes and F. Mayinger

Franz Mayinger · Oliver Feldmann (Eds)

Optical Measurements

Techniques and Applications

Second corrected and revised Edition

With 238 Figures

Springer

Series Editors
Prof. -Dr.-Ing. Dieter Mewes
Universität Hannover
Institut für Verfahrenstechnik
Callinstr. 36
30167 Hannover, Germany

Prof. em. Dr.-Ing. E.h. Franz Mayinger
Technische Universität München
Lehrstuhl für Thermodynamik
Boltzmannstr. 15
85748 Garching, Germany

Editors
Prof. em. Dr.-Ing. E.h. Franz Mayinger
Dr.-Ing. Oliver Feldmann
Technische Universität München
Lehrstuhl für Thermodynamik
Boltzmannstr. 15
85748 Garching, Germany

Library of Congress Control Number: 2001020640

Additional material to this book can be downloaded from http://extras.springer.com.

ISBN 978-3-642-63079-8 ISBN 978-3-642-56443-7 (eBook)
DOI 10.1007/978-3-642-56443-7

Originally published by Springer-Verlag Berlin Heidelberg New York in 2001
Softcover reprint of the hardcover 2nd edition 2001

Typesetting: Digital data supplied by editors
Final processing: PTP-Berlin Protago-TEX-Production GmbH, Berlin (www.ptp-berlin.com)
Cover-Design: deblik Berlin
Printed on acid-free paper 62/3141/Yu – 5 4 3 2 1 0

Preface

Increasing possibilities of computer–aided data processing have caused a new revival of optical techniques in many areas of mechanical and chemical engineering. Optical methods have a long tradition in heat and mass transfer and in fluid dynamics. Global experimental information is not sufficient for developing constitution equations to describe complicated phenomena in fluid dynamics or in transfer processes by a computer program. Furthermore, a detailed insight with high local and temporal resolution into the thermo– and fluiddynamic situations is necessary.

Sets of equations for computer program in thermo dynamics and fluid dynamics usually consist of two types of formulations: a first one derived from the conservation laws for mass, energy and momentum, and a second one mathematically modelling transport processes like laminar or turbulent diffusion. For reliably predicting the heat transfer, for example, the velocity and temperature field in the boundary layer must be known, or a physically realistic and widely valid correlation describing the turbulence must be available. For a better understanding of combustion processes it is necessary to know the local concentration and temperature just ahead of the flame and in the ignition zone.

Here optical measuring techniques provide comprehensive and detailed information. Its results also supply valuable evidence on the formation of phase interfaces, on particle movement, or on the size distribution of droplet swarms. By using the results of optical measuring techniques, not only is it possible to improve computer programs to give a better description of physical processes and a better adaption to the physical reality but also these optical techniques are very sensitive touchstones for checking the grade of reliability and the extent of general validity of computer programs.

On the other side, evaluating optical data, for example from a hologram, from an interferogram, from Raman–spectroscopy, or from laser–induced–fluo-rescence signals, has become much faster. A few years ago it took hours to evaluate an interferogram. The same work is done today by a computer within seconds. But also the huge storage capacity of modern computers – even of the PC type – was an important requirement for preparing the way for the revival of optical methods.

The book is intended to demonstrate the possibilities of optical measuring techniques – especially image–forming techniques – and to introduce the processes of recording, reprocessing and electronically evaluating the data. It is intended to inform the reader to such an extent that he can design and construct simple experimental set–ups. For more difficult and highly sophisticated techniques he is referred to the specialist literature in the field.

Munich,
April 1996

Franz Mayinger

Preface to the 2nd Edition

The first edition of the book "Optical Measurements – techniques and applications" found so much interest, that it was out of sale within a few years. The present second edition comprises the completely revised first edition and is supplemented by new contributions on the latest developements in "Phase–Doppler–Velocimetry" and "Paticle–Image–Velocimetry".

It was a special matter of concern of both the editors and the authors not only to describe and to explain the fundamentals of optical measuring techniques, but also to give guidelines for their application and to demonstrate the capability of the various methods. In combination with that effort, a CD–ROM is added as a supplement to this book, which demonstrates the applicability of visualisation methods, such as Fluorescence, Schlieren–photography, Holography and Holografic Interferometry combined with High Speed Cinematography. The high–speed–movies on the CD–ROM give vigorous impressions of fluiddynamic transport phenomena in technical processes.

The book provides comprehensive and detailed information on most of the modern optical measuring techniques being of interest for engineers and physicists in practice. It is also of interest to Master– and Ph.D.– students who want to become familiar with optical measuring techniques.

The editors would like to thank each author for his outstanding contribution to this book. We also wish to acknowledge "Springer Verlag" for its contribution towards the publication of the book.

Munich,
August 2000

Franz Mayinger
Oliver Feldmann

List of Contributors

Prof. P. Andresen
Fakultät für Physik D3
Universität Bielefeld
D - 33501 Bielefeld
andresen@physik.uni-bielefeld.de

Dr. M. Buchmann
Institut für Verfahrenstechnik
Universität Hannover
D - 30167 Hannover
dms@c36.uni-hannover.de

Dr. A. Chavez[1]
Lehrstuhl A für Thermodynamik
Technische Universität München
D - 85747 Garching
feldmann@thermo-a.mw.tum.de

B. Durst[1]
Lehrstuhl A für Thermodynamik
Technische Universität München
D - 85747 Garching
bodo.durst@web.de

Dr. V. Ebert
Physikalisch–Chemisches Institut
Universität Heidelberg
D - 69120 Heidelberg
volker.ebert@urz.uni-heidelberg.de

A. Eder
Lehrstuhl A für Thermodynamik
Technische Universität München
D - 85747 Garching
eder@thermo-a.mw.tum.de

O. Feldmann
Lehrstuhl A für Thermodynamik
Technische Universität München
D - 85747 Garching
feldmann@thermo-a.mw.tum.de

Dr. P. Gebhard[1]
Lehrstuhl A für Thermodynamik
Technische Universität München
D - 85747 Garching
feldmann@thermo-a.mw.tum.de

C. Gerlach
Lehrstuhl A für Thermodynamik
Technische Universität München
D - 85747 Garching
gerlach@thermo-a.mw.tum.de

U. Glückert
Lehrstuhl A für Thermodynamik
Technische Universität München
D - 85747 Garching
gluecker@thermo-a.mw.tum.de

Dr. M. Jordan[1]
Lehrstuhl A für Thermodynamik
Technische Universität München
D - 85747 Garching
Martin.Jordan@ri.dasa.de

Dr. B. Kruppa[1]
Lehrstuhl A für Thermodynamik
Technische Universität München
D - 85747 Garching
gerlach@thermo-a.mw.tum.de

Prof. F. Mayinger
Lehrstuhl A für Thermodynamik
Technische Universität München
D - 85747 Garching
may@thermo-a.mw.tum.de

Prof. W. Merzkirch
Lehrstuhl für Strömungslehre
Universität GH Essen
D - 45117 Essen
wolfgang.merzkirch@uni-essen.de

Prof. D. Mewes
Institut für Verfahrenstechnik
Universität Hannover
D - 30167 Hannover
dms@c36.uni-hannover.de

B. Ofner
Lehrstuhl A für Thermodynamik
Technische Universität München
D - 85747 Garching
ofner@thermo-a.mw.tum.de

M. Pitschmann
Lehrstuhl A für Thermodynamik
Technische Universität München
D - 85747 Garching
pitschmann@thermo-a.mw.tum.de

R. Schmidt
Lehrstuhl A für Thermodynamik
Technische Universität München
D - 85747 Garching
schmidt@thermo-a.mw.tum.de

Prof. J. Straub
Lehrstuhl A für Thermodynamik
Technische Universität München
D - 85747 Garching
jstraub@thermo-a.mw.tum.de

Dr. G. Strube[1]
Lehrstuhl A für Thermodynamik
Technische Universität München
D - 85747 Garching
gerlach@thermo-a.mw.tum.de

R. Tauscher
Lehrstuhl A für Thermodynamik
Technische Universität München
D - 85747 Garching
tauscher@thermo-a.mw.tum.de

Prof. J. Wolfrum
Physikalisch–Chemisches Institut
Universität Heidelberg
D - 69120 Heidelberg
wolfrum@urz.uni-heidelberg.de

[1] Authors marked with [1] formerly worked at the "Lehrstuhl A für Thermodynamik". Please refer to the given e–mail address in case of any questions.

Contents

1 Introduction

Franz Mayinger and Oliver Feldmann

Optical measurement techniques have been receiving an increasing attention during the last years. Although most of the techniques which are described in this book were developed decades ago, their application to the analysis of phenomena in heat– and mass–transfer becomes more and more common. This originates from several reasons. The most important one – common to all optical techniques – is that they work non–intrusively and therefore do not influence the investigated process. Even highly transient and/or sensitive phenomena can be analyzed with a high spatial and temporal resolution. Furthermore, in most cases it is possible to visualize directly (on–line) the processes in the test section. This yields the very high accuracies of the measurements and new insights in heat– and mass–transfer phenomena. Both are required to analyze and to improve processes in chemical and power engineering. Secondly, the rapid improvements in computer technology allow to process and to store the huge amount of data within a reasonable time that is obtained by applying optical measuring techniques. High development rates can also be found in the field of light detector, laser, and semiconductor technology.

The continuing developments in laser, detector, optical fiber and computer technology will further augment the already high applicability and versatility of optical measuring techniques. Therefore, it can be expected that optical techniques will continue to gain in importance in many fields of application.

The first point of interest for a given thermofluiddynamic measurement problem is to determine the parameters of the process that have to be measured. Such parameters may be the temperature distribution in a fluid, the concentration of a certain substance in a fluid or its local velocity. Once the parameters are determined, the following requirements have to defined to be able to select the most suitable technique:

- required accuracy of measurement
- required time resolution
- required spatial resolution
- overall dimensions of the measuring volume
- maximum duration of measurement
- optical accessibility of measurement volume

A summary of the optical techniques presented in this book, as well as the physical effect the technique is based on is given in Table 1. Additionally, common applications for each technique are listed, as well as the dimension of the measurement and whether or not a real-time application is possible. The dimension of the measurement is referred to as the resolvable spatial dimension of the obtained data. "Integ." in parenthesis indicates that the measurement is integrated over the path length of the recorded light.

Table 1.1. Overview of the optical measuring techniques discussed in this book

measuring technique	physical effect	application	dimension	real-time application
schlieren and shadow	light deflection	heat, mass-transfer	2d (integ.)	yes
holography	holography	particle size, velocity	3d	no
interferometry	change of light velocity	heat, mass-transfer	2d (integ.)	yes
laser Doppler velocimetry	Mie scattering	flow velocity	point	yes
phase Doppler velocimetry	Mie scattering	particle size, flow velocity	point	yes
dynamic light scattering	Rayleigh scattering	density, temperature	point - 2d	yes
Raman scattering	Raman scattering	mol. concentration temperature	point - 1d	no
laser induced fluorescence	fluorescence	concentration, temperature	point - 2d	no
absorption	absorption	concentration, temperature	point - 2d (integ.)	yes
pyrometry	thermal radiation	temperature	1d	yes
thermography	thermal radiation	temperature	2d (integ.)	yes
particle image velocimetry	light scattering	flow velocity	2d (3d)	yes

As shown in Table 1 several parameters can be measured by more than one optical method. For example, the measurement of particle sizes and velocities can be performed by applying either the short time holography or the phase Doppler Analysis (PDA). Short time holography records the entire flow field twice in two sequential, ultra short exposures on the same holographic plate. The sizes and the three-dimensional velocity field of the particles are obtained with a very high spatial resolution at a given time. On the other hand, PDA is capable of determining time–resolved size– and velocity–measurements at a given point. This is a good example on how different optical measuring techniques can complement one another. The combined application yields the best result: both spatial and temporal resolution in three dimensions with a high accuracy.

A striking example for the efficiency of optical measuring techniques is given in Fig. 1.1. Holographic interferometry was applied to observe a condensing bubble over its lifetime. In the images shown in Fig. 1.1 each effect connected with the process of subcooled boiling can clearly be observed: phase change, convection, and heat– and mass–transfer. The heat–transfer coefficient as a function of the time is one of the most important design parameters for heat–exchangers. Its variations with the time and and the ohase interface can be determined quantitatively from the evaluation of the images.

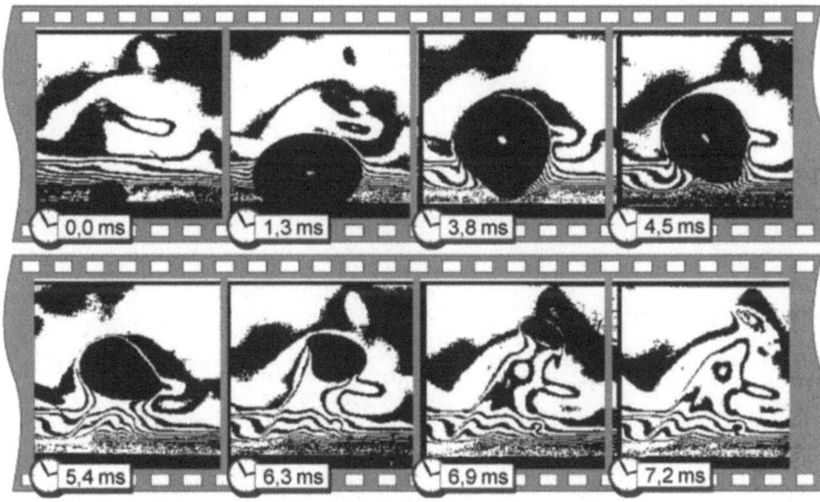

Figure 1.1. Life of a condensing bubble (s. Fig. 3.13)

Each technique listed in Table 1 is presented and discussed in a separate chapter. A brief description of the theoretical background is followed by the description of experimental set–ups which are typical for this technique. Examples of application demonstrate the wide field of application, but also the limits of the techniques are stated. In addition, the references given in each chapter provide a helpful overview of more detailed information.

Optical measuring techniques require suitable lasers as light sources. Important criteria for all lasers are wavelength, bandwidth, stability, tunability and coherence length as well as the energy output of the laser. The requirements of the laser for application in the different measuring techniques are given in each chapter.

When the image forming techniques are applied to analyze highly transient processes, they are often combined with the high-speed cinematography. The recorded sequences yield new and substantial insights into these processes. This book includes a CD–ROM that contains a collection of such sequences which demonstrate the efficiency and the versatility of the presented optical measurement techniques.

Although not every practically applied technique can be discussed in detail, the book may be used as a basis to easily understand and to penetrate into additional techniques. The book is intented to serve as a guide to select the best suitable technique for a given measuring problem. It is both, a general introduction to the most important optical measuring techniques applied in heat and mass transfer, and a guideline to select the technique and to start its application to a specific problem.

2 The Schlieren Technique

Andreas Eder and Martin Jordan

2.1 Introduction

The measuring–principle of the schlieren technique is based on the deflection of a collimated light beam crossing gradients of the index of reflection in a transparent medium. It is, therefore, suited for applications in which deviations of light are intended to be visualized as they appear e.g. at refraction–gradients due to density–discontinuities in a fluid. This classical non–invasive optical measuring technique was established by August Toepler [1,2] in 1864. It is often applied at heat– and mass–transfer phenomena, such as convection, mixing processes of gases or fluids, flame–propagation phenomena, or the investigation of (super–) sonic flows, where the density–gradients of the gas are strong enough for a sufficiently high deviation of the light.

The schlieren technique is very suitable for a first qualitative insight into the investigated physical process because its set–up is of low complexity. The main disadvantage of the schlieren technique is that only integral measurements through the whole depth of the measurement volume can be recorded, which makes a quantitative evaluation of the gradients very difficult or even impossible. Though, as first shown by Schardin in 1934 [3], it is possible for some applications to apply this technique to a quantitative determination of occurring gradients.

Using modern recording devices such as high–speed video cameras, the schlieren technique opens new perspectives for the investigation of highly transient processes by recording a series of images with a very high temporal resolution. Frame–rates of several million images per second can be achieved with sophisticated camera–systems.

2.2 Basic Principle

The basic principle of the schlieren technique is shown in Fig. 2.1. L is a confined light source of homogeneous brightness. The objective O, which should be spherical and with achromatic correction, generates an image of the light source at L'. At this point, the schlieren edge K (e.g. a knife–edge) is located. The sharp edge of K is set parallel to and apart by the distance a to the optical axis of the set–up so that the image of the light–source is partially covered by the schlieren edge. An optic behind the schlieren edge displays a sharp image of the schliere on a screen or a film F.

In the absence of any light disturbing schlieren, the screen F is uniformly illuminated. The brightness of the image on F depends on the position of the schlieren edge K. The closer the edge of K is to the optical axis, the more light will be covered by the schlieren edge and the darker is the undisturbed image. The distance a (edge of the schlieren edge to optical axis) is, therefore, a measure for the sensitivity of the set–up.

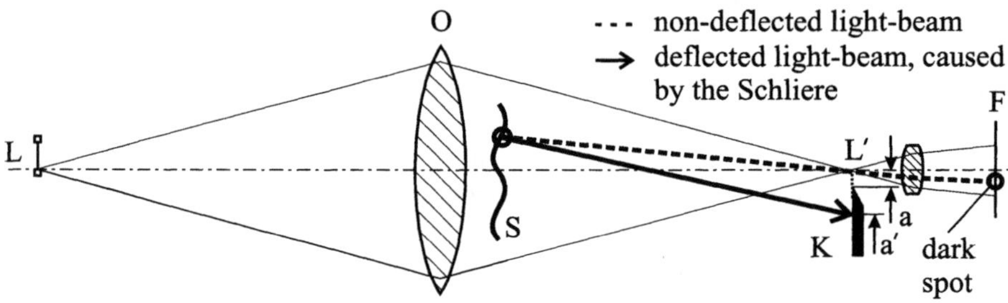

Figure 2.1. Optical arrangement for the schlieren technique due to Toepler.

If a light–beam is deflected by a schliere S, it can be refracted either in the direction of the schlieren edge or away from it, depending upon the orientation of the gradient. When refraction is in the direction of the schlieren edge the spot on the image, which corresponds to the position of the schliere within the measurement volume, will be darker in contrast to the undisturbed spots on the image. Conversely, when the beam is reflected away from the schlieren edge, the corresponding spot will be brighter.

If the density distribution is to be determined quantitatively, it is necessary to evaluate this illumination density–distribution by means of photometry. In addition, further optical adjustments within the arrangement of Fig. 2.1 must be provided to reveal a quantitative relation between the intensity distribution on F and the light deflection at S, which is a function of the density variation in the schliere.

2.3 Optical and Thermodynamic Interrelations

The deflection of light due to a change of the refractive index in the path of the light beams is the basis of the schlieren technique. For the application of this technique in the field of heat and mass transfer, correlations between thermodynamic and optical quantities have to be determined.

2.3.1 Refraction Index and Temperature Field

The refractive index n of a transparent medium applied to a definite wavelength within the domain of visible light can be expressed by:

$$n = \frac{c_v}{c}, \tag{2.1}$$

with the velocities of light in vacuum c_v and in the medium c, respectively. If we are concerned with the investigation of temperature fields in ideal gases, the equation of state reads:

$$\rho = \frac{p}{RT}, \tag{2.2}$$

where ρ is the density and R is the specific gas constant of the medium. T is its absolute temperature and p is the barometric pressure, imposed on the field. For gases the simplified Lorentz–Lorenz relation between the refractive index and the density is valid, because of $n \approx 1$:

$$n - 1 \sim \rho. \tag{2.3}$$

In the reference state ($T_0 = 273.15\ K$; $p_0 = 1013\ hPa$) we find from tables for air and for a wavelength of e.g. $\lambda = 544\ nm$ ("green" light):

$$n_0 = 1.000293. \tag{2.4}$$

With this value, (2.2) and (2.3) yield to the relation:

$$\frac{n-1}{n_0 - 1} = \frac{\rho}{\rho_0} = \frac{pT_0}{p_0 T}, \tag{2.5}$$

which can be solved for n as follows:

$$n = 1 + 7.9 \cdot 10^{-7} \left(\frac{p}{T}\right). \tag{2.6}$$

Equation (2.6) differentiated with respect to T leads to:

$$\frac{\mathrm{d}n}{\mathrm{d}T} = -7.9 \cdot 10^{-7} \left(\frac{p}{T^2}\right). \tag{2.7}$$

2.3.2 The Deflection of Light in an Inhomogeneous Medium

In Fig. 2.2, line 1 marks the front of a light wave at the time t, which propagates in a medium with a locally varying field of refractive index. Therefore, the velocity c of light in this medium can be regarded as a position function (s. (2.1)). The velocities of the light at the two points A and B on wave front 1, kept apart by the elementary distance $\mathrm{d}r$, are expressed by c and $c + \mathrm{d}c$. When the time interval $\mathrm{d}t$ has elapsed, the wave front reaches position 2, which is slightly rotated with respect to position 1. A light beam propagating through point A perpendicular to the wave front will follow a

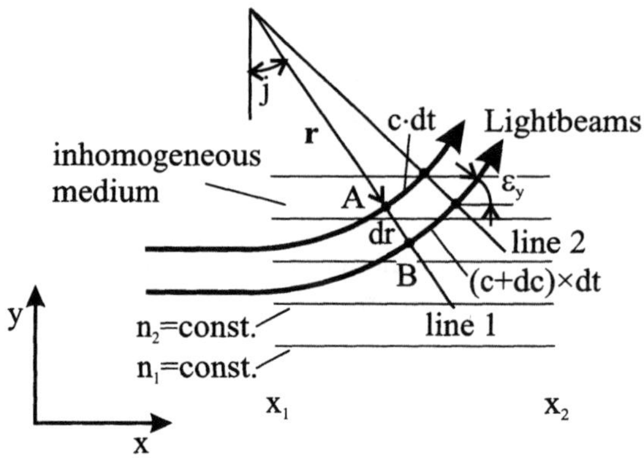

Figure 2.2. Deflection of light in an optically inhomogeneous medium.

curve with a radius of curvature of r. Basic geometric considerations lead to an equation for r:

$$\frac{1}{r} = \frac{1}{c}\frac{\mathrm{d}c}{\mathrm{d}r}. \tag{2.8}$$

As (2.1) states that c is proportional to $1/n$ and taking into account that n is only a function of the field variable T, (2.8) can be expressed by:

$$\frac{1}{r} = -\frac{1}{n}\frac{\mathrm{d}n}{\mathrm{d}r} = -\frac{1}{n}\frac{\mathrm{d}n}{\mathrm{d}T}\frac{\mathrm{d}T}{\mathrm{d}r}. \tag{2.9}$$

In Fig. 2.2 the angle between the coordinate y and the radius $\boldsymbol{r}$ – which is equal to the angle between the coordinate x and the light–beams – is given by φ. We, therefore, find:

$$\frac{1}{\mathrm{d}r} = \frac{1}{\mathrm{d}x}\sin(\varphi) - \frac{1}{\mathrm{d}y}\cos(\varphi), \tag{2.10}$$

with $tan(\varphi) = \mathrm{d}y/\mathrm{d}x = y'$. Regarding a variation of the temperature along T, we can define:

$$\frac{\mathrm{d}T}{\mathrm{d}r} = \frac{\partial T}{\partial x}\sin(\varphi) - \frac{\partial T}{\partial y}\cos(\varphi). \tag{2.11}$$

Expressing $\sin(\varphi)$ and $\cos(\varphi)$ with y' yields to the the following equation:

$$\frac{\mathrm{d}T}{\mathrm{d}r} = \frac{y'\partial T/\partial x - \partial T/\partial y}{\sqrt{1+y'^2}}. \tag{2.12}$$

The differential geometry leads to a relation for r:

$$\frac{1}{r} = \frac{y''}{(1+y'^2)^{3/2}}. \tag{2.13}$$

Inserting (2.12) and (2.13) into (2.9) we can define a differential equation for the light path:

$$\frac{y''}{1+y'^2} = -\frac{1}{n}\frac{\mathrm{d}n}{\mathrm{d}T}\left(y'\frac{\partial T}{\partial x} - \frac{\partial T}{\partial y}\right). \tag{2.14}$$

For experimental conditions with a temperature field varying only in y–direction ($\partial T/\partial x = 0$) and surmising further that only small deflections of the light beams in the test section occur, we can set $1 + y'^2 \approx 1$. This results in a considerably simplified differential equation for the light path:

$$y'' = \frac{1}{n}\frac{\mathrm{d}n}{\mathrm{d}T}\frac{\partial T}{\partial y}. \tag{2.15}$$

A similar expression can be found for density gradients when solving and differentiating (2.5) with respect to ρ and following the lines shown above.

Assuming that the light follows the path of a straight line, we can evaluate the angle ε between an undeflected and a deflected light beam in the x–y–plane by integrating (2.15) along x:

$$\varepsilon_y = (y')_{x_2} = \frac{1}{n}\frac{\mathrm{d}n}{\mathrm{d}T}\int_{x_1}^{x_2}\left(\frac{\partial T}{\partial y}\right)dx, \tag{2.16}$$

where x_1 and x_2 mark the beginning and the end of the inhomogeneous medium in x–direction.

Together with a correlation between ε and the deviation of the light beam at the schlieren edge (s. (2.17) in Sect. 2.4) it is possible to correlate the measurable variable a' with a density or temperature gradient.

2.4 Application of the Schlieren Technique

The optical arrangement of the schlieren technique as shown in Fig. 2.1 is only one of several possibilities. Although the basic optical principle is always the same, four main praxis–orientated optical arrangements can be distinguished, each more or less suitable for a specific application. Instead of the objective O in Fig. 2.1, consisting of one collecting lens with a long focal length, set–ups with the following may be applied:

- two collecting lenses with a parallel light–beam in between,
- one concave mirror (s. Fig. 2.3), and
- two concave mirrors with a parallel light–beam in between (s. Fig. 2.5).

The advantage of set–ups with only one optical component is their compactness, but they can only be applied if the measurement volume is close to the collecting lens or the concave mirror. When using two optical components, the measurement volume can be placed at any axial position of the parallel light beam. In general, there is no difference when using either lenses or mirrors. Nevertheless it is worth noting that it is easier and less expensive to obtain mirrors of the highest quality as opposed to merely adequate lenses for applications with larger measurement volumes, in order to avoid deflection mistakes or schlieren from the lens glass.

The application of only one concave mirror leads to two possible set–ups. The optical arrangement may be constructed to allow the light beam to pass the schliere either once or twice. In the former, the light source is outside the optical axis of the concave mirror (s. Fig. 2.3). Due to the requirement

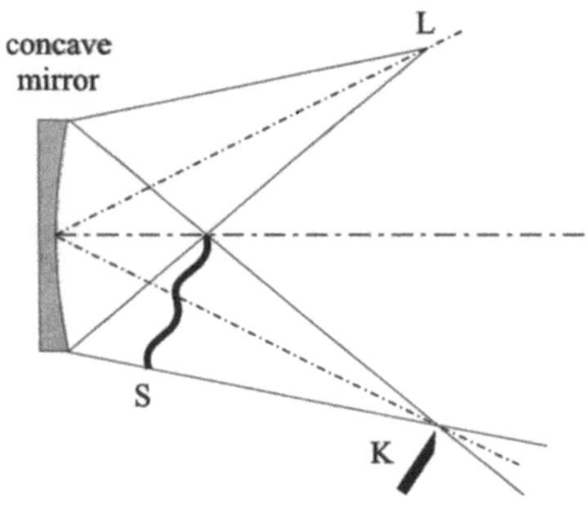

Figure 2.3. Optical arrangement with one concave mirror.

that the schliere should be as close as possible to the mirror, the practical application of this arrangement is limited. In addition the illumination of the mirror from the side leads to astigmatic errors: two focal points originate from the two main layers of the system. Assuming in Fig. 2.3 that the drawing layer is the tangential layer and perpendicular to it the sagital layer, the two focal points as shown in Fig. 2.4 occur. Depending on the alignment of the schlieren

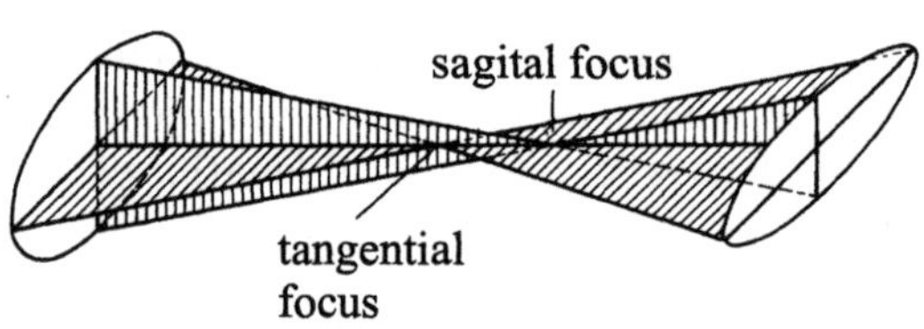

Figure 2.4. Tangential and sagital focal point, originating from applying concave mirrors [3].

edge (the alignment in Fig. 2.3 is vertical), it must be either in the tangential or the sagital focal point. Using a small angle between the incoming and outgoing light beams (by using a large focal distance of the concave mirror), the distance between the two focal points can be minimized.

In order to increase the sensitivity of the schlieren set–up with one concave mirror, the light beam can pass the schliere twice or even more often. This can be realized with a so called "coincidence set–up "in which the light–source is on the optical axis of the concave mirror. This arrangement should only be applied for small deviations. In larger deviations, the reflected light beam might pass the schliere at a different position. The interpretation of these images is, therefore, very difficult as the angle of deviation of the beam no longer depends on the respective position on the schliere alone.

A so called "Z set–up "incorporating two concave mirrors (s. Fig. 2.5) is a good compromise, as compared to the set–up with two collecting lenses, between compactness of set–up and high flexibility. The astigmatic errors which

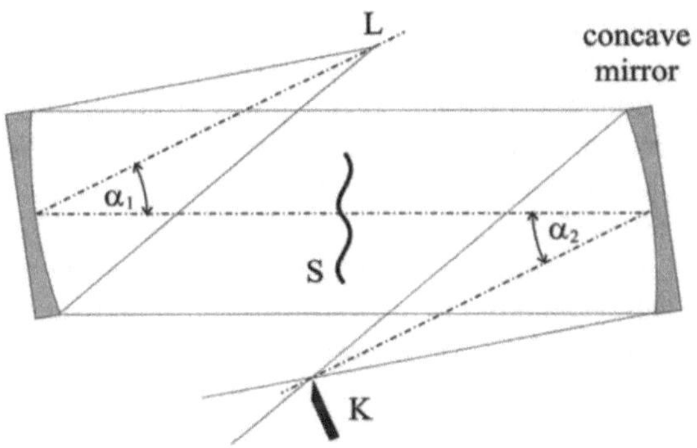

Figure 2.5. Optical arrangement with two concave mirrors, "Z set–up ".

appear when using one concave mirror, are compensated, if the focal distance of the two concave mirrors and the angles α_1 and α_2 between incoming as well as outgoing light are identical and the light source is on the opposite side of the measurement volume from the schlieren edge [3].

The directions of incoming and outgoing light at the second mirror must be contrary to the directions at the first mirror. In cases where a surface mirror is used within the parallel light beam in order to change its direction, the directions of incoming and outgoing light at both mirrors must be identical.

Light Source

The correct choice of light source is highly intrinsic to the quality of the measurements and to the sensitivity of the entire set–up. In general, sensitivity increases with a decreasing wavelength of the light. Argon–Ion and similar cw–lasers can be used only for a limited number of applications with high gradients as the requirement for a light source with a finite extent and a homogeneous brightness is not met. In many cases it is necessary to generate an intermediate image of the light–source and to confine its size by an adequate aperture. In general, mercury and sodium discharge lamps, and other arc–lights are adequate light–sources to be used for schlieren measurements.

Schlieren Edge
Two different types of schlieren edges can be distinguished. If the deviations are small the best choice is to use a schlieren edge, which can be either a knife–edge for rectangular light sources or a round aperture in cases of using a circular light–source. The latter opens furthermore the opportunity to visualize gradients of all directions within the measurement volume (without distinguishing the directions), whereas the application of a knife–edge limits the optical arrangement to visualize gradients only in one direction. It is essential for the quality of the visualized gradients that the schlieren edge is located right in the focal point of the mirror or the lens, respectively.

The exact position of the schlieren edge can be found by applying the following procedure: After the approximate position has been determined, the schlieren edge is moved from one side into the beam path. If the image on the screen becomes darker from the same side as the schlieren edge is moved into the beam path, the schlieren edge must be moved away from the screen. The contrary darkening occurs if the schlieren edge is too close to the schlieren optics. The exact position is found if the image on the screen gets homogeneously darker when the schlieren edge is moved into the light beam.

If the gradients are stronger, a gray–key or a color–slide can be used instead of a schlieren edge. The gray–value on the image or the color respectively facilitates the determination of both the relative strength and the direction of the gradient within the measurement volume (s. Sect. 2.4.1).

It is worth mention that the deviations of the light beams are a function of the focal length of either the concave mirror or the collecting lens. The light–beam, deviated at the schliere with an angle of ε causes an deviation at the schlieren edge a' as follows:

$$a' = \varepsilon f, \tag{2.17}$$

with the focal length f of the concave mirror or the collecting lens. Equation 2.17 is only valid for an optical arrangement with a parallel light beam. The deviation a does not dependent on the position of the axial schliere within the parallel light beam.

2.4.1 Application to Transient Combustion Research

The schlieren technique may be used within the field of transient combustion research to record the following:

- mixing processes of fuel and oxidizer [4–8],
- the global flame propagation process by means of visualizing the density–gradient between the unburnt mixture and the exhaust gas [9–11], and
- the propagation and structure of precursor pressure– and shock–waves [12,13].

Although the reaction–zone of the flame can not be distinguished from the hot exhaust gases, this measurement technique can give a deep insight into the physical–processes of transient combustion phenomena and opens new perspectives when combining it to modern recording devices.

The Schlieren Cinematography

The time–resolved visualization of changes in the location and shape of density gradients, referred to as schlieren cinematography, delivers essential results for the investigation of fast processes such as the propagation of flames. An exemplary optical set–up consisting of two concave mirrors ("Z set–up") as applied in practice is shown in Fig. 2.6. The semi–permeable mirrors facilitate a very exact arrangement of the optical components and avoid astigmatic errors.

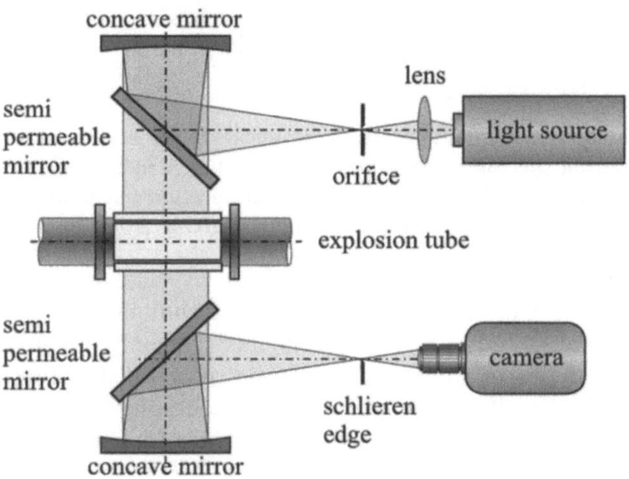

Figure 2.6. Optical arrangement of the schlieren cinematography in combustion research [10].

A mercury discharge lamp confined by a rectangular aperture is used as a light–source. The schlieren edge consists of a vertical knife–edge which facilitates a visualization of gradients in the horizontal direction. The temporal changes of the density gradients in the test chamber are recorded by a high–speed video–camera.

The sequence shown in Fig. 2.7 gives a good impression of the capability of the applied set–up. During the combustion process a flow–field arises in front of the flame–front due to the expansion of the hot, burnt gas. By the presence of obstacles in the path of a flame, such as the horizontal tube in Fig. 2.7, the flow gets more and more turbulent, which leads to an increase of heat and mass transfer between burnt and unburnt gas and, therefore, to a significant modification of the flame–shape: the surface of the flame increases and, therefore, more fuel is consumed per unit of time, which leads to significant pressure rises.

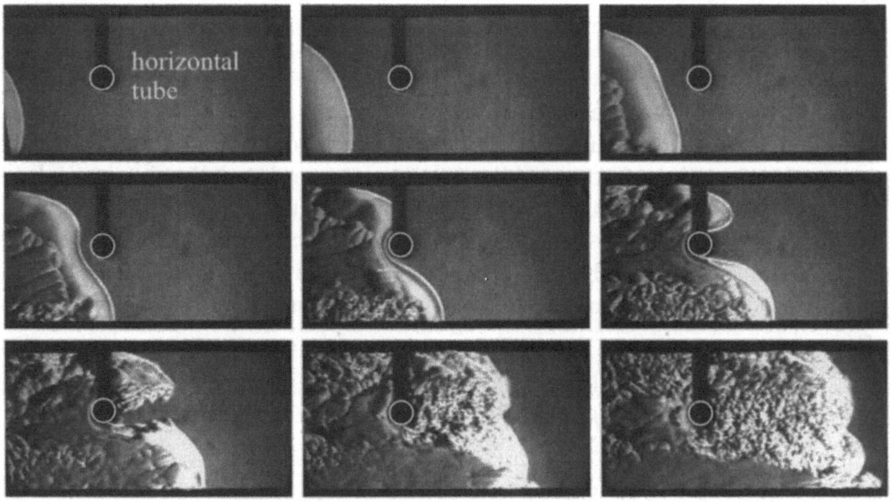

Figure 2.7. High–speed schlieren cinematography, showing the flame–acceleration due to a horizontal tube during the combustion process of a 8 Vol.% hydrogen–in–air flame, time between two images: 2.7 ms. Various flame–acceleration phenomena, recorded by means of this technique, can be found on the *CD–ROM* [10].

The main disadvantage of the schlieren measurement technique in combustion research, as can be observed in Fig. 2.7, is the fact that only integral images through the entire depth of the chamber can be recorded. In order to visualize the reaction–zone itself with a very high spatial resolution, light scattering techniques such as the Laser–induced Fluorescence (s. Chap. 12) have to be applied.

The Color Schlieren Technique

The color–distribution in a schlieren image proviedes important information for the interpretation of physical phenomena as the relative strength of the density gradients can be determined also for applications with large deviations at the schlieren edge which can not be determined by using a knife edge. This measuring technique is very useful e.g. when the interaction of shock–waves and flame–fronts is to be investigated simultaneously. Replacing the schlieren edge with a transparent color–filter transforms the black–and white system into a color schlieren system. Furthermore, the light source in color schlieren set–ups should have a homogenous spectral distribution curve in order not to be limited in assigning a specific color to a certain class of gradients. For an optical set–up, adequate color–filters, and exemplary experimental results refer to the *CD–ROM*. Further information on the color schlieren technique can be found in [14].

2.4.2 Application to Fuel–Injection Systems

The development of new fuel–injection systems for a more efficient reduction of the exhaust–emission of internal combustion engines requires a detailed understanding of the physical processes in the combustion chamber. Schlieren cinematography facilitates the visualization of the fuel injection process, the evaporation in case of using liquid fuels, and the combustion and is, therefore, a first important step in the optimization–procedure of injection–systems.

In Fig. 2.8, an optical arrangement for schlieren cinematography applied to a single stroke rapid compression machine (bore: 220 *mm*) is shown [7]. Optical access to the combustion chamber is provided by means of a glass piston. In this experiment, a coincidence set–up with one concave mirror is

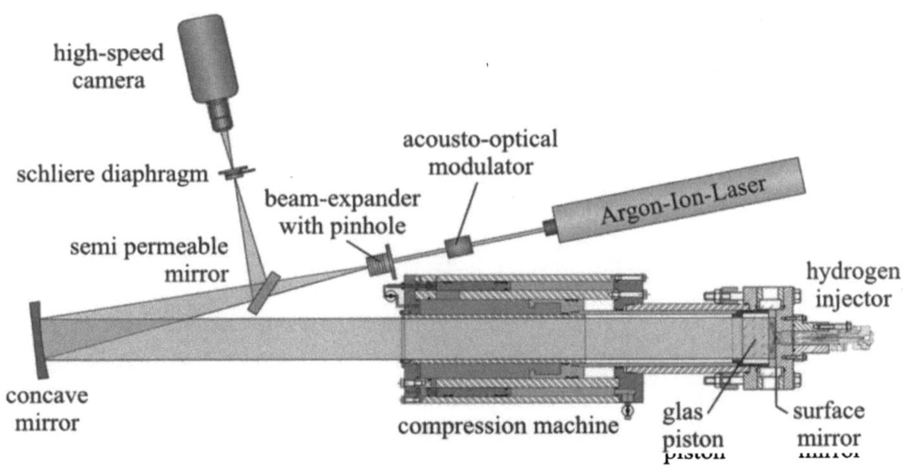

Figure 2.8. Optical set–up for schlieren cinematography, applied to a single stroke rapid compression machine.

applied. The parallel light is reflected by a surface–mirror that is mounted on the cylinder top and which doubles the deviation of the light at the schliere. As a light source, an Argon–Ion laser is used. Using the internal shutter of the camera running at a frame–rate of 40.000 images/sec, the relatively long exposure time would lead to a movement blur. Therefore, the light of the cw–laser is chopped by means of an acousto–optical modulator, which generates laser–flashes with a pulse–duration of some microseconds. The laser beam is expanded by means of a beam expander including a pinhole. A semi–permeable mirror allows use of the same beam path for incoming and outgoing light of the measurement volume.

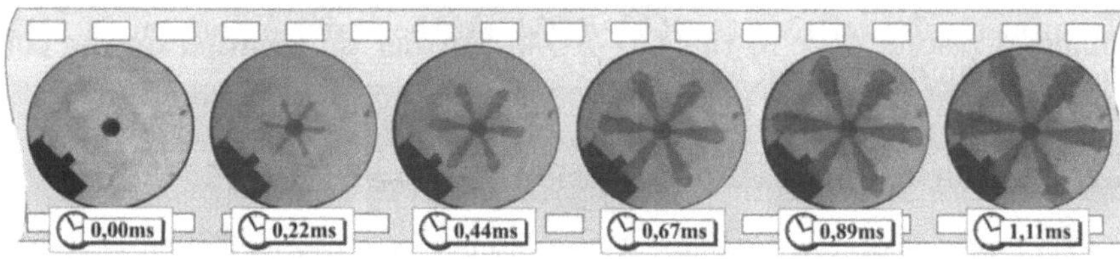

Figure 2.9. Series of schlieren images showing the injection of 30 MPa hydrogen into 1 MPa cold pressurized air [7]

The potential of this technique can exemplary be seen in Fig. 2.9, showing the injection of 30 MPa hydrogen into 1 MPa cold pressurized air [7]. By means of this technique it is possible to determine important variables of the mixing process, described above, which are prerequisites for the development of clean combustion engines.

3 Fundamentals of Holography and Interferometry

Franz Mayinger

3.1 Abstract

Holography allows for the use of various interferometric methods for measuring processes of heat and mass transfer. In this chapter the principles of optical arrangements for such experiments are shown. The techniques of data acquisition and evaluation are discussed. Examples demonstrate the advantages of the techniques used. Holographic interferometry has displaced Mach–Zehnder interferometry completely due to its greater cost effectiveness, simplicity of operation and convenience. Holographic interferometry does not require machining or manufacturing of test section windows, mirrors or lenses with special precision or accuracy, because imperfections are automatically balanced by the holographic two step procedure.

3.2 Introduction

Optical methods have been used for many years in heat and mass transfer because of their advantages as measuring techniques. They do not influence the process under examination and because of their inertia free response they can also be used for very fast and highly transient phenomena.

Usually one distinguishes between image forming and non–image forming methods. The former give information about the situation in a larger volume or area whereas the latter are usually restricted to a narrowly sized volumetric element–usually below one cubic millimeter–and therefore are called local methods. Image–forming methods mostly utilize photography to record the information of interest.

Fifteen years ago, interferometric methods using Mach–Zehnder, Michelson, Schlieren or shadowgraph methods were used in heat and mass transfer and in gas dynamics. They were used, for example, in combustion research. The optical arrangement, data evaluation and theoretical fundamentals of these methods are described in detail by Hauf and Grigull [21]. In the middle of this century Gabor [20] invented a new method for recording and storing optical information which was called holography.

Unlike photography which can record only the two–dimensional distribution of the radiation emitted by an object, holography can store and reconstruct three–dimensional pictures. The name holography is derived from the ability of the method to record the totality (holos) of the light information as regards the wave–front, namely the amplitude (as brightness), the wavelength (as color) and the phase position of the light. Using these possibilities, completely new interference methods can be developed. Holography demands, however, a source which emits coherent light. Therefore, holography, as invented in 1949 by Gabor, could not be used until the laser had been developed ten years later.

Today, lasers are available for almost every wavelength, even in the infrared and ultraviolet regions, and they can emit considerable power. They work by continuously emitting light of a monochromatic nature, for example, as in gas or diode lasers. Pulsed lasers emitting flashes of ultra–short duration down to a few nanoseconds have also been designed.

3.3 Principle of Holography

The general theory of holography is so comprehensive that one has to refer here to the literature [16,17,23,43] for a detailed description. This chapter will treat only those simple fundamentals which are necessary to understand holographic interferometry. In Fig. 3.1 the holographic process is described as used for recording and reconstructing an arbitrary wave–front. The object is illuminated by a light source emitting monochromatic light. This light, reflected and scattered from the object, directly meets a photographic plate. The light originating from the object and scattered and reflected by it forms a very complicated wave–front which is called an object wave. According to the principle reported by Huygens we can regard this wave–front as a superposition of many elementary and simple spherical waves. To simplify the description only one spherical wave of the light reflected from this object is drawn in Fig. 3.1.

Now a second wave, called a reference wave or reference beam, is superimposed to this spherical wave. If both waves are coherent they form a stable interference pattern when landing on the photographic plate. This system of fringes or spots can be stored by the light sensitive emulsion of a photographic plate. After chemical treatment, developing and fixing, this plate is called a "hologram". The resulting pattern, consisting of microscopic small elements, contains all the information about the wave. This pattern usually consists of from 1000 to 5000 fringes or spots per millimeter. The information density is a function of the angle formed by the two light beams, namely the object and reference waves, and of the wavelength of the light emitted by the laser. The amplitude of the waves is stored in the form of the various brightnesses and contrasts of the fringes. The phase position is fixed in the form of various distances between the fringes or the spots of the pattern.

If one illuminates the photographic plate by using a light beam which is similar to the original reference wave, the microscopic pattern now acts like a defractive grating with variable grating constant. The light leaving the

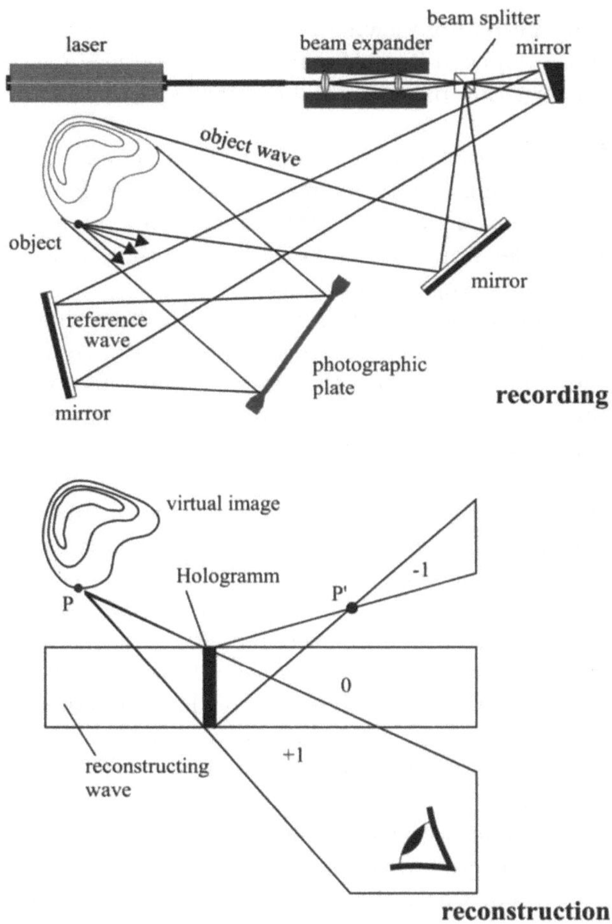

Fig. 3.1. Principle of recording and reconstructing a wave–front

plate on its rear side consists of three waves: one wave of so–called zero order which travels exactly in the direction of the incoming illumination wave; and two waves of first order. One of these waves of first order travels in the same direction as the original object wave of the recording process with the same amplitude and phase distribution. This wave mediates a virtual image of the original object. The second wave of first order reconstructs a real image of the object. The latter mostly seems to be very unfocused and distorted. There is, however, a method to obtain an undistorted and well–focused real picture which will be described later.

Due to the fact that the wave originating from the object point P is stored in every part of the hologram, this point can be seen on the hologram, even if only a very small spot of the plate is illuminated. Thus by using holographic plates of large extension this point can be regarded from any direction and the magnitude of the viewing angle is limited only by the size of the holographic plate. This method of viewing is not restricted to a single point of the object. Though the superposition of many waves produces a very complicated interference pattern this pattern can be recorded without any difficulty.

With this approach a very extended object can also be seen via each part of the hologram when reconstructing the object wave. Human eyes look at the holographic plate from different angles, so a three–dimensional picture is reconstructed which yields a lifelike impression. An observer can thus envision grasping the object behind the holographic plate.

By using these recording capabilities, different waves, even those shifted in time and in series, can be stored on the same holographic plate. By illuminating the developed holographic plate with the reference wave all object waves are reconstructed simultaneously. Where they only differ slightly from each other they interfere with each other. These are the fundamentals of holographic interferometry.

This new possibility of superimposing diffuse reflected wave–fronts has found many applications, including a new method of non–destructive material testing and also in vibration analysis. Such investigations are usually performed with so called on light holography, as shown in Fig. 3.1 by creating two or more exposures which interfere with each other during reconstruction. The fringe pattern resulting from this interference gives information about the movement, with respect to magnitude and direction, of the object during the time difference between the two exposures. In heat and mass transfer investigations, however, the temperature and the concentration distribution in a fluid is of special interest.

In heat and mass transfer investigations a modified method yields superior results. Instead of recording the reflected wave, i.e. the on light method, a so–called through light method is used, where the object wave radiates through a volume or a space in which the heat or mass transport takes place. This "through–light wave" is then recorded.

3.4 Simple Holographic Arrangement

Before we can discuss the various holographic interferometric methods it is necessary to give a brief description of the optical arrangement used in holography. An arrangement quite frequently used for through–light holography is shown in Fig. 3.2. A beam splitter divides the laser beam into two waves: an object wave and a reference or comparison wave. It is recommended to use a beam splitter with a variable splitting action, because the intensities of the object and the reference wave can then be varied. Both waves are expanded to parallel wave bundles behind the beam splitter via lenses, usually consisting of an arrangement of an microscope lens and a collecting lens. The expanded and parallel organized object wave travels through the space with the research object of interest–called a test section–in which the distribution of temperature or concentration may be measured. The reference wave, however, bypasses the test section and falls directly onto the photographic plate.

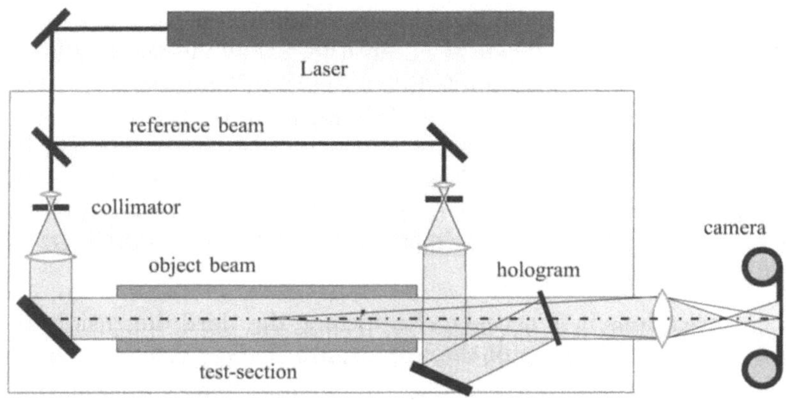

Fig. 3.2. Holographic arrangement for the examination of transparent media

As previously mentioned, holography requires a coherent light source. Therefore, only lasers are used today. Continuous measurements can be performed only by using lasers which emit continuous light. Several kinds of lasers are now available for this purpose, such as He–Ne, argon, krypton, and diode lasers. For highly transient processes pulsed lasers are needed, for example, those with rubies as a light source which allow very short exposure times.

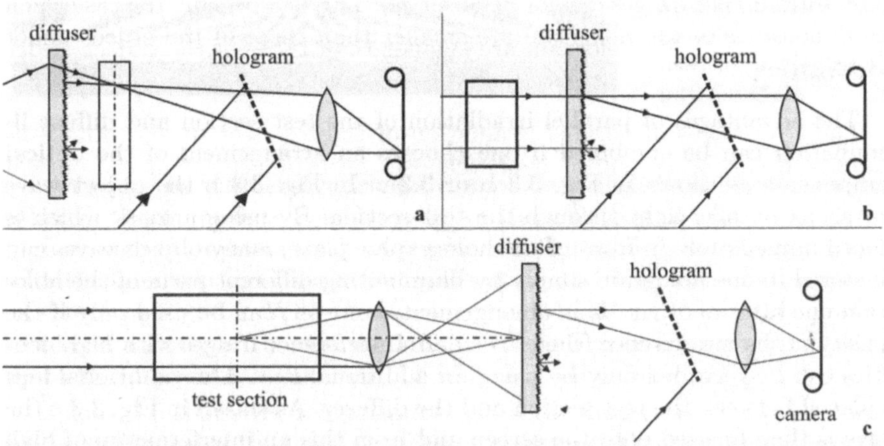

Fig. 3.3. Holographic set–up with diffuse illumination

To obtain a well focused picture of a hologram the interference pattern between the object and the reference wave must be completely stable during the illumination or exposure time of the plate. Therefore, the optical components must be mounted on vibration–free tables or scaffoldings. in some cases even the cooling of the laser can cause vibrations. It is therefore advisable, especially with water cooled lasers of high power, to line up the laser separately from the optical components. There is no need for a high optical quality of the optical components, as for example in conventional interferometry. This is because only relative changes of the object wave are recorded, and optical errors are automatically compensated by the interferometric method. In many cases only a two–dimensional temperature distribution over a test object is of interest. It is then sufficient to record the change of the refractive index in a two–dimensional way. For such cases object waves with parallel light provide pictures of high quality.

For some applications it is necessary to include the third dimension. In these cases a diffuse illumination of the test section is advantageous or even necessary. As shown in Fig. 3.3 it is easy to produce diffuse illumination. A diffusor, for example in the form of a disk of frosted glass, can simply be placed into the object beam. The test section can now be regarded from various angles and additional information about the change of the refractive index in the test section may be obtained.

Diffuse illumination is also chosen if the diameter of the object is larger than the dimensions of the available photographic plates or lenses because parallel beams cannot record a complete picture on the plate. The fact that the various light beams are travelling via different angles through the test section, however, also has disadvantages. The capability to form interference becomes weaker, and evaluation of the interference pattern formed by diffuse light is very complicated. Especially in the immediate neighborhood of the walls, interference fringes cannot be observed. This is especially the case when the dimensions of the hologram are smaller than those of the object under investigation.

The advantages of parallel irradiation of the test section and diffuse illumination can be combined if one chooses an arrangement of the optical components as shown in Fig. 3.3 b or 3.3 c. In Fig. 3.3 b the object wave travels as parallel light through the test section. By using a mask which is placed immediately in front of the holographic plate, many object waves can be stored in one hologram simply by illuminating different parts of the hologram one after another. This arrangement, however, can be used only if the density of the interference fringes is small. Interference fringes with high densities can be recorded only by using an additional lens. This additional lens is placed between the test section and the diffuser. As shown in Fig. 3.3 c the wave is then focused onto the screen and from this an interferogram of high quality is produced.

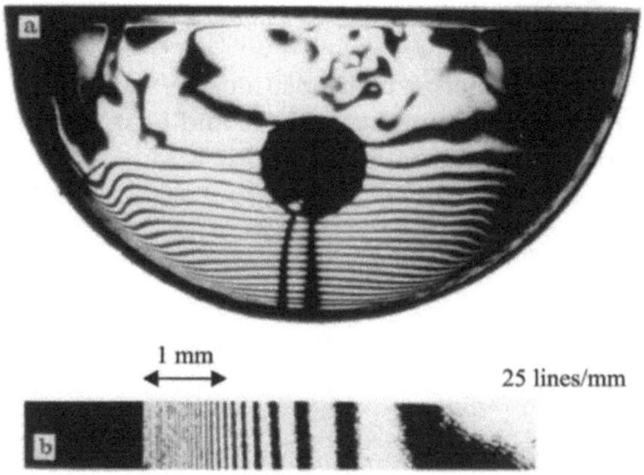

Fig. 3.4. Temperature field in a semicircular cavity. **a**, Diffuse and **b**, parallel object beams

Though the production of a holographic interferogram has not yet been explained it seems advantageous to illustrate a few examples of the possibilities of parallel and diffuse illumination. Fig.s 3.4 and 3.5 show temperature fields in fluids under free convection, heated by internal heat sources [22]. The walls of the semi–cylindrical space in Fig. 3.4 a were uniformly cooled. The diameter of this test section was relatively large and, therefore, diffuse light was used to obtain a picture of the whole temperature field.

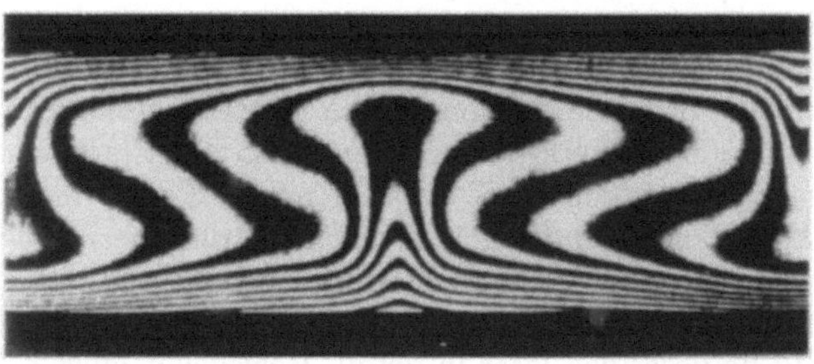

Fig. 3.5. Temperature field in a liquid layer heated at the bottom. Parallel object beam used for recording the interferogram

As seen, the interference fringes cannot be well distinguished from each other in the immediate neighborhood of the wall and the resolution is not high enough to evaluate the heat transfer coefficient. In a second experiment with the same test section a wave–front was used with an optical arrangement as shown in Fig. 3.2. With this arrangement, however, it was only possible to record a sector of the test section. Now interference fringes with densities up to 25 lines per millimeter can be well resolved as seen in Fig. 3.4 b.

Figure 3.5 shows the temperature field if the fluid (water) is heated from below and cooled at the top of the rectangular vessel. One can see a convection type known as Benard convection in the references.

3.5 Holographic Interferometry

As noted earlier, in the 1970s interferometric methods, mostly according to Mach–Zehnder, were used to investigate temperature fields and heat transfer. Such a Mach–Zehnder interferometer is shown in Fig. 3.6.

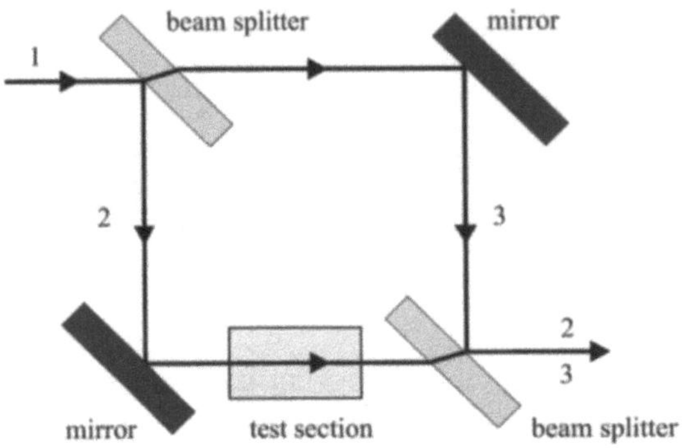

Fig. 3.6. Mach–Zehnder interferometer

A light wave coming from the left side is split into two parts by a semi–permeable mirror, in which one half travels through the test section and the second half circumvents the test section. This first section of the parallel light experiences a phase shift in the test section due to temperature fields. The wave–front changed in this way is then superimposed with the second half of the wave in a semi–permeable mirror, and the interference develops. The interference shows the difference in the optical path between both waves. From this difference the temperature or concentration field in the test section can be evaluated. In holographic interferometry two waves are also superimposed. The difference is that they travel through the test section at different times. Therefore, holographic interferometry and Mach–Zehnder interferometry differ from each other in that the former is a two–step method and the latter is a two–way method. The superposition of the two object waves can occur in various ways, which will be described later in detail. However it must be pointed out here that macroscopic interference caused by the superposition of two object waves has to be clearly distinguished from the microscopic interference which takes place when the object and the reference wave are superimposed during the recording of the hologram. In our discussion of these interferometric methods these two object waves will be called comparison and measuring waves respectively.

3.5.1 Double Exposure Technique

As previously mentioned, several object waves, in series, can be recorded on one and the same hologram. When illuminating the hologram with the reference wave all are reconstructed simultaneously. If the reconstructed object waves differ only a little from each other they can macroscopically interfere. From the interference pattern one can determine the differences between both or several object waves. This principle is used for the double exposure technique and explained graphically in Fig. 3.7.

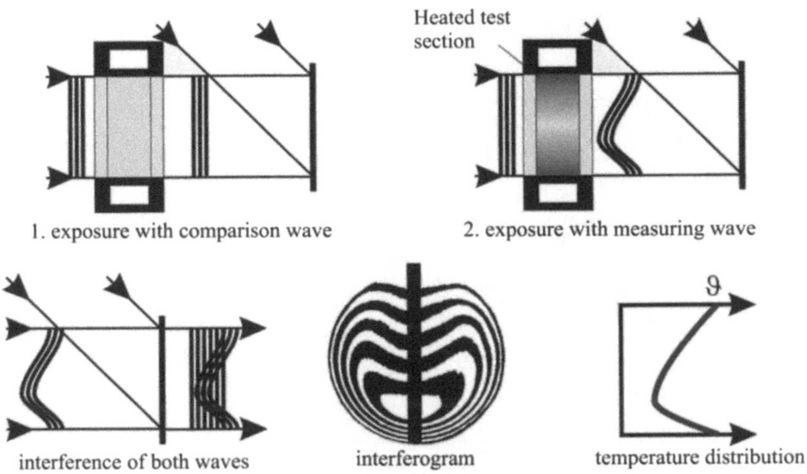

Fig. 3.7. Principle of double exposure technique

The temperature distribution around a heated tube was chosen as an example. In a first exposure the light wave travels through the test section in which the process of interest – the heat or mass transfer from the wall to the fluid – has not yet begun. The wave, however, can already be influenced or distorted by irregularities in the glass of the windows, or by bending of the windows due to pressure in the vessel of the test section. After this wave – called the comparison wave – has been recorded on the holographic plate, the process of interest in the experiment, for example the heat transfer, is started. In the example in Fig. 3.7 a temperature field is formed in the fluid due to heat transfer from the wall. If now a new wave is sent through the tube – the so–called measuring wave – it experiences a continuous additional phase shift due to the new temperature field created by the heat transfer. This wave is also stored on the holographic plate. The holographic plate is now chemically treated, i.e. developed and fixed, and afterwards will be illuminated by the reference wave. As a result of this illumination, both object waves (the first without and the second with heat transfer) are simultaneously reconstructed, i.e. released from the holographic plate, and they interfere with each other. The interference picture can be observed with the eye or stored photographically.

The main difference between classical interferometry and holographic interferometry is that in holographic interferometry the object beam is compared with itself. This allows plenty of possibilities for additional applications when compared to classical interferometry. Experiments performed with holographic interferometric methods are also less expensive and simpler than those using the previous procedures.

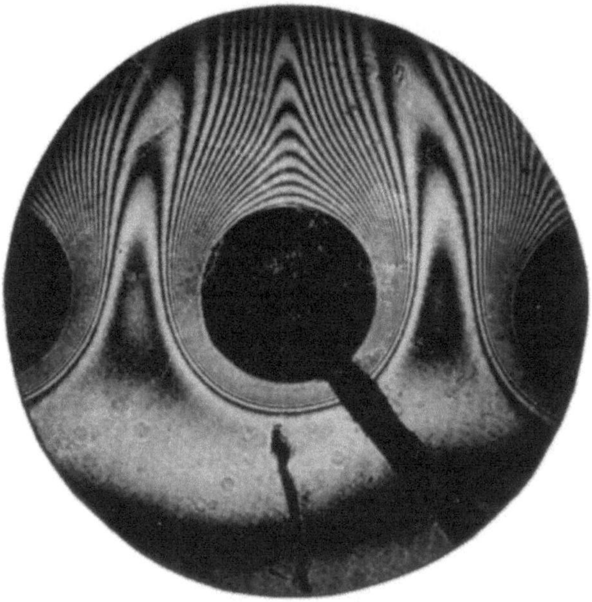

Fig. 3.8. Temperature field around a tube bundle at free convection

An example of a holographic interferogram, produced with the double exposure technique is shown in Fig. 3.8. The interference pattern represents the field of isotherms around horizontally oriented heated tubes with natural convective flow at the outside. The temperature difference between two isotherms is 2.3 K in this example. In areas of high heat transfer – at the front stagnation point – the densely packed isotherms reveal a high temperature gradient whereas at the rear stagnation point they demonstrate much lower heat transfer. The advantages of these double exposure techniques are simplicity, high reliability and low experimental cost.

However, the method also has disadvantages. The most favorable moment for the exposure – i.e. for recording the measuring wave – cannot be deliberately and knowingly chosen, because the interference pattern only appears after the chemical treatment of the plate when the hologram is reconstructed. Transient processes cannot be continuously observed or recorded. Furthermore, the stored measuring wave cannot be reconstructed separately from the comparison wave.

When the principles of holography were initially explained in this chapter it was mentioned that when reconstructing the hologram not only a virtual but also a real picture is produced. This real picture can be used for holographic interferometry when small objects are investigated. Figure 3.8 shows what an interferogram looks like if the virtual picture of the holographic process is used in recording. In this case both waves, recorded at different times, are superimposed during the reconstruction, and a well focused undistorted interference pattern can be gained only if the waves are focused on the film or the camera by using lenses. In general the lens system is focused to the middle plane of the object. If constant temperature exists in the test section, which is the case when recording the comparison wave, the light wave travels through the test section in straight lines. During the recording of the measuring wave, however, a temperature gradient exists over the cross–section of the test section due to the heat transfer process, and the measuring beam is deflected especially in the boundary layer. One has to take care that this wave is well focused onto the film of the camera by the camera lenses. When focusing the camera to the mid–plane of the object, one has the impression that the light beam comes from this mid–plane.

This procedure can be simplified by using the real picture of the hologram. A real, undistorted picture can be produced by using a parallel reference beam during recording and by using a conjugated reference beam during reconstruction of the hologram. The latter can be easily done by turning back the direction of the reference wave, or even more simply by inverting the hologram for reconstruction as shown in Fig. 3.9.

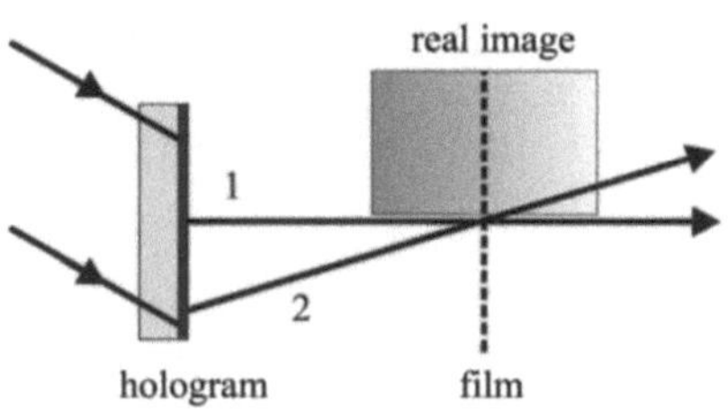

Fig. 3.9. Arrangement for reconstructing a real image

The corresponding beams of both waves stored on the holographic plate then form a real picture of the interference pattern, and the beams intersect on the focused plane. At this point the recording tool, a photographic plate or a light sensitive film for example, must be in position. There are no additional lenses necessary as this could distort the picture. As mentioned, this method also has the advantage that very small objects can be investigated, i.e. by placing a microscope lens on the focused plane instead of a light sensitive film or a photographic plate. This microscope lens can enlarge the real picture. In this case, however, one has to take into account weak distortions by the lens system.

In Fig. 3.10 this possibility of recording interference patterns around small objects is illustrated. The interference pattern in Fig. 3.10 mediates the impression of the temperature field around a vapor bubble condensing in a

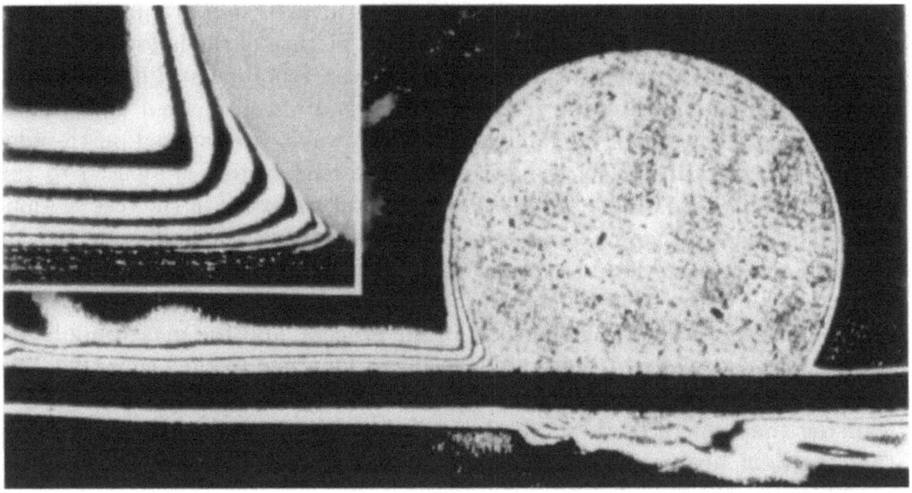

Fig. 3.10. Temperature field around a bubble

subcooled liquid. The interference fringes, however, do not represent lines of constant temperature, because complicated corrections have to be applied due to the bending of the phase interface and of the temperature field. If the bent temperature field is symmetrical, for example with spherical bubbles, a method proposed by Abel can be applied for this correction. This method is described in detail in [21].

3.5.2 Real–Time Method

As mentioned before, the method described up to now has the disadvantage that the process under investigation cannot be observed continuously. Compared with classical interference methods like Mach–Zehnder interferometry this is a great disadvantage, because such classical methods allow continuous observation. However, a slightly modified method also based on holographic interferometry which is called the real–time method avoids this disadvantage. This method is briefly outlined in Fig. 3.11. There are two modifications of this method.

After the first illumination in which the comparison wave is recorded on the hologram the photographic plate, now called the hologram, is fixed and developed.

After chemical treatment and drying it is necessary to replace the holographic plate in the mounting with an accuracy of half a wavelength, less than 300 *nm*. This accurate replacement is not as difficult as one may believe. An approximation of the original position with an accuracy of a few micrometers can be done by mechanical means. This is possible, for example, by using two microscrews in the mounting device. The final adaptation

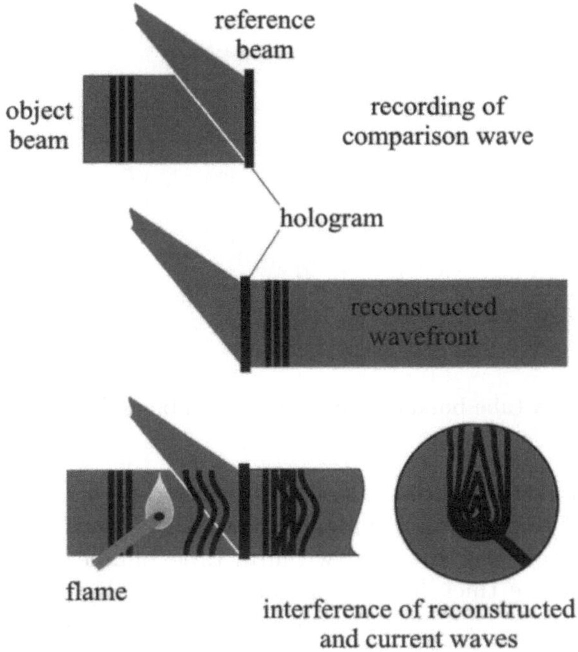

Fig. 3.11. Principle of real–time method

is then made by a piezo–quartz device by applying electrical voltage to the two quartz crystals. During this adaptation, the hologram is illuminated by the reference and the object wave, the test section is returned to the same condition as during the first illumination, and the adaptation is continued until all interference patterns originating out of the hologram due to misalignment disappear. Mounting devices working with piezo–quartz crystals can be purchased easily. For the second exposure or any following exposures the object wave of the unheated test section is reconstructed by illuminating the repositioned hologram with the reference wave. The test section is then simultaneously irradiated by the measuring wave. However, this is now undertaken with the heat or mass transfer process switched on.

The new measuring wave also penetrates the hologram and meets the old object wave which is released by the reference wave and which has all the information concerning the test section even when the heat or mass transfer process is not yet going on. So both object waves, the new and the old, interfere behind the hologram and the interference pattern is a result of the temperature or concentration field produced by the heat or mass transfer process. In Fig. 3.11 this procedure is represented schematically by a burning match. It is now possible to observe the resulting interference pattern continuously or to photograph it with a still or movie camera. By illuminating it with a high powered laser, high–speed cinematography can also be applied, taking several thousand pictures per second.

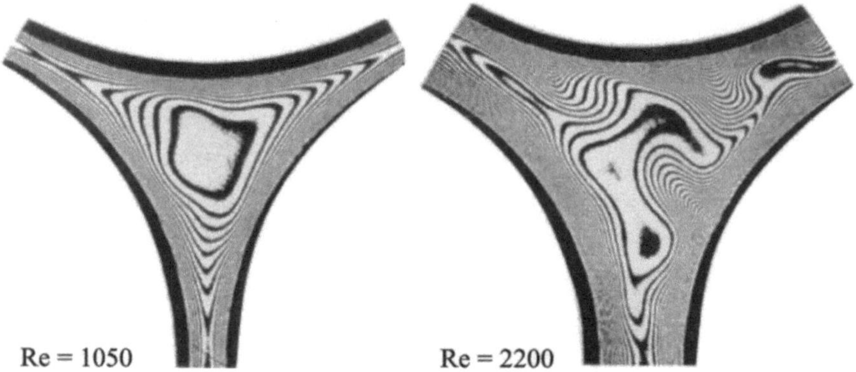

Fig. 3.12. Longitudinal flow in a tube bundle: Heat transfer and boundary layer

Examples of interferograms taken with this real–time method are shown in Fig.s 3.12 and 3.13. Figure 3.12 shows the temperature field and, specifically, the formation of the boundary layer in a rod within a tube–bundle which has longitudinal flow. The three thick lines at the outside are segments of the tube walls. Figure 3.12 (right side) was recorded during a flow situation where the change from laminar to turbulent flow (Re = 2200) occurred. The fluctuations in the boundary layer can be clearly seen. With predominant laminar flow (Re = 1050) the boundary layer is much more uniform and also smoother at its surface, as one can see in Fig. 3.12 (left side).

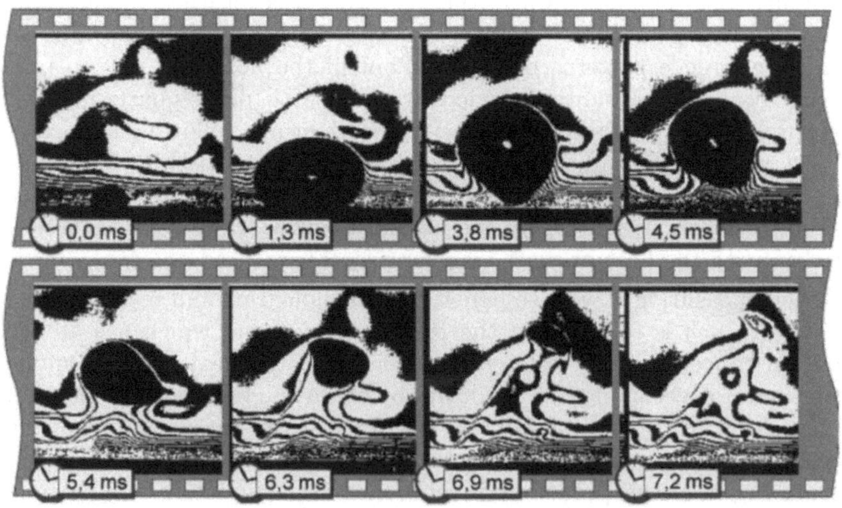

Fig. 3.13. Growth and condensation of a steam bubble (pressure $p = 1$ *bar*, water temperature 8 K below saturation temperature, flow velocity $w = 0.25$ m/s heat flux $\dot{q} = 9$ W/cm^2)

Figure 3.13 provides evidence that it is also possible with this method to record situations of an extremely transient character. In this figure the history of life of a steam bubble and its surrounding temperature field is shown. The steam bubble is produced at a heated horizontal plate and grows into the surrounding water, which has a temperature below the saturation temperature. The steam bubble initially grows in the superheated boundary layer. However, with its increasing volume, it enters the subcooled bulk flow where it is again rapidly condensed. The high–speed exposures prove that processes can be demonstrated by this method, even those having a total lifetime of only a few milliseconds.

3.5.3 Evaluation of the Interferograms

This chapter presents only a very simple explanation and basic hints for the evaluation of interferograms produced using the holographic methods described above. For more detailed information and instructions reference is made to chapters five, six and eight of this book.

Generally the evaluation of holographic interferograms is the same as that of Mach–Zehnder interferograms [21]. As mentioned above, the two object waves penetrate the test section at different times. The object wave taken without heat transfer in the test section is superimposed upon the second where the heat transfer process (or any other process of interest) is going on. By this method the differences in the optical paths between the waves are superimposed upon each other. This difference in the optical path expressed in the multiple S of a wavelength is calculated as follows:

$$S(x,y)\lambda = l\,[n(x,y)_2 - n(x,y)_1] \tag{3.1}$$

In (3.1) l is the length of the test section in which the refractive index n is changed due to varying temperatures or concentrations. For a simplified evaluation it is assumed that the distribution of the refractive index $n(x,y)$ is two–dimensional during the recording of both waves, i.e. the refractive index is assumed to be constant along the path of each light beam through the test section. Equation (3.1) shows that only local changes can be detected at first. Only if the distribution of the refractive index $n(x,y)_1$ during the recording of the comparison wave is known is it possible to derive absolute values of the temperature distribution from the interferogram. Therefore, constant temperature is usually applied during the recording of the reference wave and this results in a constant refractive index over the whole volume of the test section.

$$S(x,y)\lambda = l\,[n(x,y)_2 - n_\infty] \tag{3.2}$$

To obtain the absolute values of the temperature field the local temperature at one position in the test section must be measured with another method; for example, with a thermocouple. This position is usually chosen in an area which is not or only slightly affected by the heat transfer. However, the surface of the wall emitting or absorbing heat can also be used as a measuring position. Equation (3.2) represents the correlation of the ideal interferometry. It is assumed therein that the light beam travels along a straight line through the test section. However, since it penetrates a thermal boundary layer, this is not the case. The light beams are deflected due to the gradients in the refractive index of the fluid. This deflection or defraction is used as a measuring signal in the Schlieren and shadowgraph methods.

This deflection of the light can be recalculated in an additional phase shift ΔS by assuming an approximately linear distribution of the refractive index within this small region:

$$\Delta S = \frac{n_0 \lambda l}{12 b^2} \tag{3.3}$$

In (3.3) b represents the fringe distance and n_0 is the refractive index averaged over this region.

In many applications of this method an ideal two–dimensional temperature field does not exist. Often the boundary layer is expanded over the ends or edges of the heated wall or temperature changes occur along the path of the light beam. Therefore, only integral values can be measured. Possibilities to correct these influences are explained in Chap. 4.

A variation of the refractive index, represented by the interference fringes, can be recalculated in a field of density distribution. The correlation for doing this is given by the Lorentz–Lorenz equation

$$\frac{n^2 - 1}{n^2 + 2} \, \frac{1}{\rho} = \frac{N}{M} \tag{3.4}$$

in which N stands for the molar refractivity and M for the molar mass.

This equation can be simplified for gases having a refractive index n very near to unity and one obtains the Gladstone–Dale equation.

$$\frac{3}{2}(n - 1)\frac{1}{\rho} = \frac{N}{M} \tag{3.5}$$

If there is a pure substance in the test section and if the pressure is nearly constant then the changes in the density are only caused by temperature variations. In this case the density is a function only of the temperature. If, in addition, the fluid in the test section is a gas, one can use the equation of state for ideal gases at moderate pressures, and one then obtains (3.6) which correlates the phase shift – or alternatively the field of interference fringes – with the temperature distribution

$$T(x, y) = \left(\frac{2S(x, y)\lambda R}{3Nlp} + \frac{1}{T_\infty} \right)^{-1} \tag{3.6}$$

For other fluids such as liquids other correlations for n(T) are available (e.g. the equation by Tilton and Taylor for water as a test fluid, s. Chap. 4.3.3).

Usually the temperature distribution h is of interest. In this case the temperature gradient at the wall has to be determined from the fringe field in the immediate neighborhood of the wall. This can be done by approximating the interferomatically measured temperature distribution at the wall with a simple equation, for example by an exponential function or a polynomial, and by differentiating this equation. The velocity of the fluid is zero at the wall, which indicates pure heat conduction there. With the well–known definition of the heat transfer coefficient h, the Fourier equation and with the condition that the heat fluxes predicted by both correlations must be equal one obtains

$$h = \frac{-k_{fluid}(\frac{\mathrm{d}T}{\mathrm{d}y})_{wall}}{T_w - T_\infty} \tag{3.7}$$

In many cases the additional phase shift due to light deflection, as formulated in (3.3), is small. However, with very high temperature gradients in the boundary layer this phase shift can have comprehensive values. Large measuring errors can also occur if the test object is not correctly brought into line with the object beam. Special care has to be taken that during the recording of both object waves, before and after initiating the heat transfer process, the same global thermodynamic states are prevalent. This means that the test section is under pressure and that the flow is circulating during the first exposure. Otherwise additional interference fringes are produced, for example, by bending of the windows due to pressure which has nothing to do with the heat transfer but which can be misinterpreted.

3.5.4 Finite Fringe Method

As mentioned above, in temperature fields with very high gradients, the deflection of the laser beam cannot be neglected, as it is done in the evaluating procedure described above. High temperature gradients are for example found in the liquid boundary layer around condensing vapour bubbles. In such a case a complicated correction procedure for this deflection has to be used which is described by Nordmann and Mayinger [487], by Chen [488] and by Chen and Mayinger [489]. With the equations and corrections described there, even temperature fields around very tiny bubbles can be evaluated.

With high heat transfer coefficients the boundary layers at the phase-interface are usually very thin – in the down to a few hundredths of a millimeter – and it is difficult to investigate them with the interferometric procedure described up to now, because only a few interference fringes would be observed within this narrow area. Therefore here another interferometric method has to be used, the so called "finite fringe method".

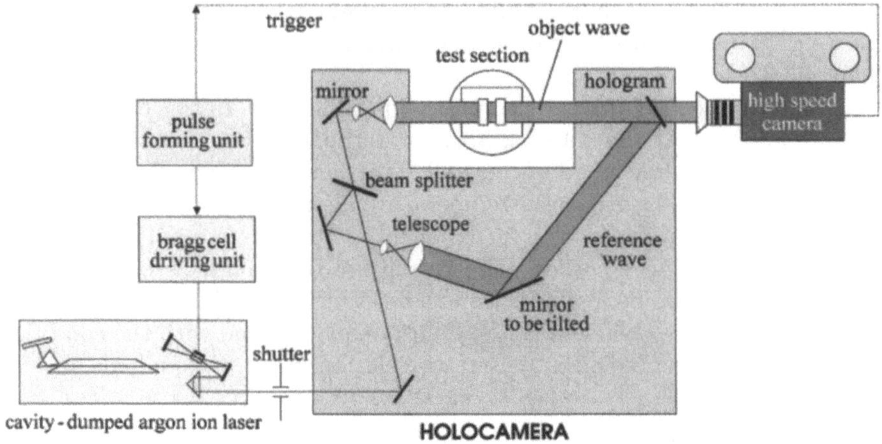

Fig. 3.14. Arrangement for holographic interferometry with finite fringe method

In this method, after the reference hologram was produced, a pattern of parallel interference fringes is created by tilting the mirror in the reference wave in Fig. 3.14, or by moving the hologram there within a few wavelength. The direction of the pattern can be selected as one likes and is only depending on the direction of the movement of the mirror or of the holographic plate.

This pattern of parallel interference fringes is then distorted by the temperature field due to the heat transfer process. The distortion or deflection of each fringe from its original – parallel – direction is, in a first rough approximation the temperature gradient and gives, by using (3.7), the heat transfer coefficient. An imagination how these interference fringes look like gives Fig. 3.15. This figure demonstrates the temperature field around a burning flame. Depending on the orientation of the parallel fringes, the distortion varies. To get most informative pattern, the parallel fringes should be orientated in the direction of the largest temperature gradients. In Fig. 3.15 the interferogram a) presents an infinite fringe pattern, whilst the interferograms b) – d) are of finite fringe nature with vertical, diagonal and horizontal orientation of the parallel fringe pattern.

Analyzing non planar temperature fields, like the above mentioned boundary layer around a condensing bubble, also with the finite fringe method the before mentioned Abel-correction has to be applied, which was described in detail by Hauff and Grigull [21]. This correction considers the different lengths of the path of the light traveling near the surface of a curved body.

The capacity of the finite fringe method is demonstrated in Fig. 3.16, which shows the boundary layer t the surface of a steam bubble created with subcooled boiling at the superheated wall and condensing in the cold

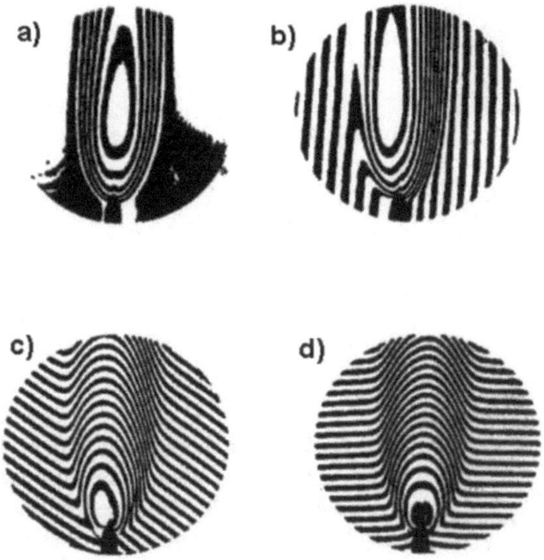

Fig. 3.15. Finite fringe interferograms (temperature field in a flame)

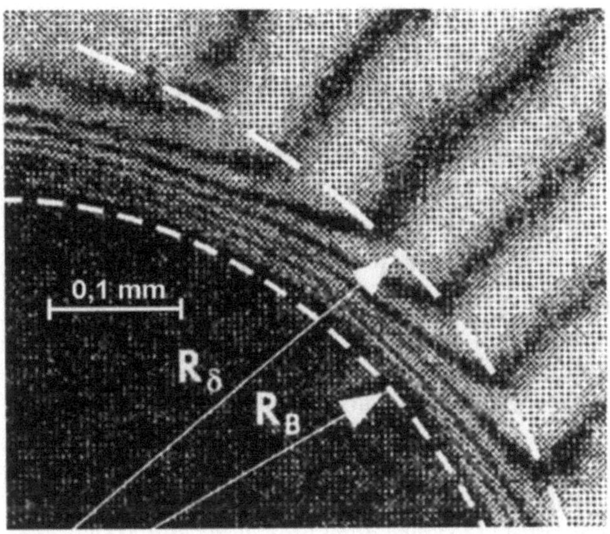

Fig. 3.16. Finite fringe interference pattern in the boundary layer of a condensing steam bubble

bulk flow of the water. In this figure R_B marks the radius and by this the true surface of the bubble and the dotted line at R_δ the outer limit of the boundary layer around the bubble.

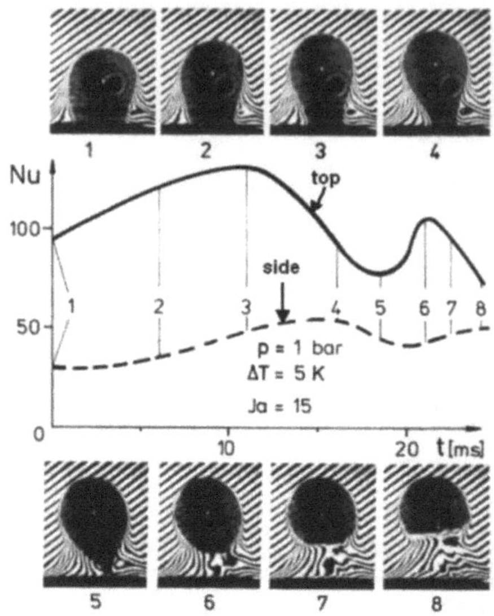

Fig. 3.17. High speed cinematography of finite fringe interferograms around a condensing bubble and the evaluated heat transfer at the phase-interface

By additionally using the high speed cinematography the temporal course of the temperature in the boundary layer can be observed. Figure 3.17. demonstrates the possibilities of using these techniques in a flow with a bubble condensing in a liquid. The heat transfer around the condensing bubble can be inertialess and noninvasive monitored by these methods. Holographic interferometry certainly can only be used if the flow situation is not too complicated and if the bubble population is not too numerous, so that individual bubbles can be identified. It is not possible to look inside the bubbles, because the light is totally reflected at the phase-interface.

3.6 An Interference Method for Simultaneous Heat and Mass Transfer

If the refractive index is simultaneously influenced by more than one parameter, for example by temperature and concentration, the interferogram cannot be evaluated directly. Therefore, combined methods are usually used for investigating simultaneous heat and mass transfer processes; for example, by combining the interferometric method with a classical measurement of the temperature, using thermocouples. This was done by Adams and McFaddens [15]. El–Wakil et al. [19] assumed that the temperature and concentration fields are similar or even identical to a large degree, which is only the case if the Lewis number is 1.

There are, however, also attempts to use the dependency of the refractive index on the wavelength of the light. This can be done by recording two interferograms originating from the light of two different wavelengths and from that to evaluate the temperature and the concentration field separately.

El–Wakil and Ross [18] first proposed this two–wavelength method and used it in a modified Mach–Zehnder interferometer to investigate the evaporation and combustion of liquid fuel. The accuracy achieved, however, was not satisfactory because both interferograms could not be superimposed accurately enough. The ability of holography to store a number of interferograms on one and the same photographic plate promised an essential improvement for this two–wavelength method. Panknin [25] took up this method and achieved very good results with a two–wavelength holographic interferometer designed and constructed by him.

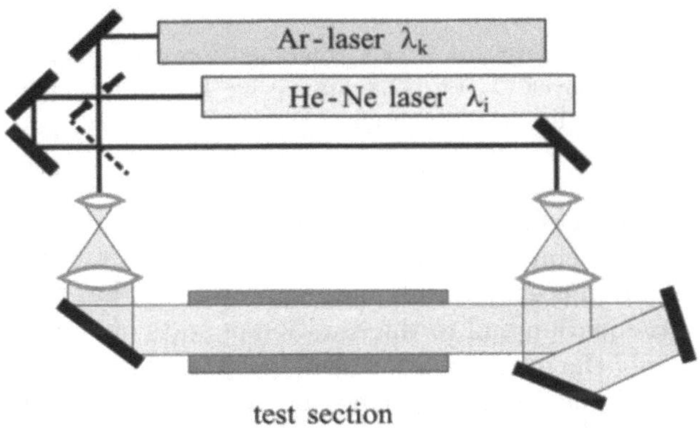

Fig. 3.18. Optical set–up for holographic two–wavelength interferometry

Fig. 3.18 shows the principle of the holographic set–up. It is similar to the arrangement shown in Fig. 3.2 to a great extent. The main difference is only that two lasers are used as light sources. The first laser is a He–Ne laser with a wavelength of $\lambda_j = 632.8\ nm$ and the other is an argon laser with a wavelength of $\lambda_k = 457.9\ nm$. The first operates in the red and the second in the blue spectrum of visible light. Both laser beams intersect with each other which makes an optical shutter necessary to guarantee exactly the same illumination time for both lasers. The beams are then superimposed and expanded in a beam expander. By doing this one gets two object and two comparison waves of different wavelengths.

It must be stated that the primary object wave λ_j is not the only wave reconstructed by the reference wave λ_j. A false secondary object wave λ_k is also reproducaed by the reference wave. This unwanted wave leaves the hologram at a different angle, however, and it can therefore be easily distinguished from the original wave.

A simple example is illustrated by the interferograms given in Fig. 3.19 (left side) as recorded from the boundary layer at a vertical heated plate whose surface was covered with naphtalin The interference fringes in the boundary layer are therefore a result of combined heat and mass transfer. Both fields of interference fringes show clear differences in the numbers and in the distances of the fringes.

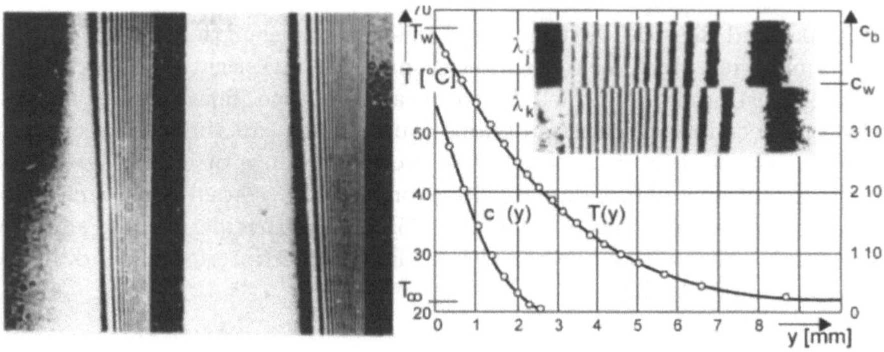

Fig. 3.19. left: Interferograms for investigation of a combined laminar temperature and concentration boundary layer ($\lambda_j = 632.8\ nm$, $\lambda_k = 457.9\ nm$), right: Temperature and concentration profiles in the boundary layer

The evaluation of these holograms is relatively easy with gases because one can use the Gladstone–Dale equation and the equation of state for ideal gases. In principle the evaluation is similar in liquids. However, one has to go back to the Lorentz–Lorenz equation and to the equation of state of the real substance. For simplicity only the evaluation for ideal gas is explained here. With the equations mentioned the following correlation between phase shift, wavelength and temperature of the mixture results in

$$S(x,y)\lambda = \frac{3pl}{2R} N_m \left[\frac{1}{T(x,y)} - \frac{1}{T_\infty} \right] \tag{3.8}$$

The molar refractivity N_n for the mixture consisting of two different gases can be calculated by using a simple mixing equation

$$N_m = N_a c_a + N_b c_b \quad \text{with } c_a + c_b = 1 \tag{3.9}$$

In (3.9) N_a and N_b stand for the molar refractivities of the pure substances and c is the concentration of the respective component in the mixture. It is assumed that during recording both interferograms, the temperature in the test section at a location far away from the heat and mass transfer process is constant and that there exists only one component, for example, the component a. By combining (3.8) and (3.9) one gets for each wavelength

$$S(x,y)\lambda = \frac{3pl}{2R} \left[\frac{1}{T(x,y)} (N_a + c_b(x,y))(N_b - N_a) - \frac{N_a}{T_\infty} \right] \tag{3.10}$$

and by eliminating the concentration $c_b(x,y)$ one can calculate the temperature $T(x,y)$

$$S_j(x,y)\frac{\lambda_j}{N_{b_j}-N_{a_j}} - S_k(x,y)\frac{\lambda_k}{N_{b_k}-N_{a_k}} \tag{3.11}$$

$$\frac{3pl}{2R}\left[\frac{1}{T(x,y)}-\frac{1}{T_\infty}\right]\left[\frac{N_{a_j}}{N_{b_j}-N_{a_j}}-\frac{N_{a_k}}{N_{b_k}-N_{a_k}}\right] \tag{3.12}$$

After determining the temperature only one interferogram is necessary to evaluate the concentration profile. Equation (3.12) shows that the difference between the phase shifts of both interferograms can be used as a measure for the temperature. This difference usually is very small and, therefore, it is advisable to use wavelengths for interferometric methods which lay as far apart as possible. Also the dependency on $N(\lambda)$ is small and reaches larger values only in the neighborhood of an absorption line which, however, is not in the region of visible light for most substances. This cuts down the application of the method with respect to the use of possible fluids. The molar refractivity of the used fluid must be sufficiently different for the wavelengths used. Some fluids, such as naphthalene, benzene and hexane, are well suited to this technique. The position of the fringes must be determined very exactly. Usually it is necessary to bring the error of reading below one micrometer. On the other hand additional phase shifting does not affect evaluation to a large extent, as discussed in the section on one–wave interferometry. The reason for this is that both waves are affected by such errors to the same extent and, therefore, the difference in the interferograms is minimal.

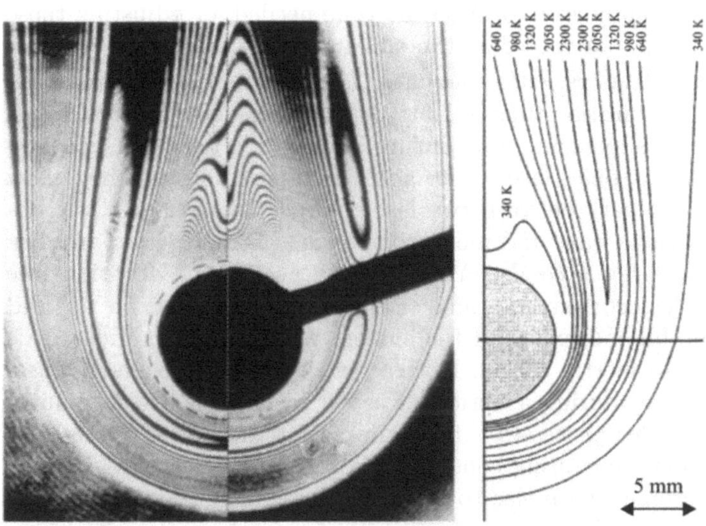

Fig. 3.20. left: Simultaneously taken interferograms (wavelength: $\lambda_k = 457,9\ nm$ (left) and $\lambda_j = 632,8\ nm$ (right)) of a flame (infinite fringe field adjustment), right: Temperature distribution (isotherms)

Figure 3.19 (right side) shows the evaluation of the combined heat and mass transfer from a heated vertical plate covered with naphthalene. To clarify the differences in the phase shift both interferograms are superimposed in this figure. This can easily be done during the recording in such a way that one fades out one wave below a certain level of interest and above that level the other one. By doing this the differences in the phase shift can be observed directly on the hologram similar to a nonius scale. In the diagram of the figure the curves for the temperature and the concentration profile near the wall are drawn as calculated by the method mentioned above. Finally, Fig. 3.16 shows an application of this two–wavelength method for combustion processes in a flame. In these figures hexane used as fuel was seeping out of a porous horizontal cylinder and evaporating on its surface. Figure 3.16 (right side) gives the temperature distribution in this flame which was evaluated from Fig. 3.20 (left side).

3.7 Comparison with Classical Methods

Contrary to conventional interferometric methods such as Mach–Zehnder or Michelson methods, holographic interferometry offers a number of essential advantages. Holographic interferometry allows a much easier experimental technique with equal accuracy yet requires less cost and experimental effort. This is because the optical components are cheaper to produce as they do not need very precise manufacturing. The glasses, lenses and mirrors do not need to be very accurately ground. Also the time needed for adjusting these optical components and for recording the holograms is greatly reduced.

The use of pulsed lasers with adequate coherence offers further possibilities. With a pulse duration time of 10 to 30 ns, ultrashort time processes, for example those in supersonic flow during boiling or cavitation, can also be investigated. Double pulse lasers offer additional possibilities. They allow the time difference to vary between two laser pulses, thus between two object waves – comparison wave and measuring wave – from a period of a few microseconds up to a few seconds. From this the need for a vibration–free support of the optical components and of the test section can be reduced because a considerable movement of the system within a few microseconds is not possible due to inertia effects. If there are any pulsed lasers giving a great number of pulses during a period of a few seconds or a few minutes very rapid heat transfer processes – for example, at the phase interface of bubbles, during boiling or during condensing – can be investigated. Finally, holographic interferometry enables to record a three–dimensional temperature and concentration field by an arrangement of optical components where the test section is irradiated from different directions. To evaluate such holograms holographic interferometry has to be combined with tomography [24] (s. Chap. 15).

Electronic data evaluation and numerical methods for integrating even very complicated systems of differential equations have reached a very high degree of development. Correlations for describing heat and mass transfer usually consist of two sets of equations having a different nature, namely balance equations formulating the laws of conservation mathematically and constitution equations describing the transport processes more or less empirically. Especially when formulating constitution equations holographic interferometry provides not only very precise and useful information but also represents a very sensitive touchstone for the physical access of such constitution equations. The interferogram allows a detailed insight into the boundary layer and its temperature distribution and a computer program describing this temperature distribution will also precisely predict the heat transport. The image–forming method of interferometry provides immediately comprehensible idea of the physical process to the theorist, and therefore facilitates to choose the right mathematical tools.

4 Holographic Interferometry

Robert Tauscher

4.1 Introduction

Holography can be used to store the amplitude and phase distribution of wave fronts on a photographic plate. In order to take a hologram the object is illuminated by coherent light. The diffuse reflected light can be stored on the hologram and reconstructed by illuminating the hologram with the reference beam.

Two coherent waves propagating through a common medium are able to interfere with each other,ie their amplitude vectors are added. As a result of this superimposition their amplitudes are amplified or extinguished depending on the phases of both waves. In practice there is a distinction between interferometry where one wave is reflected by an opaque object allowing a measurement of path changes (Michelson interferometry) and interferometry where the object wave penetrates a transparent test medium with a changing refractive index distribution (Mach–Zehnder interferometry).

Holographic interferometry is a combination of the techniques of interferometry and holography. The distribution of phase and amplitude of the undistorted object wave, called a comparison wave in the following, is stored on a hologram and can be reconstructed after its development by an illumination with the reference beam. The heat or mass transfer is then activated, so that the momentary object beam experiences an additional phase shift.The object and the comparison wave beam then interfere behind the hologram describing the investigated physical phenomenon.

4.2 Components of a Holographic Interferometer

In this chapter advice is given on the selection of the components required to build a holographic interferometer The optical set–up of a commonly used holographic interferometer for examination of transparent media is shown in Fig. 4.1.

4.2.1 Light Source

Lasers are the most commonly used light source for holographic interferometry. Lasers make possible an interference of the reference and object beam.

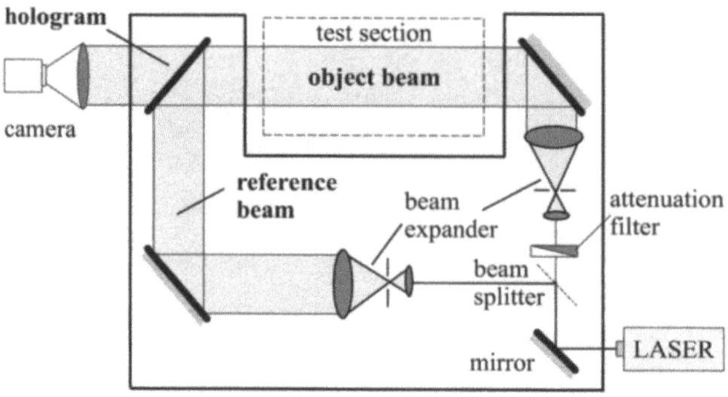

Fig. 4.1. Optical set–up of a holographic interferometer for the investigation of transparent media

Gas lasers are the most suitable light source for continuous measurements because of their long coherent length. Helium neon lasers, for example, are simple constructions, but with a comparable low power. Argon ion lasers operate with a higher light power and thus are are more expensive to use. A frequency stabilizing etalon should be integrated to enlarge the coherence length. As a result of the high light power, up to several watts, a water–cooling of the plasma tube is necessary. In practice a laser wavelength with the highest light power is chosen. The required power depends on the optical equipment, i.e. the diameter of the expanded light beams; and on the time dependence of the investigated physical phenomenon, as a high light power leads to a low exposure time. It is adventageous to have power over the light source, since the exposure time can then be adapted to the optical set–up and photographic plates. The laser should be fixed on a separate table in order to avoid vibrations of the optical set–up that may be caused by the cooling system of the laser.

4.2.2 Optical Table

The mechanical stability of the optical set–up is essential to the quality of a hologram. During the exposure the difference between both optical paths should change less than $\lambda/10$ at the hologram plane. Experience shows that neither the laser nor the test facility should be mounted on the optical table. Vibrations of the optical table introduced by vibrations of the ground should also be minimized. One possibility is to use a heavy rack or table plate with a high damping factor and a low eigenfrequency. Additionally, the table should be positioned on air filled damping elements. The optical components must be fixed on the table, so that the interferometer can be adjusted to the test facility. One method for fixing components is to use a magnetic plate. A second method is to drill a grid of holes in the table, where the components can be tightly screwed in place.

4.2.3 Shutter

A shutter provides a precise control of the exposure time of the holographic plate. Shutters commonly used for photography can also be used for lasers. The shutter time must be variable in a range between a few hundredths and a few seconds. Shutter times can be determined by measuring the light intensity in the hologram plane with a common light meter.

4.2.4 Beam Splitter

The laser beam is divided into two beams, the object beam and the reference beam, by a (variable) beam splitter, a prism or a semipermeable mirror. Usually the ratio of the reflected and transmitted beam is adjusted by a sheet of glass that is coated with a reflective film with a changing thickness. Another possibility is to apply a polarization beam splitter. In this case the direction of the polarization of one beam must be retarded by a quarter wave plate so that both waves are able to interfere.

4.2.5 Attenuation Filter

A (variable) attenuation filter can be placed in the object and/or in the reference beam so that its intensity can be weakened during the measurements. The intensity of the stored comparison wave is less than that of the object beam. By varying the transmission of the attenuation filter the intensity of both waves should be brought into line with each other in order to obtain a high contrast.

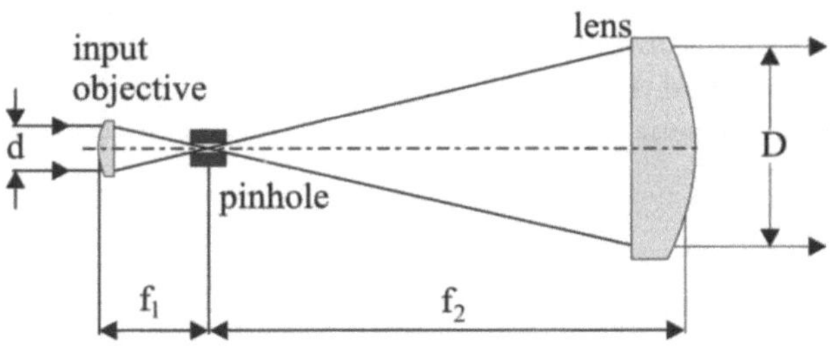

Fig. 4.2. Beam expander

4.2.6 Beam Expander

Due to the small laser beam diameter both the object and the reference beam must be expanded to allow an illumination of larger objects. A beam expander or collimator consists of an input objective, a pinhole and a lens (s. Fig. 4.2). The pinhole (diameter approx. $10-30$ μm) is placed at the focus of the objective. This spatial filter provides a simple way to block out random fluctuations (spatial noise) from the intensity profile of the laser beam.If the focus plane of the input objective and the lens is the same, the diverging beam is brought parallel. The effect of a spatial filter is shown in Fig. 4.3.

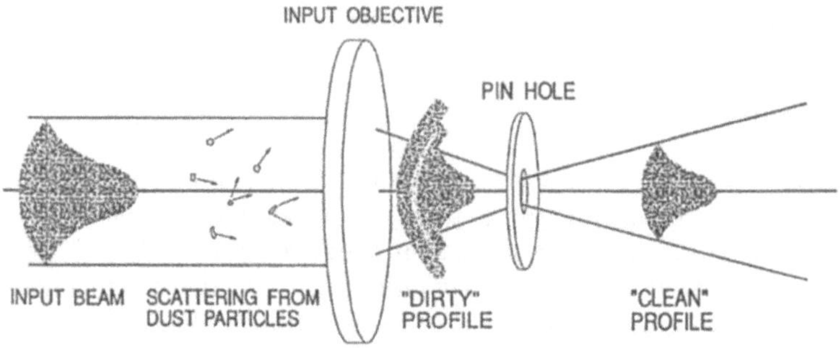

Fig. 4.3. Blocking out noise by a pinhole

4.2.7 Mirrors, Lenses

Aberrations caused by inhomogenities of the optical components introduce distortions in both the measuring and the comparison wave which are eliminated, since interference patterns result only from a difference of their optical paths. For the test section streak free sheet glass is recommended. Acrylic glass with a large coefficient of volume expansion should be avoided because a change of temperature would cause additional interference fringes.

Generally as few mirrors as possible should be used in order to avoid light power losses through reflection and absorption; and to enhance mechanical stability. A problem free adjustment of the laser beams in relation to the collimators should be also possible.

4.2.8 Recording Materials

The photographic layers must be be sensitive to the wavelength of the laser. Also the recording material must be suitable for high spatial frequencies which require a very fine grain. The carrier frequency of the microscopic interference pattern is dependent on the wavelength and on the angles β_o of the object and β_r of the reference beam measured from the perpendicular of the hologram plate. According to Fig. 4.4 it can be calculated by:

$$f_r = \frac{\sin\beta_o - \sin\beta_r}{\lambda} \tag{4.1}$$

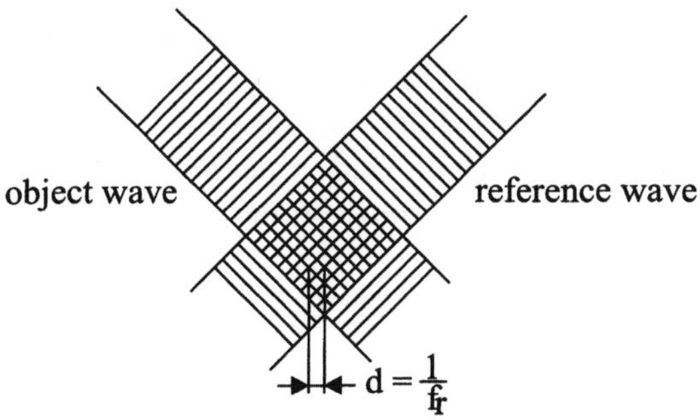

Fig. 4.4. Calculation of the carrier frequency

This leads to a line density of 2750 *lines/mm* for a simple but commonly used arrangement with $\beta_o = 45°, \beta_r = 45°$ and $\lambda = 514\ nm$. As a result of the high resolution the sensitivity is too low.

Photographic materials with silver halide emulsions are commonly used for holography. Photographic plates should be selected for film materials because of their mechanical stability.

The processing procedure for phase holograms consists of development, fixation, rinsing, bleaching, rinsing and, in order to avoid water sediments, wetting. A wide number of recommended baths exist, especially for bleaching. The user should refer to the manufacturer's instructions according to the type of holgraphic plates used.

Reflections at the back of the hologram plate cause an interference of the reflected light with the incoming light and result in additional interference fringes.These reflections are also reconstructed by the reference beam and thus disturb the macroscopic interference pattern. This can be avoided by an anti–reflection coating of the rear glass surface or by rotating the hologram by the Bragg angle. Deviations of the reconstructed object beam due to a shrinking of the hologram during drying is minimized by an adjustment of the plate normal to the bisector of the angle between the object and the reference beam [40].

Other materials such as thermoplastic materials or ferroelectric crystals can be used instead of silver halide emulsions. They allow a fast and dry development without chemicals; but the maximum spatial frequency and therefore the contrasts, are very low.

4.2.9 Piezo Mirror

In order to avoid complicated *in situ* development, the hologram plate must be repositioned on to the plate holder after development. Due to the fact that an exact reposition in a range of half of the wavelength is nearly impossible, it is necessary to rotate the reference beam by a fine adjustable mirror which can be positioned by a Piezo–Quartz device. As a result the direction of the reconstructed object beam is also changed. The exact position of the reconstructed object beam is identical to the comparison wave and is reached when no interference line can be seen behind the hologram plate (infinite fringe field adjustment). Straight interference lines can be preadjusted by rotating the mirror by a small angle (finite fringe field adjustment). In this case the local heat transfer can be determined from the derivations of the interference lines.

4.2.10 Test Facility

The interference field is a result of the phase shift of the object wave caused by a change of the optical refractivity, the path length, or both. The length of the test section influences the number of the interference lines. A low number leads to a large evaluation error. A test section which is too long results in a high deviation of the laser beam according to Snell's law. In the area of interest, for example the boundary layer, approximately 5 to 10 lines should exist.

4.3 Evaluation of Interferograms

4.3.1 Theoretical Principles

In Fig. 4.5 the interferogram of a boundary layer with an infinite fringe field can be seen as left–handed, whereas its origin is shown as right–handed. The deformed wavefront of the measuring beam (only one wave is shown) intersects with the plane wavefronts (distance λ) of the undistorted comparison wave. In this case the interference pattern can easily be interpreted as a field of constant density, and therefore approximated as lines of constant temperature or concentration. As shown later, the heat flux in the vicinity of walls can be calculated from the distance of the interference lines. For this reason a valuable evaluation is possible only if enough lines exist.

In order to generate an interferogram with an overlaid finite fringe field the planes of the comparison wave are rotated by an angle relative to the momentary object wave (Fig. 4.6). The interference order is now a function of the deflection of the preadjusted interference line. An advantage of the finite fringe field adjustment is the easy evaluation of small or irregular objects, for example condensing bubbles.

Determining the temperature or concentration distribution the phase shift can be calculated as a multiple S of the wavelength λ of the laser light:

$$S \cdot \lambda = \int_0^l (n - n_c) \mathrm{d}z \tag{4.2}$$

Passing the temperature or concentration field, the light beam is deflected according to Snell's law. On the condition that the derivative of the refractive

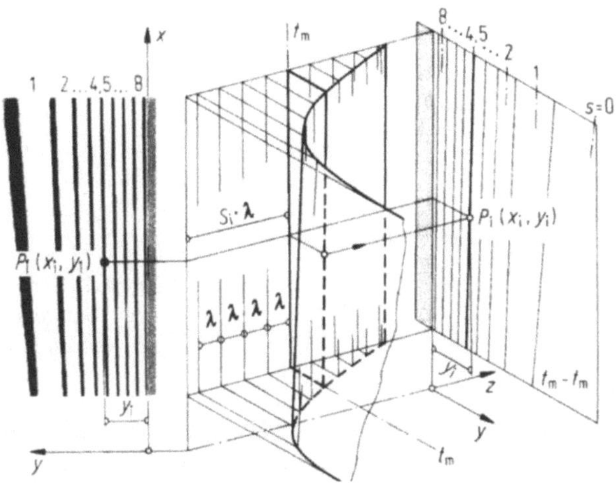

Fig. 4.5. Origin of a interferogram with a infinite fringe field adjustment

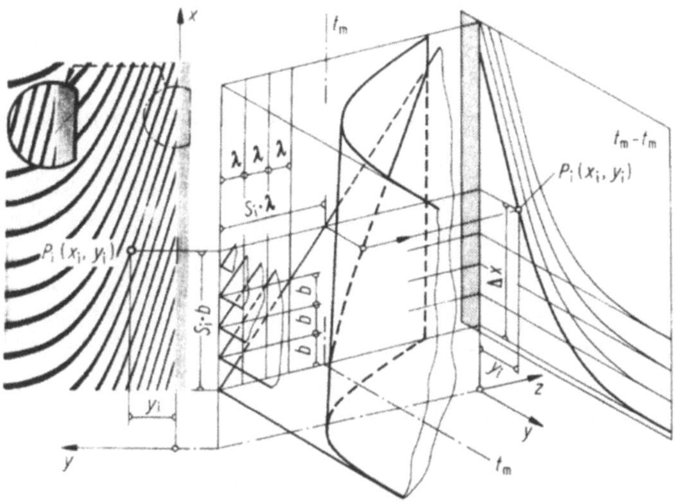

Fig. 4.6. Origin of a interferogram with a finite fringe field adjustment

index is constant perpendicular to the wall, the light beam is following a parabolic curve [34].

In Fig. 4.7 the paths of comparison and object beam are drawn. They enter the test section under an angle of ϵ and ϵ_0 (s. Fig. 4.8), respectively, and interfere on the photographic plate. The comparison wave passes straight through the unheated test facility because of the constant refractive index n_c. The object beam experiences a derivation as a result of the refractive index distribution. The straight object beam intersects the comparison beam

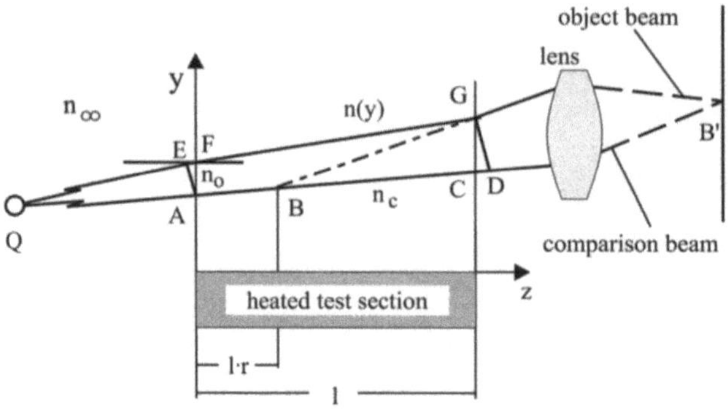

Fig. 4.7. Phase difference between object and reference beam

at point B. The optical paths of both beams from B to B′ and therefore the optical paths $\overline{GB}'$ and $\overline{DB}'$ have the same length assuming an ideal lens. The distances of points E and A to the light source Q are the same as well. Now (4.2) can be written as:

$$S\lambda = \int_F^G n(s)\mathrm{d}s - \overline{AC}n_v + (\overline{EF} - \overline{CD})n_\infty \tag{4.3}$$

For small deflections of the light beam the refractive index $n(y)$ is approximated by:

$$n = n_0 + n'(y - y_0) \quad \text{for} \quad 0 \leq z \leq l \tag{4.4}$$

with:

$$n' = \frac{\mathrm{d}n}{\mathrm{d}y} = const \quad \text{and} \quad n_0 = n_F \quad \text{for} \quad y_0 = y_F \tag{4.5}$$

so that the integral in (4.4) is calculated by:

$$\int_F^G n(s)\mathrm{d}s = \int_{z=0}^{z=l} (n_0 + n'(y - y_0))\sqrt{1 + y'^2}\mathrm{d}z \tag{4.6}$$

with:

$$y' = \frac{\mathrm{d}y}{\mathrm{d}z} \tag{4.7}$$

A relation between the y–coordinate and its derivation y' is received from Snell's law according to Fig. 4.8:

$$\frac{n}{n_0} = \frac{\cos\epsilon_0}{\cos\epsilon} = \sqrt{\frac{1 + \tan^2\epsilon}{1 + \tan^2\epsilon_0}} = \sqrt{\frac{1 + y'^2}{1 + \epsilon_0^2}} \tag{4.8}$$

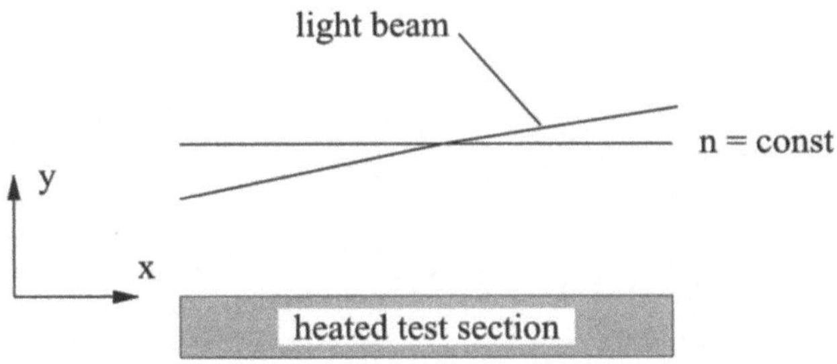

Fig. 4.8. Deflection of the laser beam in the temperature field

with which the differential dz in (4.6) can be determined:

$$\mathrm{d}z = \frac{\mathrm{d}y}{y'} = \frac{\mathrm{d}y}{\sqrt{(1+\epsilon_0^2)(1+\frac{n'}{n_0}(y-y_0))^2 - 1}} \tag{4.9}$$

With a Taylor row:

$$\left(1+\frac{n'}{n_0}(y-y_0)\right)^2 \approx 1 + 2\frac{n'}{n_0}(y-y_0) \tag{4.10}$$

the following integral can be derived:

$$\int_0^z \mathrm{d}z = \int_{y_0}^{y_z} \frac{\mathrm{d}y}{(1+\epsilon_0{}^2)(1+2\frac{n'}{n_0}(y-y_0)) - 1} \tag{4.11}$$

After integration the following relation is received:

$$y(z) - y_0 = z\epsilon_0 + \frac{n'}{n_0}\frac{z^2}{2} \tag{4.12}$$

and

$$y'(z) = \epsilon_0 + \frac{n'}{n_0} z \tag{4.13}$$

In (4.6) the square root $\sqrt{1+y'^2}$ is now approximated by a Taylor row:

$$\sqrt{1+y'^2} \approx 1 + \frac{1}{2}y'^2 \tag{4.14}$$

so that the following equation for the integral in (4.6) is received:

$$\int_F^G n(s)\mathrm{d}s = n_0 l + \frac{1}{2}n_0 l \epsilon_0{}^2 + n_0\frac{n'}{n_0} l^2 \epsilon_0 + \frac{1}{3} n_0 \frac{n'^2}{n_0^2} l^3 \tag{4.15}$$

The optical path of the comparison beam is now calculated on the condition that the refractive index outside the test chamber n_∞ is equal to the one in the test section n_c during the exposure

$$\overline{AC}n_c + (\overline{CD} - \overline{EF})n_c = (\overline{AC} + \overline{CD} - \overline{EF})n_c \tag{4.16}$$

with:

$$\overline{AC} = \frac{l}{\cos\epsilon_0} = l(1+\frac{1}{2}\epsilon_0^2) \tag{4.17}$$

$$\overline{CD} = l(1-r)(\frac{1}{\cos\epsilon_1} - \frac{1}{\cos\epsilon_0}) = l(1-r)\frac{1}{2}(\epsilon_1^2 - \epsilon_0^2) \tag{4.18}$$

$$\overline{EF} = \sin\epsilon(y_F - y_A) = \epsilon(y_F - y_A) \tag{4.19}$$

The cosine of the very small angle ϵ is substituted by a Taylor row with two elements and the sine by the angle itself. The different coordinates y are calculated by (4.12) and set into (4.16):

$$(\overline{AC}+\overline{CD}-\overline{EF})n_c = n_c l + \frac{1}{2}n_c l {\epsilon_0}^2 + \frac{1}{2}n_c\frac{n'}{n_0}l^2{\epsilon_0}^2 + n_c\frac{n'^2}{n_0^2}l^3(\frac{1}{2}-\frac{r}{2}) \tag{4.20}$$

This relation is set into (4.3). Terms of less orders of the factor $(n_0 - n_c)$ are neglected, leading to the following relation for the phase shift:

$$S\lambda = (n_0 - n_c)l + \frac{1}{2}n_c\frac{n'}{n_0}\epsilon_0 + \frac{1}{3}n_c\frac{n'^2}{n_0^2}l^3 \tag{4.21}$$

The phase shift is seen at y_B, so that (4.21) must be calculated as a function of the refractive index at point B:

$$(y_F - y_B) = (y_0 - y_B) = \frac{n'}{n_0}l^2(\frac{1}{2} - r) - lr\epsilon_0 \tag{4.22}$$

The refractive index n_B at y_B is determined by (4.3),

$$n_0 = n_B + n'(y_0 - y_B) \tag{4.23}$$

so that the refractive index n_B is given by:

$$n_B = n_0 - n_0\frac{n'^2}{n_0^2}l^2\left(\frac{1}{2} - r\right) + n'lr\epsilon_0 \tag{4.24}$$

The entrance angle ϵ_0 is calculated by Snell's law:

$$\epsilon_0 = \epsilon\frac{n_\infty}{n_c} \tag{4.25}$$

so that the following relation for the phase shift can be derived:

$$S\lambda = (n_B - n_c)l + \frac{n'}{n_c}l^2\epsilon\left(\frac{1}{2} - r\right) + \frac{n'^2}{n_c}l^3\left(\frac{1}{3} - \frac{r}{2}\right) \tag{4.26}$$

4.3.2 Conclusions

The effect of the entrance angle ϵ is taken into consideration by the second term of (4.26). It can be seen that a parallel adjustment of the laser beam to the test section is necessary in order to eliminate this term. It is useful in practice to place an additional mirror between the collimator and the test section. A rough preadjustment of the test facility should be possible as well.

This adjustment is often very difficult, especially for long test sections. For parallel object waves an oblique entrance angle results in an additional phase shift:

$$\Delta y = \epsilon l(\frac{1}{2} - r) \tag{4.27}$$

which disappears when focussing at the middle of the test section ($r = \frac{1}{2}$).

The third term of (4.26) stands for the phase shift caused by the light deflection. Its influence also depends on focusing and vanishes when focusing at $r = \frac{2}{3}$. Other cases concerning the fault, for example a wrong adjustment of the collimators, are described by [40].

With (4.26) the assumptions of ideal holography can be shown:

1. The laser beam enters the test section parallelat a parallel, and is not deflected.
2. The refractive index distribution is only two–dimensional $n(x, y)$ and does not change along its path through the test section.

On this condition the second and the third term of the equation vanish so that (4.26) is simplified to the equation of ideal holography:

$$S\lambda = \Delta n(x, y)l \tag{4.28}$$

4.3.3 Calculation of Temperature and Concentration Distributions

With a infinite fringe field adjustment interference lines can be interpreted as lines of constant refractive index. A relation between the refractive index n and the density ρ is given by the Lorentz–Lorenz equation:

$$N(\lambda) = \frac{n(\lambda)^2 - 1}{n(\lambda)^2 + 2}\,\frac{M}{\rho} \tag{4.29}$$

The molecular refractivity N is a substance specific constant, independent from pressure and temperature. For a refractive index near 1, i.e. for most gases, (4.29) can be simplified to the Gladstone–Dale equation:

$$N(\lambda) = \frac{2}{3}(n(\lambda) - 1)\frac{M}{\rho} \tag{4.30}$$

For ideal gases the density ρ is determined by the equation of the state of air so that the refractive index is calculated from the temperature:

$$N(\lambda) = \frac{2RT}{3p}(n(\lambda) - 1) \tag{4.31}$$

Putting this equation in the equation of ideal holography, the temperature distribution is given as a function of the phase shift:

$$T(x,y) = \left(\frac{2S(x,y)\lambda R}{3Npl} + \frac{1}{T_\infty}\right)^{-1} \tag{4.32}$$

with: $S < 0$ heating $S > 0$ cooling

If a temperature at any location is known it can be measured, for example by a thermocouple. The temperature field in the test section can be determined from the interferogram.

For water as test fluid the equation by Tilton and Taylor [44] can be used:

$$n(T) = 1,33711 - 9,3784 \cdot 10^{-6}T - 2,1726 \cdot 10^{-6}T^2 + 5,9387 \cdot 10^{-9}T^3 \tag{4.33}$$

in order to achieve a relation between the temperature and the refractive index.

4.3.4 Determination of the Local Heat Transfer Coefficient

In the vicinity of a wall where a laminar boundary layer or a phase boundary is assumed the local heat transfer coefficient is calculated from the distance of the interference lines $\mathrm{d}y$ and the corresponding temperature difference $\mathrm{d}T$:

$$h = \frac{-k_{fluid}\left(\frac{\mathrm{d}T}{\mathrm{d}y}\right)_{wall}}{T_{wall} - T_\infty} \tag{4.34}$$

It is assumed, that the heat transfer in the boundary layer results only from molecular heat conduction and not from convection. The local mass transfer coefficient can be defined as being analogous.

4.4 Examples

4.4.1 Determination of the Temperature Distribution in a Compact Plate Heat Exchanger with Plain Fins

Description of the Test Section

Compact heat exchangers are key components for the development of future technical devices. An enhancement of the heat transfer rate results in a decrease of the heat exchanger size and thus in lower weight and also lower investment costs. An exact knowledge of the temperature distribution in the

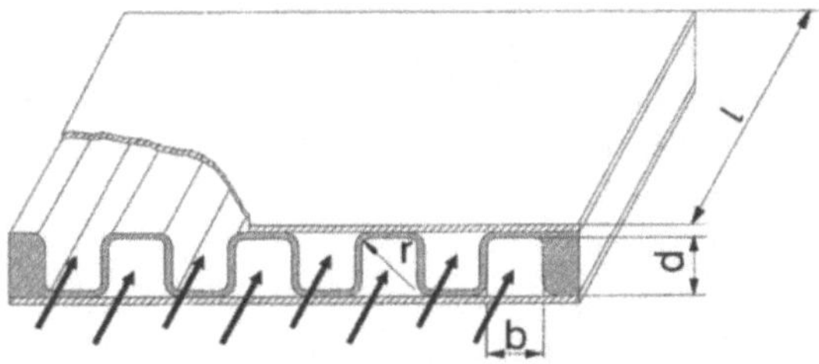

Fig. 4.9. Heat exchanger with plain fin arrangement

boundary layer is necessary for a specific improvement of the heat transfer. In this example the heat transfer in a plain fin arrangement of plate–fin heat exchangers (Fig. 4.9) is investigated in order to optimize the cross section area.

The longitudinal section of the test channel is shown in Fig. 4.10. The test matrix is heated by hot water flowing through six heating plates, three at the bottom and three at the top. The water temperature of each plate can be measured by thermocouples so that a uniform heating temperature can be obtained by regulating the mass flow rate of the heating water. Cooling of the test facility is also possible. The inlet has a length of 50 *cm* and is 10 *mm* wider than the measuring section due to optical reasons. The entrance and the transition to the matrix were shaped carefully, in order to leave the flow pattern undisturbed. At the air outlet the test section was closed by an ordinary sheet of glass. The air temperature at the entrance and at the exit of the test matrix is measured with six thermocouples so that the logarithmic mean temperature difference can be calculated.

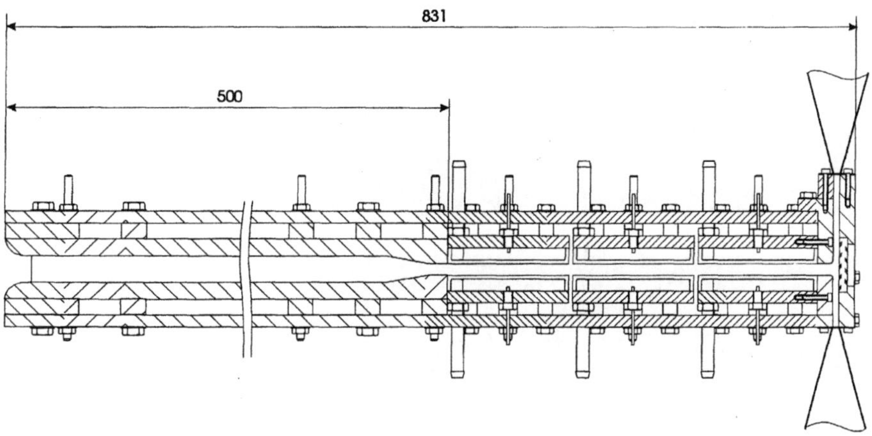

Fig. 4.10. Longitudinal section of the test section

The air is sucked from the ambient into the inlet by a compressor and flows through the test matrix. The mass flow rate can be measured by orifices in order to obtain the Reynolds number in the test section. The flow velocity can be regulated with a throttle at the outlet.

Generally matrices with an overall length up to thirty centimeters can be investigated. Different hydraulic diameters of the channels can be examined with several test matrices. For this investigation a channel length of thirty centimetres was chosen incorporating square ducts of ten millimeters height and width. The corner radii of the flow channels were varied between one and five millimeters. The ducts were milled out of tweleve millimeter thick sheets of aluminum. The minimum fin thickness is two millimeters.

Description of the Interferograms

In Figs. 4.11 and 4.12 the interferograms of channels with different corner radii for laminar (Re = 500) and low turbulent flow (Re = 2500) are shown. The wall temperature was uniform $50°C$ for all measurements, whereas the air inlet temperature at the inlet was equal to the ambient temperature ($20°C$ to $21°C$). The temperature difference between two neighbouring isotherms is $2.3\ K$, approximately. The evaluation of the interferograms was carried out

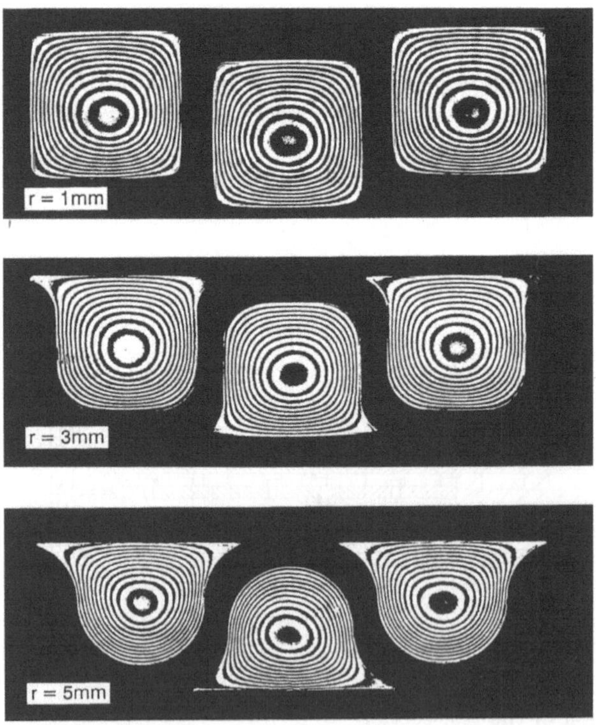

Fig. 4.11. Interferogram of channels (Re = 500)

according to the formalism described above. The distance between adjacent interference lines and thus the distribution of the local heat transfer coefficient was determined using the digital image processing system and the procedures explained in Sect. 6.4.

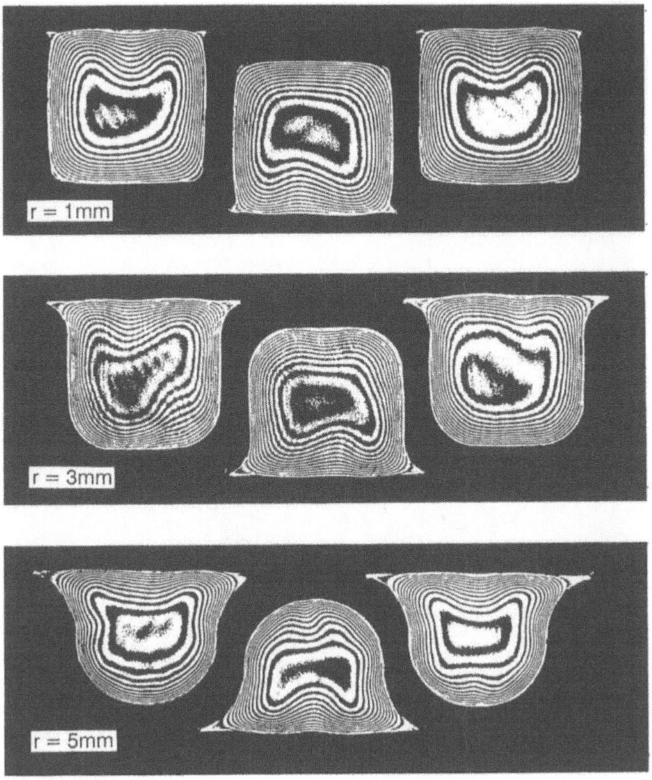

Fig. 4.12. Interferogram of channels (Re = 2500)

5 Short Time Holography

Oliver Feldmann, Peter Gebhard, and Anselmo Chavez

5.1 Introduction

Holography is an imaging method based on the interference ability of coherent light. Apart from artistic and commercial applications, holograms are often used as storage media in databank systems and in devices for optical measurement.

After a brief review of holography's historical development and a general physical description of the process, this chapter will focus on short time holography. The choice of proper optical components and the procedures for recording and reconstructing a hologram are explained. A representative application of this measuring technique is then described.

5.1.1 Historical development of holography

- 1948 Production of the first optical hologram, Gabor [59].
- 1952 – 1960 Lack of suitable light sources and photographic materials. No important progress.
- 1962 First holocamera with a separate reference beam. The coherence length of the light becomes more important, Leith & Upatnieks [66].
- 1964 Application of upgraded gas laser as light source, production of the first clear holograms. Beginning of commercial holography, Leith & Upatnieks [67].
- 1965 Discovery of holographic interferometry (Brooks et al. [51]; Burch [53]; Collier, Dothertyfi & Pennington [57]).
- 1965 – 1969 Multiple applications (non destructive mechanical tests, particle sizing [64], displays). Improved materials for holography [58].
- 1969 Production of the first rainbow hologram, Benton [48].
- 1970 – 1980 Consolidation of holographic methods as a measuring technique [65,77,79,81]. Realization of previous ideas about interferometric studies on heat and mass transfer [69,73,75]. First application of digital image processing to evaluate holograms [50,63].
- 1981 – *present* New optical components – glass fiber, holo–optical elements, dielectrical surface coatings, etc. – permit holographic evaluation of difficult problems in the fields of heat and mass transfer [54,56,80,70], particle sizing [55,60,61,76], fluid mechanics [71,74,78], combustion [72], etc.

5.1.2 The holographic image

Conventional photography yields two–dimensional (2D) records of three–dimensional (3D) scenes. When shooting a photograph, the distribution of the light intensity (i.e. the square of the light wave amplitude) going through or reflected by the imaged objects is stored on a photosensitive surface as illustrated in Fig. 5.1. The phase distribution information of the light is not required.

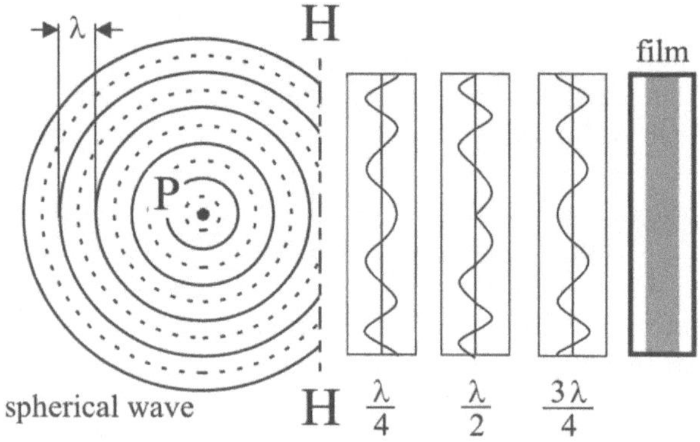

Fig. 5.1. Recording of a spherical wave on a photo plate

In contrast to this, holography records the entire wave field. Both the amplitudes and the phases of the light waves are recorded. If a photosensitive material can store the entire wave field reflected by or transmitted through a three–dimensional scene, it is also possible to reconstruct the scene in three dimensions. This is the characteristic feature of holography as an imaging method.

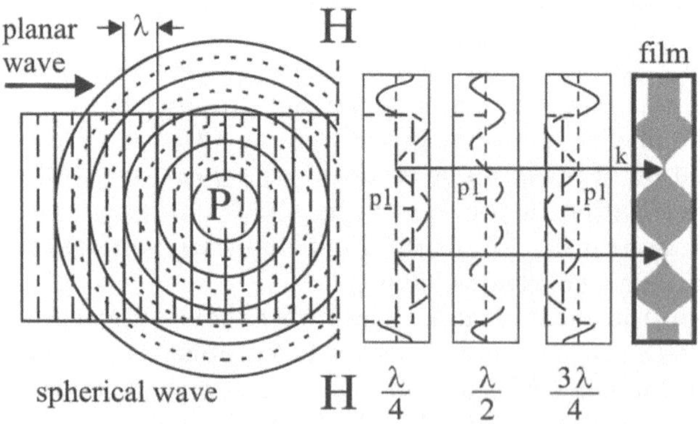

Fig. 5.2. A spherical object wave and a planar reference wave of monochromatic light illuminate a photo plate. The phase distribution information is also recorded

Recording materials consist of a uniform sheet of photographic emulsion bonded onto a celluloid film (photo film) or a glass plate (photo plate). The materials are sensitive only to the intensity of light. Therefore, it is necessary to transform the phase information into a light intensity code which is then recorded. This is performed by illuminating the object with coherent light (object wave) and by superimposing a reference wave from the same light source (Fig. 5.2). The superposition of these two waves results in an interference figure in which the local fringe frequency describes a function of the phase distribution. The density of the fringes is proportional to the intensity of the object wave.

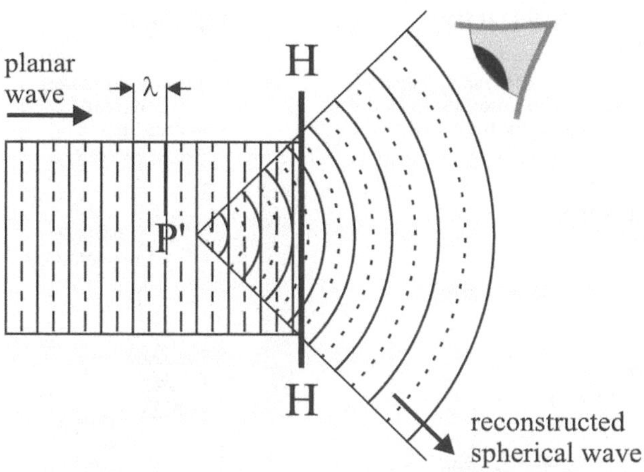

Fig. 5.3. Reconstruction of the spherical object wave by illuminating the hologram with a planar wave. The reconstructed spherical wave is a 3D image of the originally recorded wave

After photographic development the photo plate is called a hologram. The recorded scene is reconstructed by illuminating the hologram with a wave referred to as the reconstruction beam. This laser beam simulates the reference beam used for exposure of the hologram with respect to its geometry and to its the angle of incidence. The laser beam is diffracted at the stored interference figure. As a result, the object beam containing the entire optical information of the imaged object is reconstructed. A stationary image, the real image of the recorded scene, is formed without distortion in three dimensions in front of the holographic plate (s. Fig. 5.3). The image corresponds exactly to the recorded scene, including the characteristics of perspective and depth of field.

5.1.3 Holography as an optical measurement method

As an optical measuring technique short time holography is mainly applied to record three–dimensional fields of rapidly moving particles. It is a non–invasive measuring technique and therefore highly suited to the analysis of transport phenomena (e.g. heat and mass transfer) in dispersed transparent flows such as dust particles or sprays in a transparent continuous medium or gas or vapor bubbles in liquids. The 3D optical information about the flow field in the test section is stored on the holographic plates.

Various arrangements of the optical components for holography are reported in the literature. Holograms can be classified either according to the applied components in the optical set–up or according to the procedure used to develop the plates. These two classification criteria are briefly reflected in Table 5.1. A classification of holography by optical methods is given in Fig. 5.4.

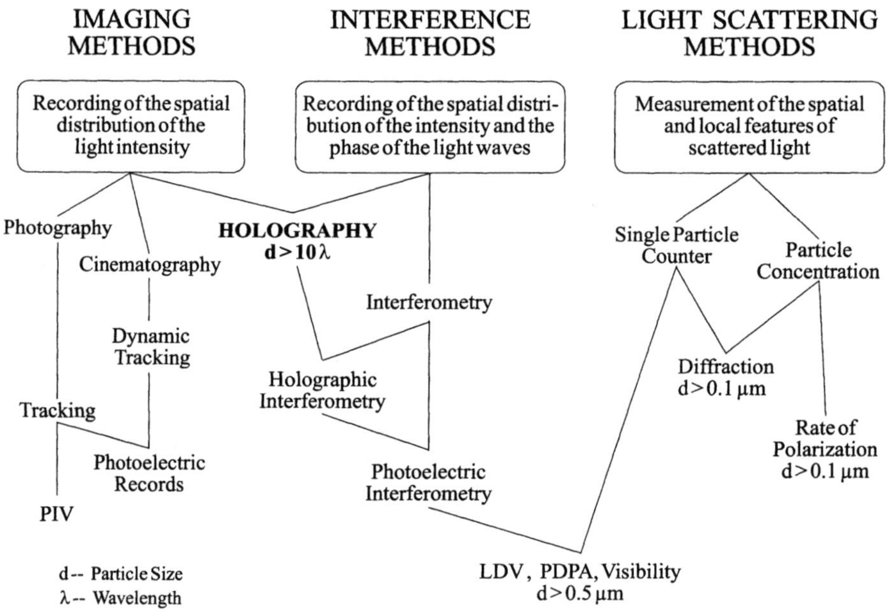

Fig. 5.4. Classification of optical measurement methods

5.2 Elements of holography

As in conventional photography, film material, a camera and a light source are required to record a hologram. In this case the camera is called a "holocamera", and the light source should generate monochromatic beams.

5.2.1 Recording materials

When recording a hologram an interference pattern, formed by superimposing the reference beam onto the object beam is preserved on the photographic plate (or film). The smallest grading cells of this interference pattern range

Table 5.1. Types of holograms

Classification by optical setup

in-line
A planar wave illuminates a particle field. A fraction of this light is diffracted at the particles and forms the object beam which interferes with the undisturbed wave (reference beam). In the holographic reconstruction the particles are identified by applying the Frauenhofer far-field theory.

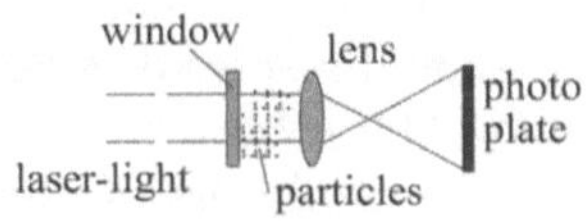

off-axis (counter light)
A light beam illuminates the objects. The resulting wave field (object beam) is superimposed with a reference beam and stored on a photo plate. This method is also suited to application in holographic interferometry.

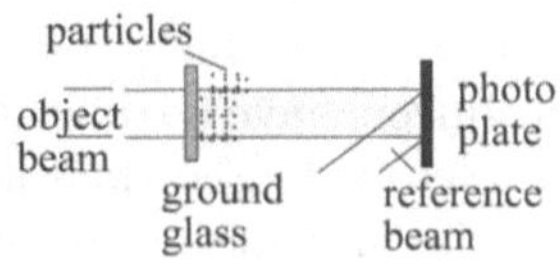

off-axis
An illuminated object reflects light (object beam) towards a photo plate. Simultaneously, the plate is exposed to a reference beam. The length of both optical paths is almost equal. This method is used to analyse opaque objects.

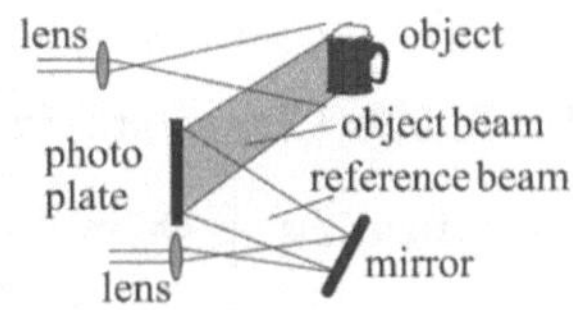

reflection
The object beam meets the photo plate at the coated side while the reference beam incides on its glass side. Such a hologram can be reconstructed with white light. Suitable objects are diffuse reflecting bodies

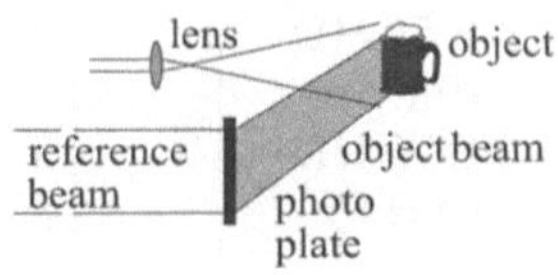

rainbow
With the aid of two cylindrical lenses a laser light sheet is generated and sharply focussed onto a ground glass plate in order to obtain a diffuse illumination of the object. The light travels through the partially transparent object towards the photo plate. A planar wave is used as a reference beam. The reconstructed image exhibits optical parallax and can be reconstructed with white light.

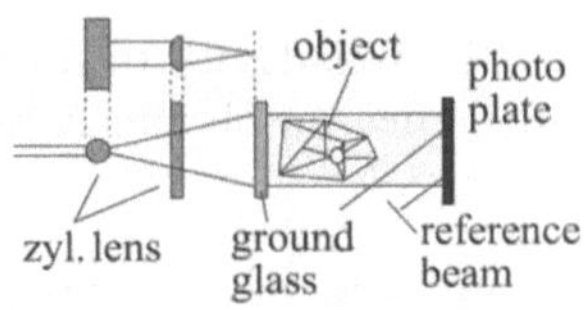

Classification by developing process

amplitude
The exposed photo plate is developed and fixed. The photo emulsion shrinks slightly but remains uniformly thick. The amplitude of the reconstructed wave is a function of the darkness distribution in the coating. Its maximum value is about 25% of its initial value.

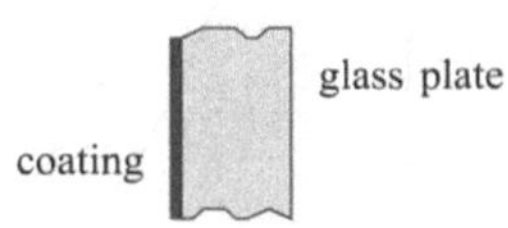

phase
Instead of fixing the developed emulsion, it is bleached to complete transparency. The silver ions are removed from the photo emulsion. This results in an irregular thickness distribution of the coating, depending on the recorded intensity distribution. The reconstruction wave behind the plate **is** phase modulated. Its maximum amplitude is about 60% of its initial value.

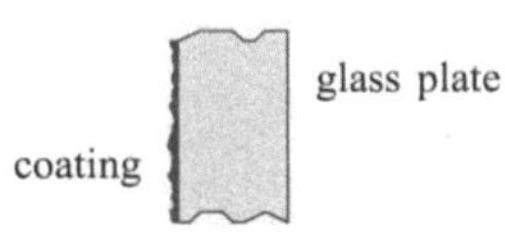

in the order of one half the laser's wavelength (typical values between 200 and 350 nm). The recording materials must therefore possess a very high resolution. In addition, their sensitivity has to be high enough to allow short exposures at reasonable light intensities ($0.2 \sim 2\ W$). Suitable photo materials should resolve at least 2000 line couples per millimeter at light intensities below 200 $\mu J/cm^2$.

The producers of photographic materials for holography have developed photographic emulsions adapted to selected wavelengths of laser light. The specifications of the coatings given in Table 5.2 refer to the holographic plates that were produced by AGFA until 1997 [47]. Since then, eastern European producers have begun manufacturing holographic plates, but the authors have not yet evaluated their products. The details given in Table 5.2 should therefore be understood as an example of what specifications the material should have.

Table 5.2. Characteristics of some selected photo materials [47]

specification	sheet thickness [mm]	resolution [lines/mm]	wavelength [nm]	exposure [mJ/cm²]
8 E 56 HD	5 - 7	5000	400 - 550	30
10 E 56	5 - 7	3000	400 - 550	2
8 E 75 HD	5 - 7	5000	633 - 694	10
10 E 75	5 - 7	3000	633 - 694	3

The sensitivity of a coating is generally given as an amplitude–transmittance curve with the light intensity expressed in J/m^2 (s. Fig. 5.5).

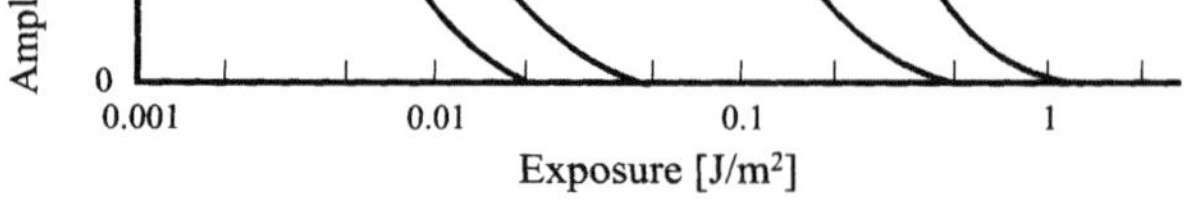

Fig. 5.5. Amplitude transmittance of some selected materials [47]

The definition of the amplitude transmittance is derived from the measurement of the light intensity behind the hologram. That is

$$\boldsymbol{T}_i = \boldsymbol{T}_a \cdot \boldsymbol{T}_a^* \tag{5.1}$$

where $\boldsymbol{T}_i$ is the intensity transmittance, $\boldsymbol{T}_a$ is the amplitude–transmittance, and $\boldsymbol{T}_a^*$ is the conjugated amplitude–transmittance. This is due to the diffractive function of the hologram on the reconstruction beam. The amplitude of the diffracted waves, which reconstruct the object image, changes as a function of the spatial amplitude transmittance of the hologram.

5.2.2 The pulsed laser

Lasers with wavelengths λ in the visible range ($400 \sim 750\ nm$) are the most common light source for holographic applications. Several types of laser are suitable. However, to record fast processes, a pulsed laser should be used. Although almost any laser can be driven in pulsed mode, only a few generate pulses in the energy range of several joules. The most suitable for holography are the solid state lasers. They can easily be driven in pulsed mode by using flash lamps as a light pump. Xenon lamps transform electrical power of up to $1\ MW$ into light. The laser crystal stores this light as stimulation energy and, through induced emission and by means of a so–called Q–Switch, releases the energy as a coherent light beam. This leads to very short and intense light pulses of energy sufficient for holographic applications.

Particles with a diameter greater than ten times the wavelength of the laser light are imaged sharply. Due to the small sizes (several microns) and high velocities of the particles the exposure time must be short enough to ensure a sharply imaged hologram. Because of the relatively low resolution of the human eye a wide unsharpness limit can be set: a particle is imaged sharply if it does not change its position for more than 10% of its size during the exposure. Although this limit is wide, it is difficult to find a laser which can achieve it. If for example a sphere $10\ \mu m$ in diameter which is moving at a velocity of $10\ m/s$ has to be imaged sharply, the exposure time must be as low as $10^{-7}\ s$. In this short interval the light source must produce light beams with a total energy of approximately $0.1\ J$ (corresponding to a power of $1\ MW$!) in order to properly illuminate a photo plate of $10 \times 12.5\ cm$.

High powered pulsed lasers such as the ruby laser and the Nd:YAG laser (Neodym/Yttrium–Aluminum–Granat) produce light pulses that meet these difficult requirements.

The ruby laser is the standard light source in holographic applications. It was the first laser capable of producing light pulses in the energy range of $0.1 \simeq 10\ J$, and its output remains stable over long periods. The wavelength of $\lambda = 694\ nm$ is in a convenient region, since specially sensitized photo materials (Table 5.2) exist. The ruby laser can be operated in both single and double pulsed mode at repetition rates of up to four pulses (or double pulses) per minute. When operated in double pulsed mode, two pulses of the same duration and energy are generated in an interval of $1 \simeq 800\ \mu s$. Two holograms are recorded on a single photo plate and information about high speed phenomena is obtained.

The Nd:YAG laser generates light pulses in the infrared region with a wavelength of $\lambda = 1064\ nm$. By means of a frequency doubler these pulses can be transformed into light with the wavelength of $\lambda = 532\ nm$, which is suitable for holography. This laser has the advantages of higher efficiency and a higher repetition rate (up to 50 Hz) when compared to the ruby laser. It can be driven in both single and double pulse modes. However, the Nd:YAG laser has the drawback that temperature oscillations significantly affect the coherence length of the emitted light and hence the hologram quality.

Comprehensive descriptions of the operating principles of solid state lasers can be found in standard textbooks about quantum optics (Bergmann & Schäfer [49]).

5.2.3 Optical set–up

When applying a pulsed laser as a light source a laser beam with a high energy density ($0.1 \simeq 10\ J/cm^2$ beam section) travels in a very short time through the optical components of the holocamera. The power density of the laser beam amounts to several MW/cm^2 which results in a high potential for damage or injury to both operator and optical components. The latter must completely reflect or transmit the light in order not to be damaged by absorbing too much energy. For this reason, the optical components, such as mirrors and beam splitters, have to be selected carefully. Dielectrical coatings with absorption ratios of about 0.05% should be installed. At power densities below 1 MW/cm^2 less expensive highly reflective metal coatings can also be applied.

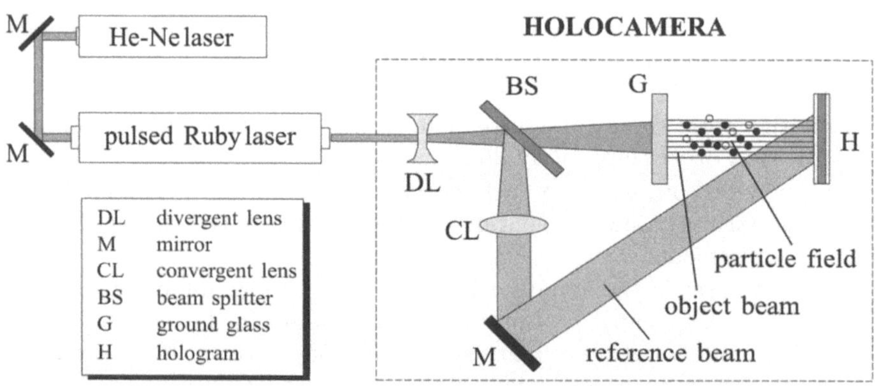

Fig. 5.6. Optical arrangement for short time holography

A simple set–up for "off–axis" short time holography is shown in Fig. 5.6. The laser produces light pulses with an energy density of 1 J/cm^2 and a pulse duration of 30 ns. The biconcave lens (DL) expands the beam which reduces the energy density by a factor of four at the plane of the beam splitter (BS) (40% transmission/60% reflection). The power density still remains over 1 MW/cm^2, so the beam splitter is dielectrically coated. The transmitted beam is the object beam and the reflected one is the reference beam. The latter is expanded until it is paralleled at the convergent lens (CL) and then directed at the mirror (M) towards the photo plate. After passing the convergent lens the beam has a diameter of 5 *cm*. The power density is less than 1 MW/cm^2 and the mirror therefore has a highly reflective metal coating. In order to prevent double reflections, the front side of the mirror must be used as the reflecting surface. The mirror is mounted on micrometric holders to direct the light beam in precisely the desired direction. The object beam passes a ground glass plate (G) before entering the test section in order to ensure uniform object illumination. After exiting the test section, the object beam contains its entire optical information and is superimposed with the reference beam. The resulting interference figure is stored on the photo plate.

The use of the ground glass allows clear focusing each part of the reconstructed image; i.e. the position of each recorded object can be determined. However, the use of diffused light also has a drawback. Speckle noise appears as an image background, which must subsequently be filtered out during image processing (s. Chap. 6).

The optical components are mounted on an optical table by means of screws, magnets or clamps to allow changes of their arrangement in order to adapt the set–up to different recording situations. The table should be supported by vibration dampers. After changes in the set–up or in more complicated optical arrangements, an analysis of the energy density at each optical plane is recommended.

5.2.4 Adjusting the holographic camera

It has already been mentioned that the object beam and the reference beam have to interfere with one another for holographic records. The basic optical set–up that is required to ensure this was shown in Fig. 5.7, but some additional aspects must be considered in order to obtain holograms of a good quality.

The object beam and the reference beam must have both the same frequency and the same phase to interfere with one another. However, if the difference of the beam path length of the two beams exceeds the coherence length of the laser, the beams lose their ability to interfere. As a rule of thumb, this difference should be less than half the laser's coherence length to attain good quality holograms.

In addition, the relation between the intensity of the two beams affects the contrast and hence the quality of the resulting hologram. This intensity ratio is set by the choice of beam splitter. Although different types of beam splitters exist, it is recommended to choose one that allows adjustment of the intensity ratio

$$\varrho = (|\boldsymbol{A}_R| \,/\, |\boldsymbol{A}_O|)^2 \tag{5.2}$$

between reference beam $\boldsymbol{A}_R$ and object beam $\boldsymbol{A}_O$. For example: assuming that the interfering beams at the plane of the photo plate are polarized with the electrical vector perpendicular to the plane of incidence, an intensity ratio of $\varrho = 1$ provides the maximum contrast. This implies that a fixed beamsplitter with 50% reflection / 50% transmission ($\varrho = 1$) produces the best results. However, if the object beam consists of diffused light reflected by the holographed object, its amplitude distribution can vary significantly. Thus, the best ϱ must be found experimentally, since in many applications the needed intensity ratio is related to operating conditions and needs to be adjusted. It ranges in general between 3 and 8. Therefore, an adjustable beam splitter ensures maximum versatility.

Finally, the centers of the two laser beams must pass the optical axis of each lens or beam splitter and hit the center of the photo plate to ensure the highest possible contrast. This requires exact adjustment of the optical components. Conventional holographic applications use continuous laser beams with which this adjustment is a more or less simple task. Therefore, it is not mentioned in any textbook. In such cases, it is very simple to find the correct position of each optical component by visual adjustment. However, if a high power pulsed laser is applied, the laser light is not visible yet highly dangerous to human eyes (consult [68] and [52] or the corresponding Radiation Safety Authority).

The ruby laser crystal can transmit red light. To adjust a holocamera with a ruby laser as a light source, a continuous $He - Ne$ laser beam with power in the milliwatt range is used. It simulates the optical path of the ruby laser as shown in Fig. 5.6 and the optical components can be adjusted one after another. If an additional fine adjustment is necessary, it should be performed by using the ruby laser directly. To protect sensitive elements, a piece of exposed and developed photographic paper (burning paper) is positioned at the place of the component to be properly positioned and exposed to a pulse of the ruby laser. By analyzing the burning marks on the paper the position of the component can be corrected.

For holocameras which use a Nd:YAG laser light source, the adjustment can be performed by setting the laser at the maximum repetition rate and minimum energy output. The laser beam is now visible and the danger to the human eye reduced to a minimum. The fine adjustment can be performed by using heat sensitive fax paper instead of photographic paper and by proceeding as described above.

5.2.5 Recording, development and reconstruction of holograms

When the optical components are correctly arranged, the laser power is adjusted to properly expose the photo plates. After the photo plates are installed in complete darknesss, the laser is fired and the hologram is recorded.

An exposed photo plate is developed similarly to a photograph. There are two different developing procedures, which lead to either amplitude or phase holograms (s. Table 5.1). The developing process is performed as follows.

Amplitude hologram

Developing is carried out in full room darkness at $20°C$ bath temperature and four minutes developing time. During the developing, the bath containing the plate is rhythmically agitated to promote a better distribution of developer on the plate surface. Hereafter, the plate is rinsed for five minutes under running water. It is best to use circulating water that has the same temperature as the developing bath. The use of a stop bath is optional. The plate is then fixed for four minutes and washed again for ten minutes. A final rinse using a wetting agent is recommended. Finally, the plate is left to dry at room temperature. If necessary, the drying process can be accelerated by careful application of a dry air jet.

Phase hologram

The development and the first washing are carried out as above. The plate is then immersed in a bleach bath. After a few seconds, the light can be turned on to visually control the bleaching process. The bleaching can be stopped when the plate is transparent again. The rinsing and drying procedures are similar to those above. Additional details about developing methods can be found in Hariharan [62].

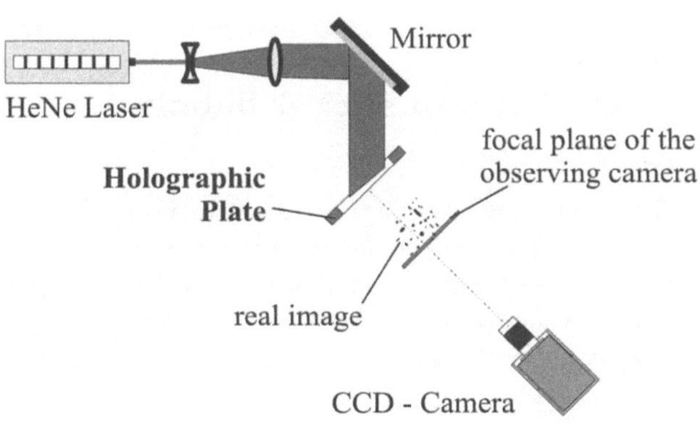

Fig. 5.7. Optical set-up to reconstruct a hologram

The developed plate – now called a hologram – is reconstructed by illumination with a continuous $He-Ne$ laser beam. This laser beam simulates the reference beam during the exposure of the hologram with respect to its geometry and to its angle of incidence. The $He - Ne$ laser beam is diffracted by the stored interference figure, and the object beam containing the entire optical information of the test section is reconstructed in front of the holographic plate. A stationary image, the real image of the recorded scene, is formed without distortion in three dimensions (s. Fig. 5.7).

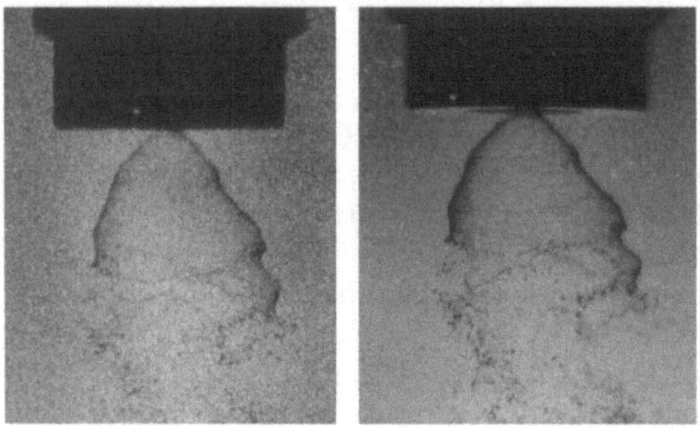

Fig. 5.8. Reconstruction of an injection jet

The reconstructed image shows exactly the same properties as the recorded scene, as far as the perspective and the depth of field are concerned. As an example, two images taken from a reconstruction of a holographed injection jet are displayed in Fig. 5.8 [54].

The spatial information of the measuring volume is recorded by means of a CCD–camera. The camera lens has a large focal length and a small depth of field, so that only a small fraction of the 3D reconstruction is in focus, as indicated by the gray bar in Fig. 5.7. By moving the camera step by step along its depth coordinate, the 3D information contained in the holographic reconstruction is transformed into a series of 2D digital images (s. Fig. 5.9).

5.3 Application example: Dispersion characteristics in stirred bubble columns

The measuring technique described above was applied to determine the dispersion characteristics of an aerated stirred vessel. The number and size of generated bubbles and their 3D velocity field was determined. The modifications in the optical set–up and the problem together with some sample results are given in the following subsections.

5.3.1 Statement of the problem

Parameters such as generated phase interface, period of dwell of the gaseous phase, and hold–up determine the optimum design of an aerated stirred ves-

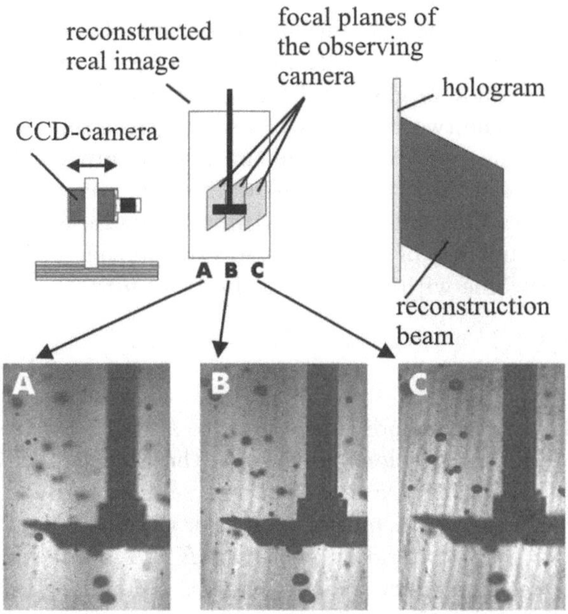

Fig. 5.9. Scanning a reconstructed hologram

sel. These parameters may be linked. Design requirements are met by the correct choice of stirrer, rotational speed, and type and location of the gas injector. Detailed knowledge is required of the number and size of bubbles generated and the velocity fields of both phases as a function of the different operating points. In this study, the dispersion characteristics of an aerated stirred vessel agitated by a Rushton turbine are analyzed. The operating points were adjusted by varying the liquid viscosity, gas flow rate, and agitator speed [60].

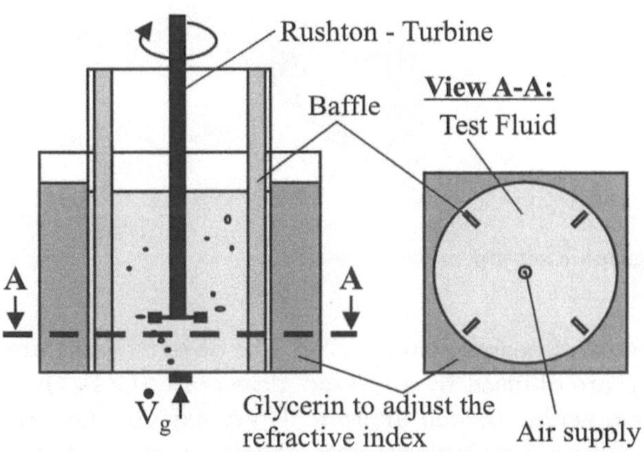

Fig. 5.10. Design of the test section

The test section that has been applied (s. Fig. 5.10) consists of a cylindrical vessel with an inner diameter of $T_i = 124\ mm$. In order to enable optical access to the vessel, it is made of glass and located in a rectangular glass vessel. The volume between the two vessels is filled with glycerin. The refractive index of the test fluid, the glass vessels, and the glycerin are equal, so laser beams can cross the test section without being diffracted. The vessel is equipped with four glass baffles. The air is supplied from a nozzle which is mounted centrally in the flat bottom of the vessel. It is dispersed in the liquid by means of a Rushton turbine with a diameter of $d = 40\ mm$.

5.3.2 Recording the holograms

The optical set–up given in Fig. 5.6 was modified in order to adapt it to the given measuring problem. These modifications resulted in the arrangement shown in Fig. 5.11. The given dimensions are related to this research project and should be understood as an example. A pulsed ruby laser generates light pulses with a wavelength of $\lambda = 694\ nm$ and an energy of $E = 1\ Joule$ for the exposure time of $t = 30\ ns$. The emitted laser beam is collimated by passing through two lenses (CL) and is then split into two object beams and

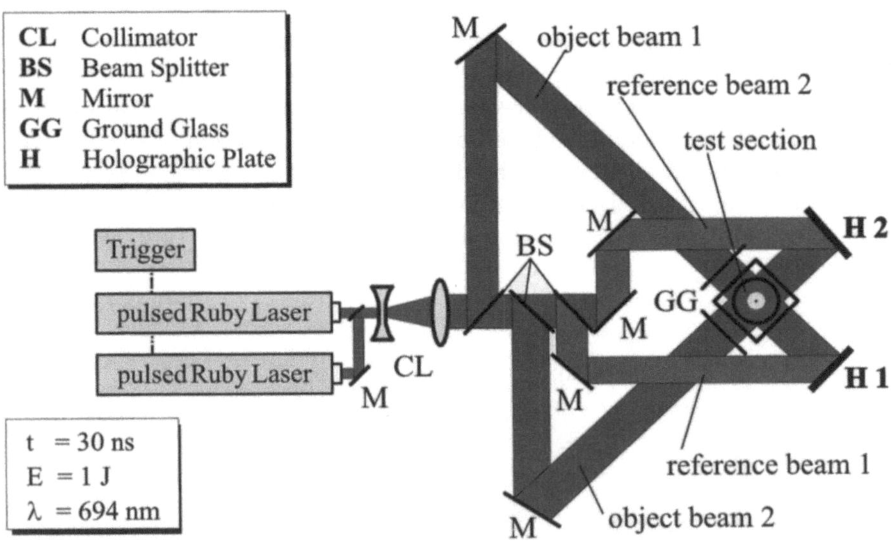

Fig. 5.11. Stereoscopic holographic set–up

two reference beams by a bank of beamsplitters (BS). The object beams are diverted at the mirrors (M), are diffused by a ground glass plate (GG), then cross the test section perpendicular to one another. Upon exiting the test section, the object beams fall perpendicularly onto the holographic plates

(H1, H2). The object beams contain the entire optical information of the test section. The reference beams are diverted directly to the holographic plates (s. Fig. 5.11).

The ruby laser can be operated in either single pulse or double pulse mode. Single pulse holograms contain both macroscopic information about the spatial distribution of the dispersed phase and microscopic information, such as the number, size, shape, and location of the generated bubbles. Double pulse mode records the bubbles' velocities and trajectories as well. In this case, two successive holograms of the flow field in the vessel are stored on each holographic plate. With knowledge of the pulse separation and the evaluated holographic reconstructions, the velocity field of the bubbles can be determined. The pulse separation of one laser can be varied from $\Delta t = 1\ \mu s$ to $\Delta t = 800\ \mu s$. To lengthen the possible pulse separation, a second ruby laser is coupled in the beam path of the first one. The second laser can be fired any time after the first laser. The pulse separation is adjusted and controlled by a trigger (s. Fig. 5.11).

It has to be mentioned that a double exposure causes a considerable increase of speckle noise in the background. This results in a decrease of the contrast obtainable and hence reduces the sharpness of the imaged particles. However, these difficulties can be overcome with advanced image processing algorithms, as will be explained in the following chapter. The holograms still provide useful velocity information.

With this stereoscopic optical set–up two holograms are simultaneously recorded. The situation inside the vessel is stored distortion free in three dimensions from two different perspectives.

5.3.3 Reconstruction and evaluation of the holograms

The holograms are reconstructed in the set–up that is shown schematically in Fig. 5.12. The spatial information of the entire measuring volume is furnished in three dimensions with a width of approximately $d = 130\ mm$ and an infinite depth of field. This 3D image is recorded by means of a CCD–camera as explained in Sect. 5.2.5.

By moving the camera step by step along its depth coordinate, the entire 3D information contained in the holographic reconstruction is scanned and transformed into a series of 2D digital video images (s. Fig. 5.10). The depth of field determines the step size and thus the resolution in the depth dimension. The images depicted in Fig. 5.10 reveal how the field of view of the camera changes with its position. Bubbles that are depicted sharply in one image are out of focus at a different camera position.

Both holograms taken in one experiment are reconstructed and scanned as described above. They represent two 3D images corresponding to one situation in the test section, so that the entire 3D information of the test section seen from two different views is provided in the form of digital video images.

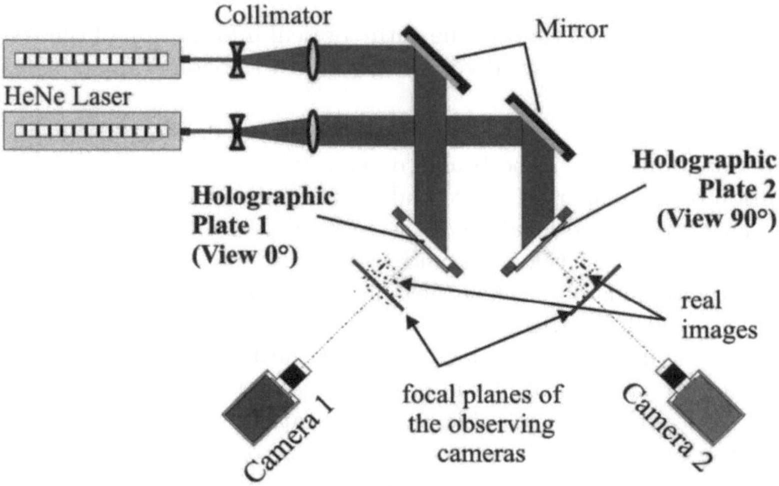

Fig. 5.12. Stereoscopic optical set–up for hologram reconstruction

5.3.4 Stereo matching of the two holograms

After both holograms taken under one set of operating conditions have been completely scanned into 2D images and evaluated, the images are numerically reconstructed. This results in two 3D reconstructions of the flow field in the vessel, displaying the same scenario from two perpendicular viewing directions. These two 3D images are correlated with one another by applying a stereo matching algorithm. The vertical and lateral coordinates of each bubble are derived from one evaluated holographic reconstruction. The depth coordinate is derived from the position of the camera, but it can only be determined within the accuracy of the depth of field. Since the two holograms were

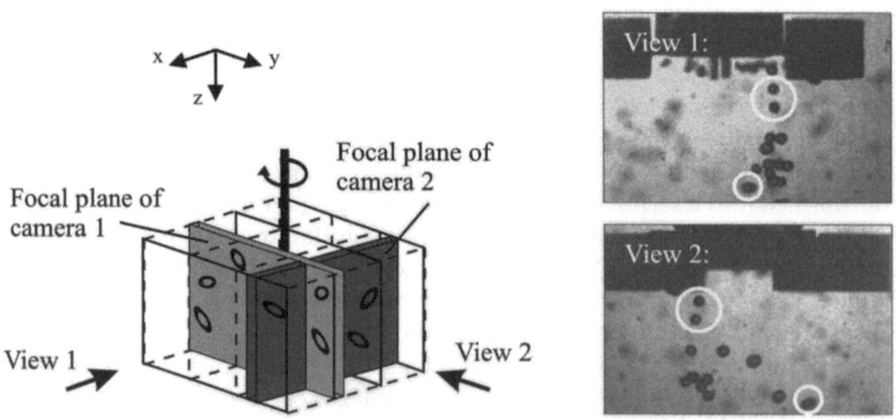

Fig. 5.13. Physical model for the stereo matching algorithm

recorded perpendicularly, the depth coordinate of a bubble in one holographic reconstruction is its lateral coordinate in the second. Thus, by considering the vertical coordinate in both views, the position of each bubble can be determined exactly. The depth of field of the cameras is adjusted in such a way that this correlation is definite. The model upon which this algorithm is based is given schematically in Fig. 5.13.

The correlation of the bubbles in both views is performed without considering any feature–based criterion. This is important because the shape and size of a bubble can vary significantly between the two views. After the two views are correlated, the sizes, spatial locations, and, in the case of a double pulsed hologram, velocity field of the bubbles are determined.

5.3.5 Results

The flow field in an aerated stirred vessel is of a highly three–dimensional character. When regarding the various operating conditions, there is no preferred flow direction for the bubbles at any place inside the vessel. In order to automatically determine the velocity of a given bubble the two images of a double pulse hologram must therefore be clearly distinguishable from one another.

A typical image taken from a reconstructed double pulse hologram is shown in Fig. 5.14 A. This image is included to facilitate the location of the image in Fig. 5.14 B. The brighter images of the bubbles belong to the first pulse (lower intensity), and the darker images belong to the second pulse (higher intensity). With this information it can be clearly derived from Fig. 5.14 B that the bubbles were moving upwards in the time interval between the two exposures to the laser beam.

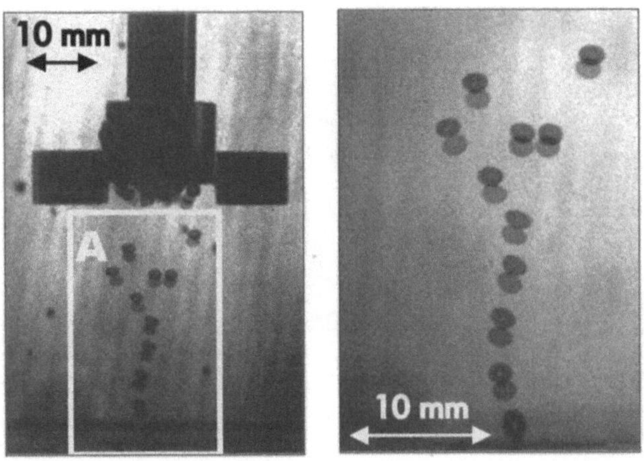

Fig. 5.14. Evaluation of a reconstructed double pulse hologram

As an example for an evaluated double pulse hologram the velocity field of bubbles generated at one operating point is given. For an agitator speed of $N = 550\ rpm$, an air flow rate of $\dot{V}_G = 3\ l/h$, a liquid viscosity of $\eta = 30\ mPas$, and a pulse separation of $\Delta t = 5\ ms$, the three–dimensional velocity field of the generated bubbles was evaluated.

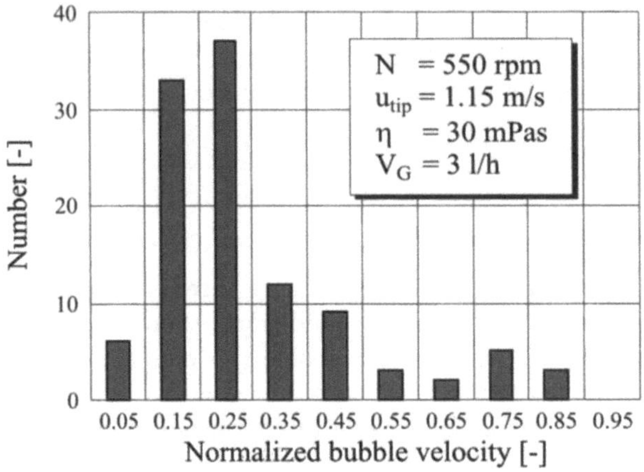

Fig. 5.15. Velocity distribution of the bubbles

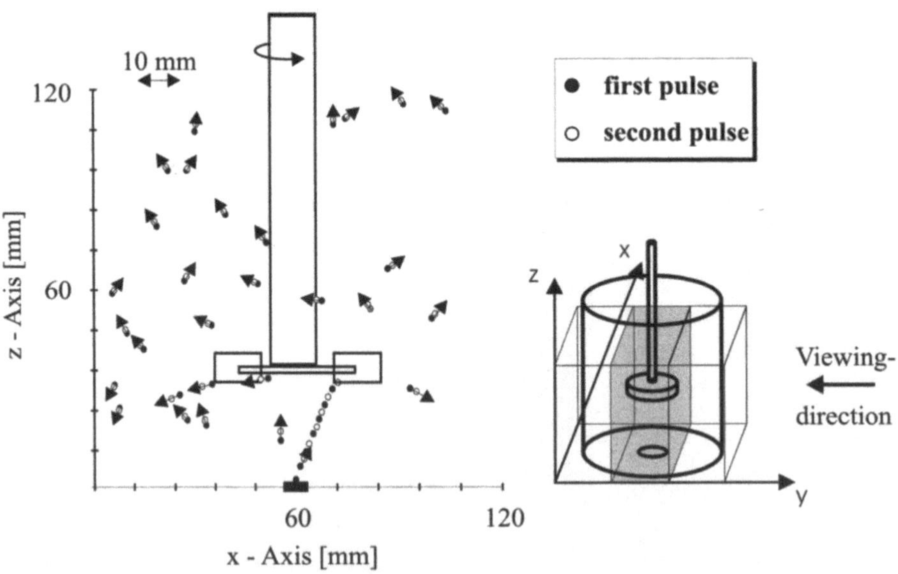

Fig. 5.16. Velocity field of the bubbles

In Fig. 5.15 both reconstructed locations of each bubble in the middle plane of the vessel are shown. Additionally, arrows are included indicating the bubble velocity and flow direction. Their length is proportional to the projected velocity in this plane. A bubble whirl below the stirrer can be identified. Above the stirrer the ascent velocity of the bubbles is almost uniform. Superimposed onto that velocity is the global swirling flow of the liquid. The upper liquid whirl is not strong enough at this impeller speed to entrain the bubbles back downwards.

The evaluation of the two double pulse holograms yielded one hundred and ten pairs of bubble images. The obtained velocity distribution of the bubbles, normalized with the circumferential agitator speed $u_{tip} = 1.15\ m/s$, is given in Fig. 5.16.

It is striking that only those bubbles adjacent to the agitator move with a velocity in the range of the circumferential agitator speed. As soon as the bubbles have left the sphere of influence of the agitator, their speed is reduced significantly due to inertial effects. Regions where the bubbles have a higher velocity can be found close to the side of the baffles facing the liquid flow and above the nozzle. The lowest bubble velocities were found in the centers of the whirls below the agitator.

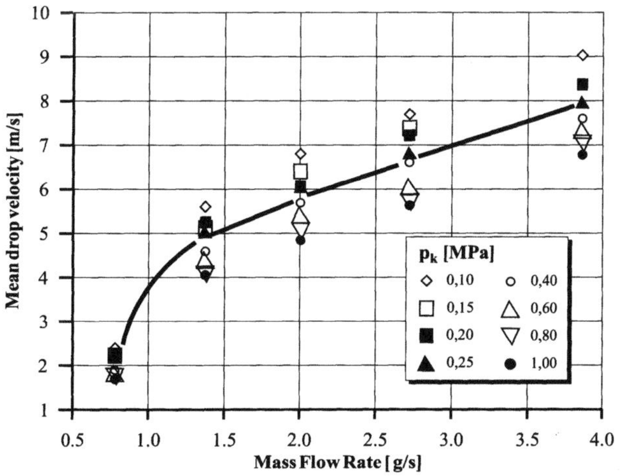

Fig. 5.17. Mean droplet velocity in a spray as a function of the flow rate at different ambient pressures [54]

Similar experimental methods may be applied successfully to a wide variety of problems. For example, analysis of the disintegration of a spray from a hollowed cone nozzle [54] (s. also Fig. 5.8) yielded the mean droplet velocity curve shown in relation to flow rate and ambient pressure in Fig. 5.17.

6 Evaluation of holograms by digital image processing

Oliver Feldmann and Robert Tauscher

6.1 Introduction

The rapid development of computer technology and the mass production of computer chips in the past decade have resulted in the application of image analysis and image processing in many technological and scientific areas. (s. Table 6.1). Numerous problems of pattern recognition, data handling of digitized pictures and computer graphics, formerly reserved for computing centers, TV stations or military organizations, can now be solved on a personal computer (PC).

Table 6.1. Selected applications of digital image processing

field	application
fluid mechanics	vortices, cavitasing bubbles, sprays, two phase flow
metallurgy	microstructures
heat transfer	temperature gradients evaluated from interferograms
medicine	pictures from tomographs, radiographs, etc.
archaeology	air photographs for reconstruction of ancient buildings
biology	microscopic pictures
art	contrast pictures for restoration and conservation of paintings and sculptures
weather forecast	satellite pictures
robotic	pattern recognition for controlling production processes and material handling
production	quality control, process control
criminology	portraits, scripts and fingerprints

The purpose of digital image processing (DIP) is to reflect the main features of a picture more clearly and informatively than in the original and to judge the contents of an image quantitatively by employing pattern recognition algorithms.

This chapter describes the application of digital image processing to the evaluation of holographic particle field reconstruction and the corresponding personal computer based image processing system configuration.

6.1.1 Digitization of a picture

Two basic steps are necessary to digitize a picture: **sampling and quantization**. These steps are presented schematically in Fig. 6.1.

When a picture is digitized, a two–dimensional grid is superimposed upon a simple black and white image to enable the spatial localization of individual picture parts. If necessary, the spatial resolution of the image is increased by choosing a finer grid. This procedure is described by the expression **sampling** and was originally used in TV technology (Fig. 6.1 a).

Depending upon the dominating color within this grid the individual picture elements or pixels are declared to be either black or white. This process is called **quantization** (Fig. 6.1 b). Eventually, the pixel pattern is transferred into a binary matrix. The value '1' represents the color white, whereas the value '0' represents the color black (Fig. 6.1 c). The obtained array is treated on a PC as a mathematical matrix.

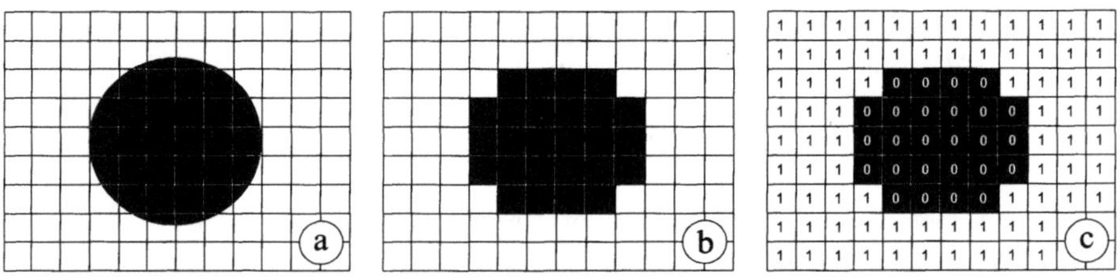

a) scanning b) quantization c) pixel matrix

Figure 6.1. The two steps necessary to digitize a picture: Sampling and Quantification

Simple pictures can be digitized by line scanners such as a telefax. More complicated images are scanned by CCD or video cameras. Specially adapted digitizer boards transform the analog signal of the video camera into a digital signal that can be read by a host computer.

6.1.2 Gray value pictures

Contrary to binary images, where grid boxes are assigned only the values, '1' or '0', gray value pictures consist of various gray values, as shown in Fig. 6.2 a. For comparison, the black and white representation of the same picture is given in Fig. 6.2 b.

The gray values are represented by bits. One bit is sufficient only (s. Fig. 6.1) to distinguish between two gray values, such as black and white. In byte oriented digitizers, gray values are represented by either one or two bytes (either 8 or 16 bits.) This provides a resolution of 256 gray values in the first case (1 byte) or 65536 gray values respectively (2 bytes). As a convention, the maximum gray value is associated with the color white and the minimum with the color black. The portrait in Fig. 6.2 a is represented by 256 gray values.

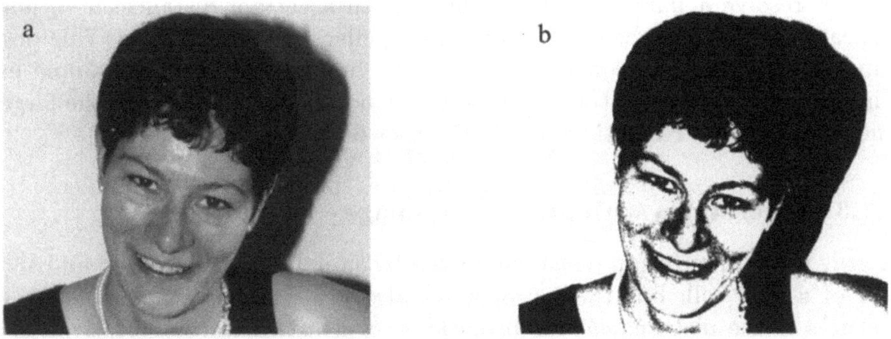

Figure 6.2. A portait: **a** gray value picture; **b** black and white representation

From Fig.s 6.1 and 6.2 it becomes clear that the quality of a digitized picture depends on the mode of digitization, i.e. on the sampling (number of grid points) and on the quantization (number of the gray values). Thus, the features of the employed digitizer are responsible for the quality of the picture. Common CCD Cameras can record images with a resolution of up to 4096 × 4096 pixels with a gray value depth of 1 or 2 byte or more. The quality of a digitized image is also influenced by the dimensions of a single pixel and the size of the entire picture. The dimensions of a grid box depend on the sampling scale of the image, when it is recorded by the video camera. They should be adapted to the size and texture of the object by choosing a suitable objective lens for the camera. For example, a detailed object of 5 × 5 mm size can be reflected well by a grid of 10 × 10 μm width while for a portrait the size of a standard photographic slide (36 × 24 mm) a grid spacing of 100 × 100 μm is sufficient to obtain a good representation of the original.

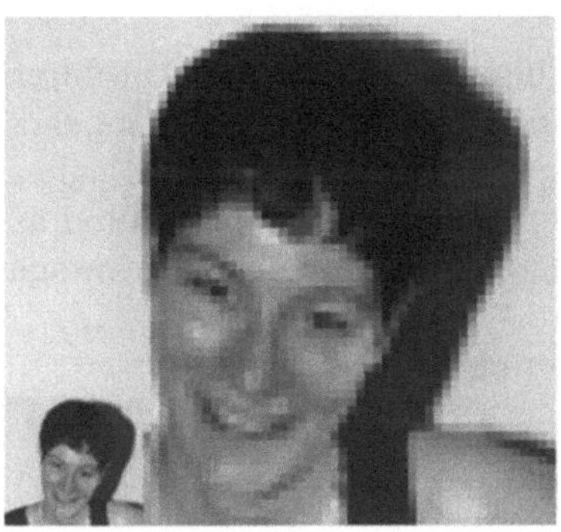

Figure 6.3. Example for the choice of grid cell and the survey sampling: 64 × 60 pixel , quantization: 256 gray values

With respect to the final size of a digitized picture one returns to the basic rule of photography that the width of a grid box should be in the magnitude of the grain of the photographic layer. Thus, the human eye should not be able to resolve a particular pixel which, from a certain distance, does not appear aesthetic or which has a disturbing effect. By obeying this rule one can get good image reconstructions for the human eye. In the example in Fig. 6.3 the sampling in the small picture is not as disturbing as in the large one, although the resolution is in both cases 64×60 pixel.

6.1.3 Operations with gray value images

A gray value picture can be characterized by sampling, quantization and the size of a grid cell, but these properties also give information about other features of the picture, such as brightness, contrast, texture and sharpness. To describe these attributes, the gray value distribution in the picture, the mean gray value, the gray value variance, and the spatial frequency of the gray values must be analyzed. A detailed description of the numerical techniques to perform these operations can be found in Gonzáles & Wirtz [82] or Pavlidis [83]. A good introduction was published by Haberäcker [84].

In this section the effects of selected operations are demonstrated. This should give an impression of how to extract the desired information from a given picture, how to suppress unnecessary and disturbing structures, and how to use the obtained results for further steps of image processing. The following definitions are assumed:

B	matrix of an image	B = [p(i,j,g)]
i	horizontal coordinate	$1 \leq$ i ≤ 512
j	vertical coordinate	$1 \leq$ j ≤ 512
g	gray value	$0 \leq$ g ≤ 255
p	pixel	p = p(i,j,g)

8–neighbor technique:	Operation with a pixel and its eight neighbors
4–neighbor technique:	Operation with a pixel and its four neighbors without considering the neighbors in the diagonal

These definitions are illustrated in Fig. 6.4 for the gray value of a pixel p with the coordinates i and j ($g = \hat{p}(i,j)$).

The application of a 3×3–average filter is illustrated in Fig. 6.5. This filter can be applied to reduce the noise level in an image. A pixel $p(i,j,g)$ appears with a gray value of $g_{ij} = 140$ (Fig. 6.5). Its neighbors clearly have higher gray values of about 240. Therefore, the pixel $p(i,j,g)$ can be interpreted as a noisy pixel. It can be eliminated by replacing its gray value by the average gray value of its eight neighbors.

The mathematical representation of an average filter is given by:

$$g'_{ij} = \frac{1}{8} \sum_{k=1}^{3} \sum_{l=1}^{3} g_{(i+2-k)(j+2-l)} \; m_{kl} \qquad (6.1)$$

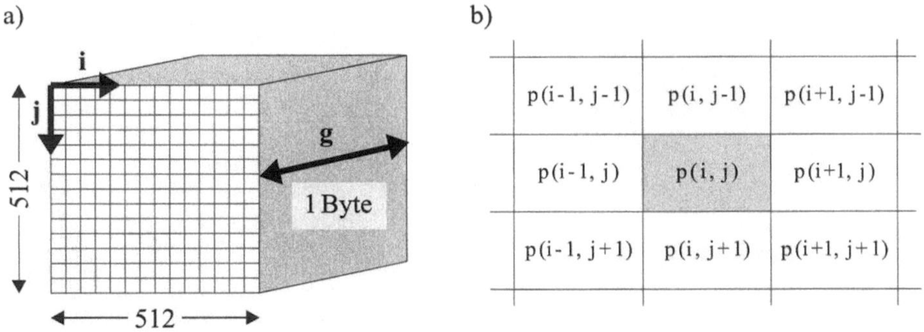

Figure 6.4. a) A matrix of pixels with 512×512 pixels of 1 byte depth; b) a pixel with its eight neighbors

with

$$[m_{kl}] = \begin{pmatrix} 1 & 1 & 1 \\ 1 & 0 & 1 \\ 1 & 1 & 1 \end{pmatrix} \tag{6.2}$$

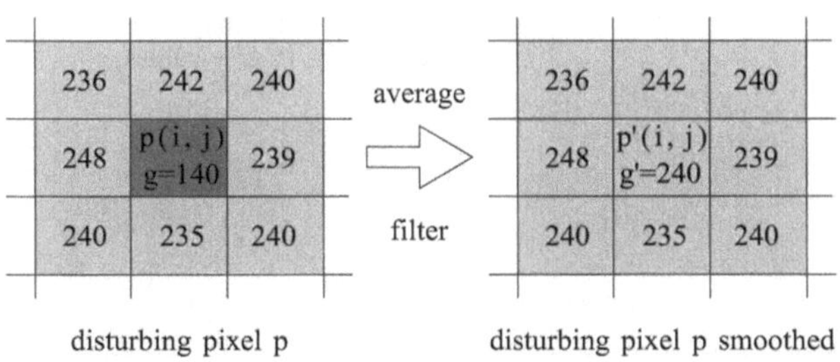

Figure 6.5. Smoothing the image with a 3×3 average filter

Morphological operations on pictures are shown in Figs. 6.6a to 6.6f. Their technical applications are described in the following. An example of the filtering operations is introduced in Fig.s 6.6.

This example shows how the different filters principally work. The gray value of a pixel is read and compared to the gray values of its neighboring pixels or to the corresponding gray value in a reference image. Depending on a certain threshold, a mathematical operation is performed between the gray value of the pixel and one or more reference pixels, and the original gray value is replaced by the result of this operation. The purpose of this operation is to highlight the wanted pixel, such as for example the edges of sharply imaged particles.

a) original image, 256 grey values

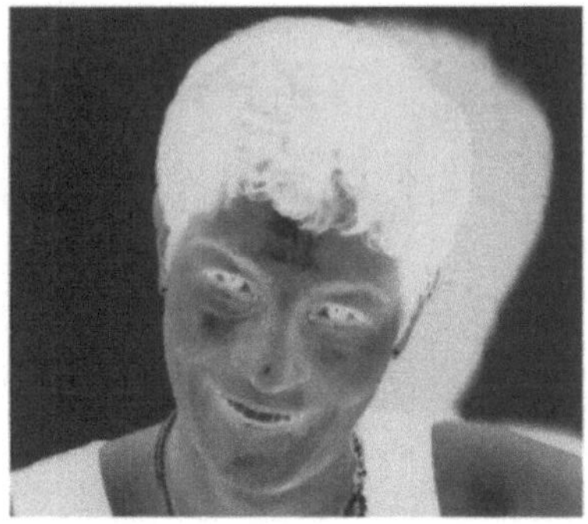

b) negative of the image

c) magnification of the image
360 * 320 pixel, 256 grey values

d) magnification of the image
360 * 320 pixel, 16 grey values

e) gradient extraction of the image

f) binarization of the gradient extraction

Figure 6.6. Operations with gray value images

6.2 A digital image processing system for the evaluation of holographic reconstructions

The optical information stored on the holographic plates is reconstructed in three dimensions as described in Chap. 5. The reconstructions are recorded by means of a camera. If macroscopic structures of some millimeters dimension have to be scanned, the positioning of the camera relative to the reconstructed object and its focusing is a fairly simple task. The CCD camera is mounted on a conventional tripod and the objective lens is focused on the holographic image. The evaluation of more detailed structures of the reconstructed object, such as the size or shape of single particles in a particle field, requires more sophisticated equipment, depending on the optical set–up of the holocamera (s. Table 5.1 in Chap. 5).

6.2.1 Evaluation of holographic images

The camera sensor can record only two dimensional images of three dimensional scenes. Therefore the focal plane of the objective is moved incrementally through the entire depth of the reconstruction. In the following, possibilities to realize this motion for the evaluation of both off–axis and of in–line holograms are given.

Scanning of in–line holograms
In the optical arrangement for recording in–line holograms, the optical axis of the object beam coincides with the reference beam (s. Chap. 5). When reconstructing such a hologram, the optical axis of the reconstructed holographic image coincides with that of the scanning camera.

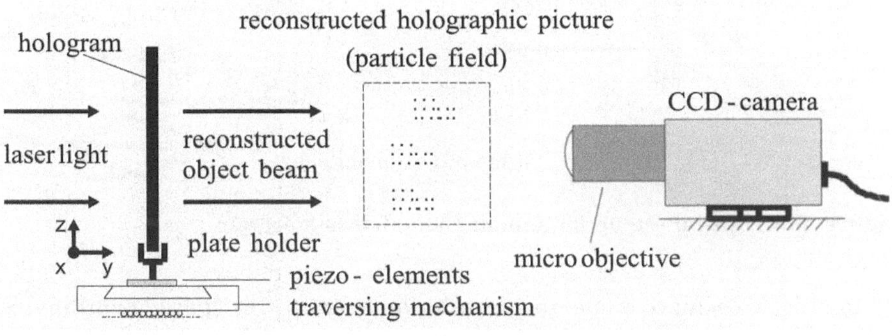

Figure 6.7. Optical set–up for scanning an in–line hologram

To resolve fine details of the imaged object the camera is equipped with a macro objective with a small depth of field. Therefore, only a small section of the entire reconstruction is imaged sharply. To completely scan the reconstruction, a relative movement between hologram and camera has to take place. Plate holders with a traversing mechanism moveable in three dimensions are commercially available. The plate holder is positioned by means of a piezo crystal, as shown in Fig. 6.7. The accuracy of the positioning lies in the range of about 10 *nm*. A second possibility is to mount the camera on a traversing mechanism. Such a traversing mechanism enables step by step movement over a large distance.

Scanning of off–axis holograms

For the reconstruction of off–axis holograms, the holograms are illuminated with a continuous laser beam (e.g.: HeNe laser) that simulates the beam geometry and angle of incidence of the reference beam during the recording of the hologram (s. Chap. 5) This illumination reproduces the object beam which contains the spatial information of the object. The reconstruction is scanned in real scale free of distortion by the CCD camera if the optical axis of the reconstructed object beam and of the camera objective coincide.

A relative movement between the photo plate and the camera is required to transform the three–dimensional image information into a series of two dimensional images. However, this relative movement can be achieved only through changing the camera position. A displacement of the plate would change the illumination conditions and thus affect the quality of the reconstruction. The set–up to scan off–axis holograms is given in Fig. 6.8.

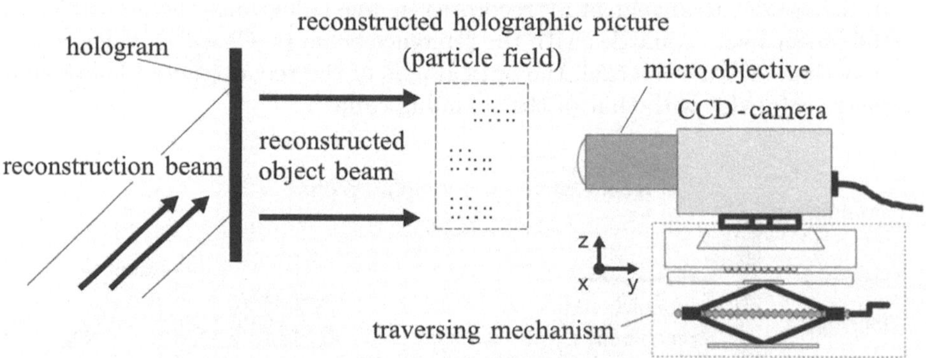

Figure 6.8. Optical set–up for scanning an off–axis hologram

In order to achieve a one–to–one relation between the spatial coordinates of the holographic image and the planar coordinates of the video pictures, it is necessary to monitor the optical distance between the picture plane in focus (focal plane) and the camera sensor, so that each picture can be correlated with a specific depth coordinate value, perpendicular to the picture plane. In order to control the position of the camera and to determine the value of this coordinate, the camera is mounted on a traversing mechanism (s. Fig. 6.8). It consists of a precision screw/steeve drive ($\phi = 12\ mm, 2\ mm$ pitch) coupled to a step motor (1000 steps/cycle). This system provides a linear resolution of 2 μm and permits the repositioning of the camera within a relative error of 10 μm over a distance of 200 mm. The movement of the step motor is monitored by the computer via a RS–232 serial port.

6.2.2 set–up of a digital image processing system

The main components of a PC–based image processing system are a camera, a digitizer, and a PC installed with image processing software. Most of the common camera systems can deliver colored (RGB; false color: red, green, blue) images. However, for the purposes required when holographic reconstructions are to be evaluated, colored pictures do not provide more useful information than do gray value images. In addition, the disc space needed to process colored images is three times greater than the space needed for gray value images. Therefore, gray value images are taken from the reconstructions and the following subsections are restricted to the processing of these images.

The CCD camera records an image of the holographic reconstruction, such as a section of the bubble column in Chap. 5, and transmits the information to the digitizer. Here, the signal is transformed from analog into digital information and is then stored in the PC in an array of 512×512 pixels with a depth of eight bits. A higher camera chip resolution is not necessarily required as it would result in either larger or more numerous images thus requiring the need for more disc space. Use of the CCD camera is preferred because the resolution is high enough to accurately determine the particle size (s. Sect. 6.3.4) and the small images produced can be processed quickly.

6.3 Image processing

The video images obtained from the holographic reconstructions show specific properties. Only a small fraction of the bubbles visible in such an image are depicted sharply, especially if the observing objective has a small depth of field. The edges of these bubbles can be blurred by unsharply depicted bubbles that are reconstructed out of focus of the observing camera. In addition, the gray value of the background varies significantly due to the Gaussian intensity profile of the laser beam and defects that are inherent in the optical parts such as unevenness of the mirrors. The gray values in the image range from 0 (= black) to 255 (= white). The disturbing noise is generated by the diffusive character of the monochromatic illumination (laser beam), which is necessary for the recording of the hologram. Tools implemented by imaging software are able to accomplish the following objectives:

- to filter noisy images to remove speckles

- to detect sharply depicted bubbles
- to create binary images
- to measure the size and the position of identified bubbles
- to determine the velocity of the bubbles

The objective of the image processing in this example is to obtain an image with the sharply depicted bubbles clearly separated from the background. The information about the shape, size, and location of the bubbles is then derived from these images. The single steps of evaluating these images are illustrated by examples obtained from holograms of the aerated stirred vessel presented in Chap. 5.

6.3.1 Evaluation of single pulsed holograms

A typical image taken from the reconstruction of a single pulse hologram is shown in Fig. 6.9. The problems mentioned above are evident in this image.

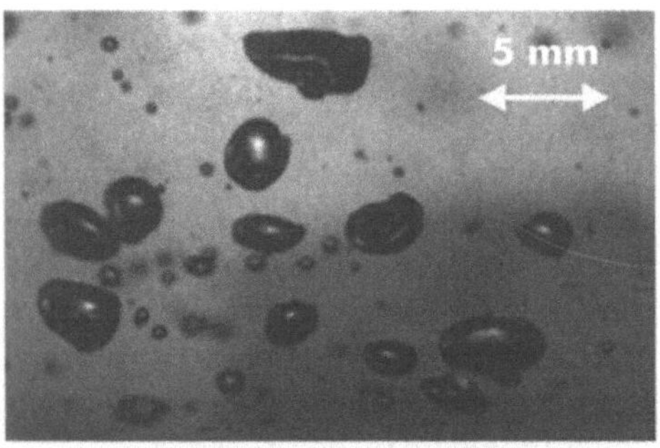

Figure 6.9. Original image from a holographic reconstruction

A reliable analysis of the bubbles that are depicted sharply in this image is impeded by the substantial variations in background gray values. The gray level difference between a sharply depicted bubble and the darker part of the background is in some cases smaller than the gray level difference between an unsharply depicted bubble and the brighter part of the background. Therefore, any common gradient filter operation which is based on a minimum gray level difference between a pixel and its neighbors followed by a threshold filter operation fails. This is because sharply depicted bubbles are undetected, or unsharply depicted bubbles are also detected. In addition, the edges of several sharply depicted bubbles are blurred by bubbles which are out of focus.

The procedure which is applied to evaluate the image starts with a common 5×5 median filter in order to reduce the high frequency noise of the image. This operation is followed by a filter that smoothes the gray level of the background. The filter is based on a very strong averaging filter operation with kernel sizes of up to a quarter of the entire image. The gray value of each pixel of the median filtered image is then divided by the gray value of the corresponding pixel in the average filtered image. The result is then multiplied by a constant factor. As a result of this filter operation, the gray values of the pixel belonging to the background scatter around this factor, and those of the pixels belonging to the bubbles have a very low gray value. The result of the application of these first two filters to the image in Fig. 6.9 is shown in Fig. 6.10. No information about the sharpness of the imaged bubbles is lost by applying this filter, but the gray level of the background becomes substantially more uniform.

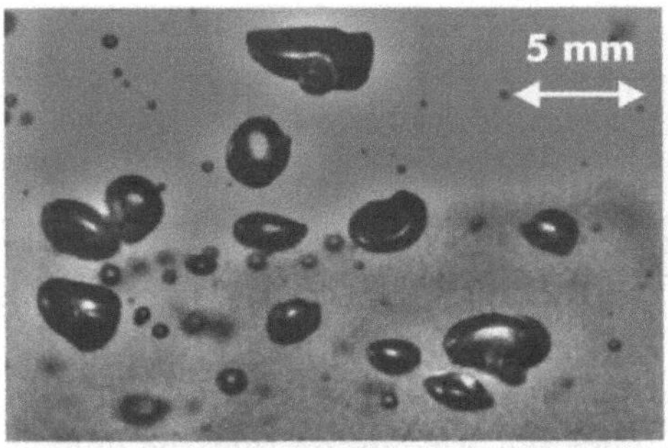

Figure 6.10. Image in Fig. 6.9 after being treated with the background filter

After the application of the background filter, the difference between the gray values of pixels belonging to the edge of a sharply depicted bubble and those of pixels belonging to the background scatter around the chosen factor. The edge detection algorithm is now applied to the image. It is based on the calculation of the resulting gradient vector for each pixel. This vector is calculated from the gray value differences between the pixel and its eight neighbors. If the gradient exceeds a given threshold, the pixel is marked. From this pixel, the edge detection algorithm searches the next points of the edge by following the normal of the gradient vector of those points along the highest gradient vector. The image is marked pixel by pixel while the direction of

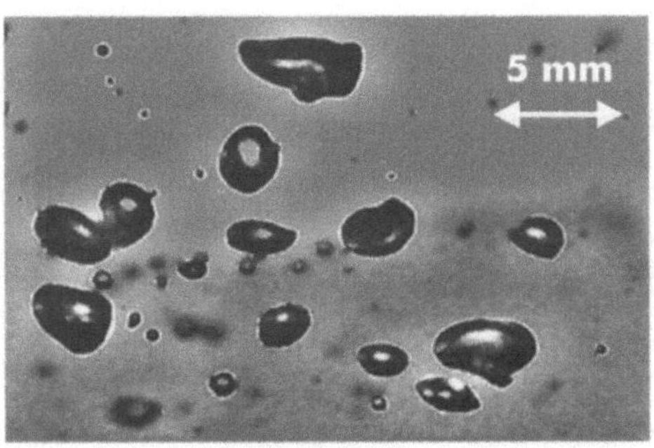

Figure 6.11. Image in Fig. 6.10 after the edge detection

movement is maintained. The condition to leave the edge detection algorithm is either a completely surrounded bubble, or that the gradient of the next pixel is below a minimum threshold value, which indicates that the edge of the bubble is blurred. As the result of the edge detection, an image in which the sharply depicted bubbles are completely or partly surrounded is obtained (s. Fig. 6.11).

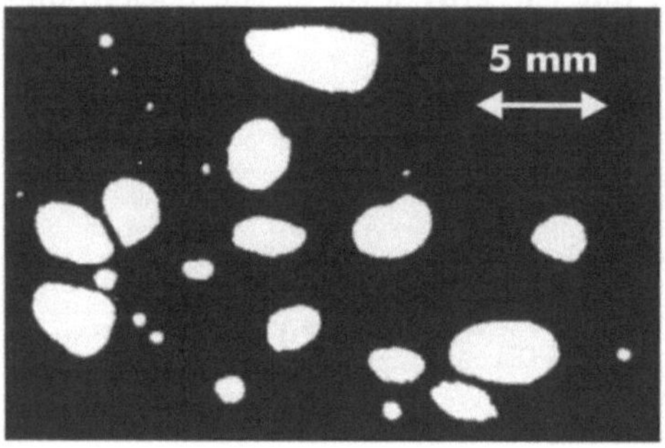

Figure 6.12. Image in Fig. 6.10 after the evaluation and binarization

In order to close the edges of only partly closed bubbles, an additional routine is applied. This is performed under the assumption that the bubble is represented by an ellipsoid. The detected part of the edge of the bubble is closed by extrapolation. When proceeding as described, every bubble sharply depicted in the image is detected and marked. The detected bubbles are filled and the image is binarized, as shown in Fig. 6.12. The detected bubbles are

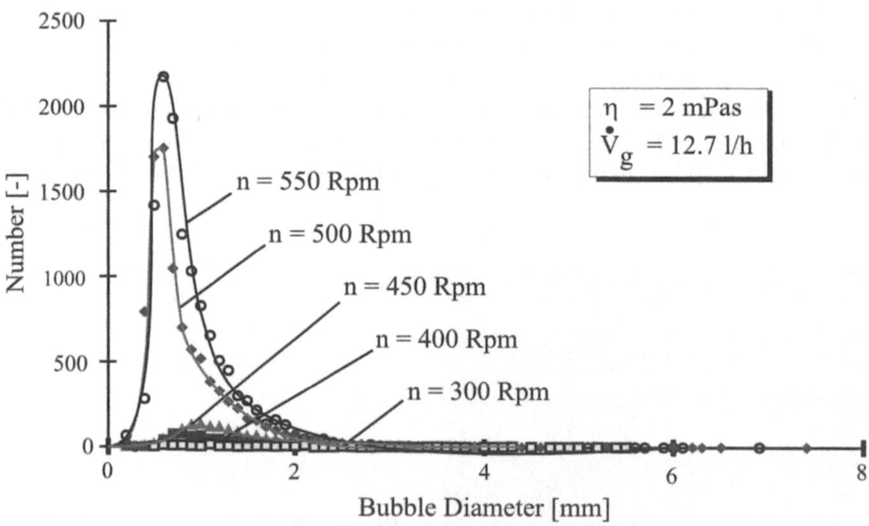

Figure 6.13. Size distribution of the generated bubbles

measured and sized. By recording an image of an object with a known size, the scale of the image is determined. The measuring unit is represented by the pixel array of the frame grabber combined with the real scaling factor after the scaling factor has been considered (pixel/length–scale). The coordinates of the center of gravity of the bubbles and their sizes are calculated using this factor from the binarized image.

The results of these procedures are stored to a file, where they serve as the input data for further data reduction. A typical result is given in Fig. 6.13. In this application example given in Sect. 5.3 for different stirrer speeds, the number of the bubbles generated is plotted versus bubble diameter. It can be derived from Fig. 6.13 that with a rising impeller speed the number of bubbles increases and their diameter decreases.

6.3.2 Evaluation of double pulsed holograms

The flow field in an aerated stirred vessel has a highly three dimensional character. When considering the various operating conditions, there is no preferred flow direction for the generated bubbles at any place inside the vessel. As the velocities of the bubbles also have to be determined, the software module maps not only the two images of one bubble against one another, but also distinguishes its first image from its second.

In order to determine the velocity vector of each bubble inside the vessel, the distance the bubble has covered in the time interval between the two exposures to the laser beam is measured. If the scale of the image and the pulse separation is known, the velocity of each bubble can be calculated from this value. From each of the two holographic reconstructions, a two dimensional velocity vector of every bubble is obtained. When correlating these two views, the real 3D velocity of each bubble is calculated. A typical image taken from a reconstructed double pulse hologram is shown in Fig. 6.14 on the left. This image is included to facilitate the location where the image at right in Fig. 6.14 is recorded. In order to optically distinguish the first pulse from the second, the two pulses were recorded with a different laser–beam intensity. The brighter images of the bubbles belong to the first pulse (lower intensity), and the darker images belong to the second pulse (higher intensity). When using this information it can clearly be derived from the image at right in Fig. 6.14 that the bubbles were moving upwards in the time interval between the two exposures to the laser–beam.

In general, for the two images of one bubble obtained from reconstructions of double pulsed holograms, three cases have to be distinguished: the images overlap partly; the images overlap completely; or the images do not overlap. The time interval between the two exposures is adjusted in such way that most of the bubble pairs partially overlap (s. Fig. 6.14). Since the region where the bubbles overlap is substantially darker than the rest of the images, it is a suitable tool to map the two images of one bubble on one another. In images depicting complete bubble overlap, the bubble had been moving

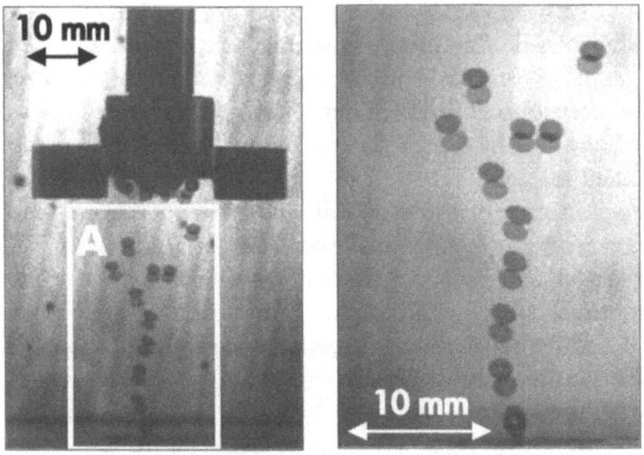

Figure 6.14. Reconstruction from a double pulsed hologram; overview (left) and detail (right)

parallel to the optical axis of the observing camera. This is detected by the software, because of the very low gray value in which the bubble is imaged. The vertical coordinate of the bubble is determined, which is the criterion needed to identify this bubble in the second view, where the bubble should have been moving horizontally. If a bubble is relatively small and moving fast, its images may cease to overlap, as was sometimes observed in the region next to the stirrer.

In this case, a search volume is being created in order to map the two images to one another. The radius of this volume is calculated from the scale of imaging, the pulse separation and the maximum speed, which is the agitator's circumferential speed. Inside this circle, the second image of the bubble is being searched. The correlation of the images to the two pulses is then performed as in the described below.

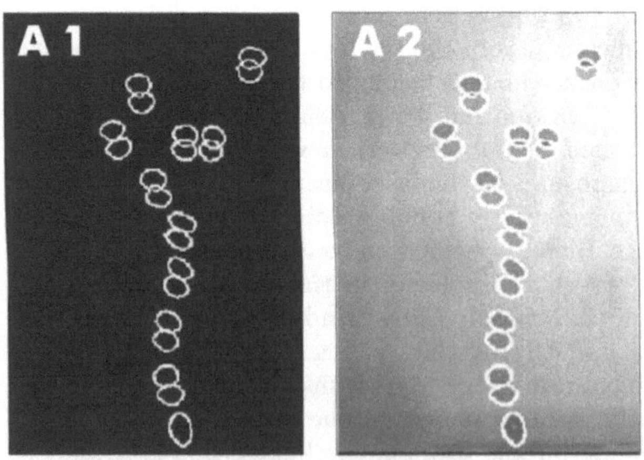

Figure 6.15. Evaluation of a reconstruction from a double pulsed hologram

Referring to the right image in Fig. 6.14, the principle features of this software module can be explained. Coming from the original image, identical procedures are performed as in the case of single–pulsed holograms, up to the point, where the edges are detected and closed, and the image is binarized (s. Fig. 6.15, left side). The following algorithm performs an ellipse–fit in order to detect the position of the two centers of gravity of one pair of bubbles. It then calculates the velocity by dividing the distance between these two points by the pulse separation. In order to determine the direction in which the bubble is moving, the binarized image is added to the original image, as shown in the image at right in Fig. 6.15.

Since the edges of the bubbles have already been detected, the average gray value of both areas is calculated. The brighter area with the higher average gray value belongs to the first pulse, whereas the darker area belongs to the second pulse. The direction of the velocity vector in this view is determined.

After this operation has been performed for each image of the two holographic reconstructions, the results are stored into a file, where they serve as the input data for the further data reduction. Results obtained by applying these filtering algorithms have been shown in Sect. 5.3.

6.3.3 Stereo matching algorithm

Two holograms were recorded simultaneously and evaluated separately from one another. The correlation of the two holograms presupposes the stereoscopic optical set–up described in Sect. 5.3.4. In this set–up the optical axis of each camera is parallel to one of the axis' of a cartesian coordinate system.

After both holograms taken at one operating point have been completely scanned into the series of 2D images and evaluated as described above, the images are reset together again numerically including the obtained results. This results in two 3D reconstructions of the flow field in the vessel, displaying the same scenario, but seen from two perpendicular viewing–directions. These two 3D images are correlated with one another by applying a stereo matching algorithm. The physical model this algorithm is based on is given schematically in Fig. 5.13 in Sect. 5.3.4.

6.3.4 Accuracy

Due to the high quality of the lenses utilized no distortion correction has to be performed. The cameras conduct only linear movements. The parallelism of the axis is ensured by alignment of the traversing slide and horizontal justification of the camera by means of a pitch plane regulating screw. A simplified calibration process can thus be applied. The projection equation is simplified and the number of unknown projection parameters is reduced.

The main source of uncertainty in this measuring technique was found to be the pixel representation of circular objects (droplets or bubbles), especially if these objects are represented by less than 10 pixel (independent of the absolute pixel size). The resolution of the area measurement tool in the developed code was set to 5 pixels, resulting in a maximum error of ±3% for areas with sizes between 6 and 40 pixel. In this work, the sizes of the bubble images ranged between 14 pixels ($\phi = 0.3\ mm$) and 200 pixel ($\phi = 1\ mm$). The error in imaging larger objects was found to be less than 1%.

The determination of the velocity of one droplet or one bubble depends on the quality of the stereo matching and the yield of the found couples. Due to the fact that thousands of droplets or bubbles are considered, the mean velocity can be calculated with the high accuracy of ±5%.

6.4 Evaluation of interferograms

The distances between interference lines next to the wall surface must be determined in order to evaluate interferograms with infinite fringe field adjustment. The local heat or mass transfer coefficient of interest can then be calculated by the form given in Chapter Five. In the past interference patterns were measured using a photometer equipped with a micrometer gauge for manual adaptations of specific directions. The gaps between the interference lines were investigated along a fixed direction and transferred to a personal computer. This action was very time–consuming. A time saving and precise digital image processing system has recently created for a personal computer. The computer program DIPS for digital image processing was written so that interferograms with any perimeter could be evaluated. Generally a rigid algorithm cannot be given because the quality and the contrast of interferograms vary over a great range depending on the test conditions. For this reason DIPS can be handled interactively, so that the application and order of the different routines can be selected freely. The most important algorithms are described below.

Generally the following method of determining local heat or mass transfer coefficients is proposed:

- Scanning and storage of an interferogram
- Enhancement of quality and contrast (elimination of incorrect pixels, filtering, smoothing etc.)
- Determination of the exact distances of the interference lines near to the wall
- Calculation of the local heat or mass transfer coefficient

When scanning interferograms, emphasis should be placed on achieving a good contrast in order to shorten the time necessary to enhance the image. In general, the user selects different procedures for image enhancing from personal experience. However the following possibility seems the most favorable among many others. The image quality is improved by spatial–domain techniques. Noise resulting from digitizing errors is eliminated using the median filter. Image disturbances caused by scattering from dust particles are reduced by a repeated application of average filters. Vicinity averaging filters with an area of 3×3 or 5×5 are suitable. Depending on the intensity of the noise, this filter needs to be applied several times to achieve a satisfactory result.

As explained in Chap. 4, the local heat or mass transfer coefficient is determined from the distance of the isotherms, i.e. approximately the distances of the interference lines. Isotherms run normal to the heat flux and thus assuming a constant wall temperature, parallel to the surface near the wall. Also for a constant heat flux boundary condition a constant wall temperature can be presupposed if the heat transfer coefficient is very small compared to the wall conductivity. Therefore, the gap between the isotherms can be measured along straight lines perpendicular to the perimeter of the interferogram. An algorithm was developed in order to achieve this distance automatically. Initially a perimeter similar to the one which is to be investigated is inputted into the computer. For this surface points of the interferogram are marked interactively using a mouse.

Two types of surface points must be distinguished: perimeter points, those describing the perimeter; and evaluation points, those where heat or mass transfer can be calculated. The perimeter points serve only to describe the surface and the perpendicular at neighboring points, whereas the heat transfer is not calculated. These points are necessary, for example, at corners with a non–continuous curvature. The distinction is realized using a variable ("flag") that is set to zero for perimeter points and to one for evaluation points. The coordinates of each point including the flag are stored in a file on the hard drive and can be applied for the examination of any interferogram similar to the stored perimeter. It should be mentioned that the number of points influences the evaluation accuracy. Depending on the interferogram between

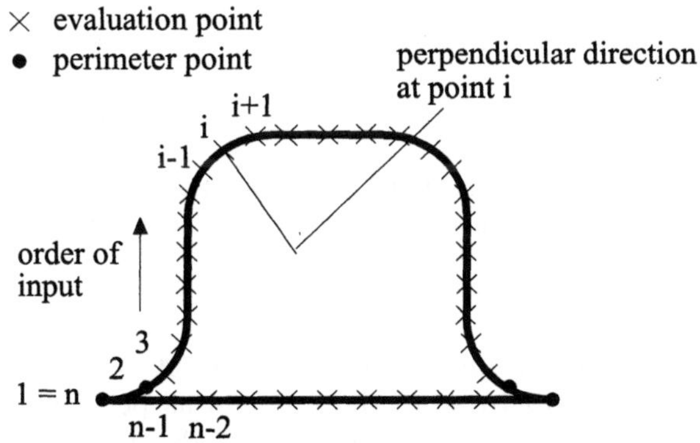

Figure 6.16. Schematic distribution of surface points along a typical wall surface

100 and 300 points are recommended. A schematic distribution of points is illustrated in Fig. 6.16.

The size and location of the interferogram in the frame memory of the image processing system vary as a result of different focusing and camera position. For this reason the interferogram is framed with a mouse, so that its location is known within the program. Now the stored points similar to the marked perimeter are read from the hard drive and converted to the interferogram surface by a central elongation.

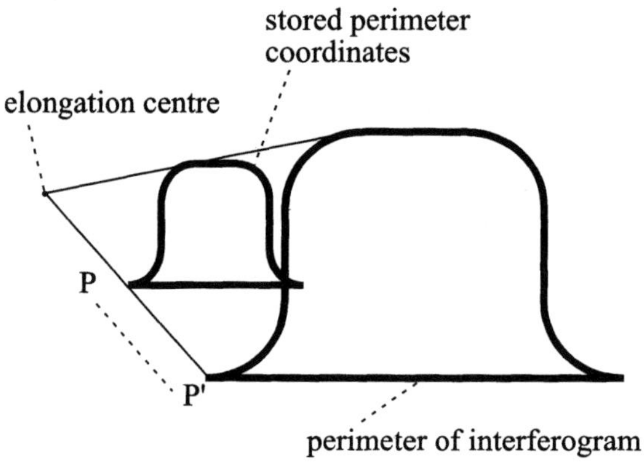

Figure 6.17. Conversion of stored surface point coordinates to the perimeter of the interferogram

In the next step the calculation of the local heat or mass transfer coefficient for each evaluation point is enforced (s. Fig.s 6.17 and 6.18). A circle is drawn through the evaluation point P and the two neighboring points in order to calculate the perpendicular at point P. The direction normal to the wall surface at point P is described by the connection of point P and the circle center. The curvature of this circle disappears, if the points are located on a straight line with a definite normal direction whose calculation does not

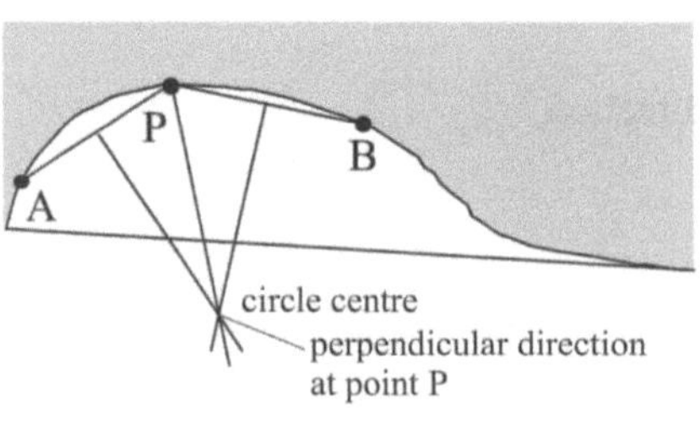

Figure 6.18. Determination of the perpendicular direction at the evaluation point P

need to be explained. Although a definite direction does not exist at corner points of the perimeter, it is necessary to store these points as perimeter points. They serve for the calculation of the curvature circle at the neighboring point and, additionally, can be used to describe the surface, especially to represent the results of separate programs.

The direction in which the gray value behavior is investigated–inside or outside the surface of the interferogram – is determined by the input direction of the surface points. It can be defined for example that points stored in a clockwise direction surround the interferogram on the outside (channel flow), and counter–clockwise stored points describe external flow.

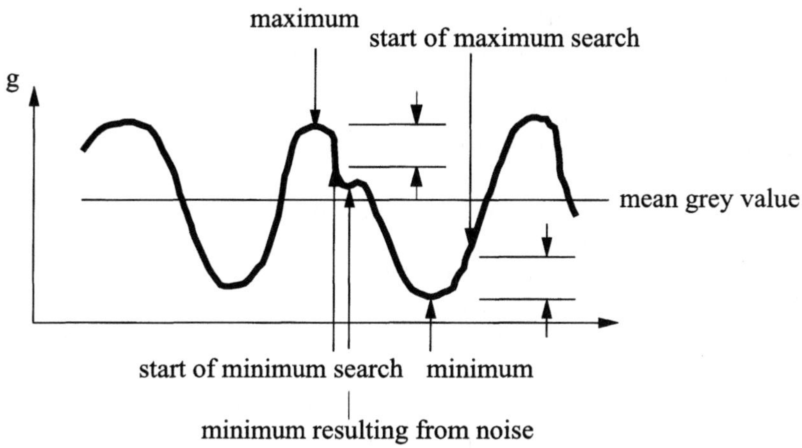

Figure 6.19. Gray value behavior during a search for extremes

The gray values along this straight line are read from the frame memory of the digitizer board and stored as a series of integer values. The detection of extreme values within this series of points follows. In principle two possibilities exist for the detection of extremes. The first is to calculate the derivative for each point considering their neighboring points. An extreme value is spotted by a change of sign. It is problematic to make a statement about the extreme itself, thus it is particularly difficult to dispose of extremes resulting from noise. For this reason an algorithm was established in order to detect the maxima and minima of absolute gray values. At the beginning the mean value and the mean variation of the series are computed. If the first gray value of the series is bigger than the mean value, the search for a maximum is initiated, otherwise a minimum is looked for. During a maximum search a value is equated to the maximum, if it is bigger than the momentary maximum. As mentioned earlier, extremes can be evoked by noise and must not be considered. Generally they are not as emphasized as those resulting from interference lines. For this reason a kind of hysteresis is introduced.

A maximum search is changed to a minimum search only if the concerned gray value is a concrete difference δ smaller than the momentary maximum. The amount of δ is equated to 0.3 times the mean variation. The coordinates of the point with the maximum gray value stored only during a switch to a minimum search. In this way the minimum in Fig. 6.19 caused by noise is detected, but not stored, because a switch to maximum search is omitted. The minimum search is continued as long as the next minimum is detected and stored after the switch to maximum search. The minimum search is realized analogously and so does not need to be described.

Now the distances between the extremes are known as numbers of pixels and must be converted to metrical units. A conversion factor is calculated from the maximum expansion of the framed interferogram whose height and width are apparent from the design of the test facility. Experience has shown that conversion factors should be determined for each coordinate direction, because distortions exist for square pixels as well. The transfer coefficient is then calculated by the formalism given in Chap. 4. It is stored together with the geometric coordinates in a file and can be plotted using a graphic program.

The evaluation of interferograms is demonstrated by the channel flow described in Chap. 4. The interferograms are evaluated using the procedures explained. For each interferogram about 250 scan points were considered. A typical distribution of the local heat transfer coefficient is shown in Fig. 6.20 for channel radii of 1 mm. In the range of low Reynolds number a typical laminar flow pattern was established resulting in a maximum of the heat transfer in the middle of the distance between two adjacent corners. Conversely the heat flux decreased at the corners, especially at the convex one. It seems that the heat flux vanished at the convex corners because of the stagnant flow.

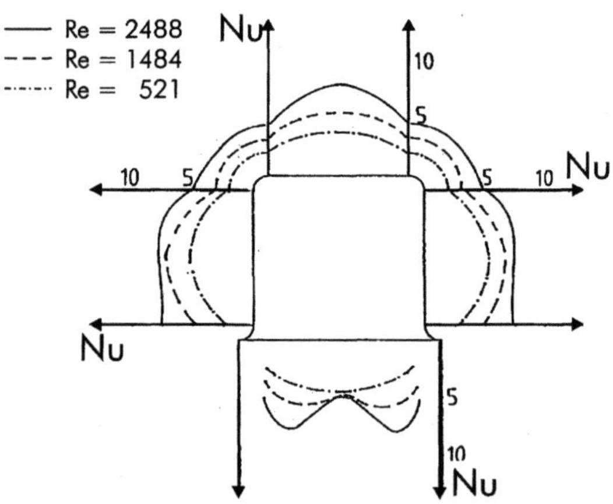

Figure 6.20. Distribution of local heat transfer coefficient

For turbulent flow the effect of the convexly curved radii in turbulence generators can be seen clearly because of the local maxima of the heat transfer coefficient in the proximity of the corners. Contrary to laminar flow the heat flux decreased only in the immediate vicinity of the convex corners.

7 Light Scattering

Boris Kruppa, Gernoth Strube, and Christof Gerlach

7.1 Introduction

The term light scattering refers to physical processes which involve the interaction of light and matter. Light that incidents on an ensemble of particles – crystals, aerosols, molecules, atoms etc. – is partially "deflected" or absorbed. In addition to a change in direction a change in frequency is also possible depending on the scattering process. The evaluation of the scattered light with regard to its intensity, wavelength, and direction often yields valuable information about the scattering matter. A set–up for scattering experiments is shown schematically in Fig. 7.1. Beside the light source and the detectors, also other optical elements, such as mirrors, filters, lenses etc., may have to be placed in the beam path to fulfil the requirements of the specific application.

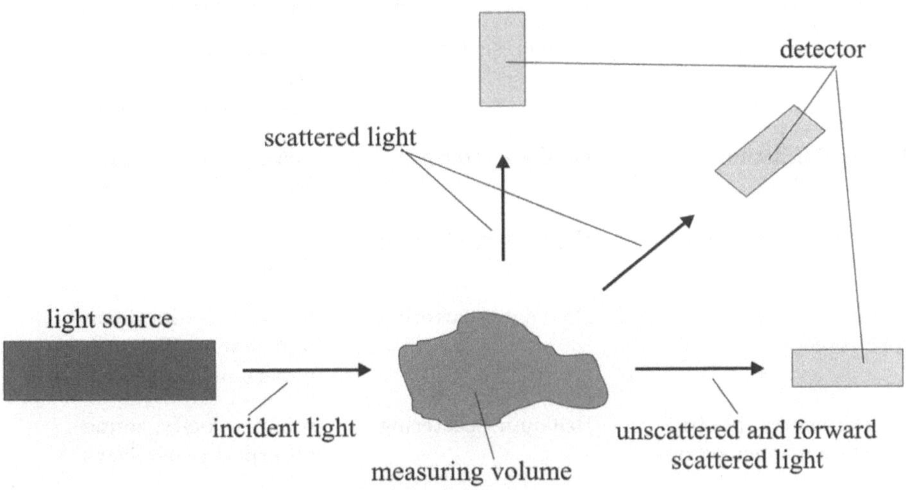

Fig. 7.1. Basic elements of a light scattering set–up

Light scattering processes gain increasing importance in heat transfer investigations, especially since their non–intrusive character prevents any disturbance on the investigated system during the measurements. The rapid development of laser technology and micro–electronics have strongly enhanced the power and versatility of measuring systems based on light scattering effects.

Most of the light scattering effects have been discovered within a period of about fifty years around the turn of the previous century (Tyndall 1869, Rayleigh 1871, Mie 1908, Brillouin 1922, Compton 1923, Raman 1928). Many different measuring techniques have been developed since then based on a few basic scattering effects. Since not all of these effects are equally important for applications in heat transfer, this section is divided into two parts. The first part provides a classification and a short description of the physical background of the various scattering processes, while the second part focuses more extensively on the technical application of these processes in the field of heat transfer. Table 7.1 gives a short overview of some measurement techniques, its corresponding scattering processes and typical applications. Some of these techniques are also described in greater detail in separate sections of this book.

Table 7.1. Typical measuring techniques based on light scattering

Measuring technique	Scattering Process	Application
Laser Doppler Velocimetry (LDV)	Mie–scattering	particle size, particle velocity
Raman spectroscopy	Raman–scattering	molecule concentration, temperature, (molecular structure)
X–Ray diffraction	Bragg–scattering	density fluctuation, structure of crystals
Rayleigh thermometry	Rayleigh–scattering	temperature, density
photon correlation spectroscopy	Rayleigh–scattering	thermal conductivity, diffusion coefficient, molecular structure
interferometry, photon correlation spectroscopy	Brillouin–scattering	sound velocity, sound absorption coefficient
fluorescence spectroscopy	fluorescence	density of atoms and molecules, temperature
absorption spectroscopy	all scattering processes	concentration of atoms and molecules

7.2 Scattering Processes

7.2.1 Interaction of Light and Matter

In order to improve the understanding of the involved processes simplified models are usually used to describe the specific effects of each physical process of interest. Scattering involves both the properties of light radiation and those of the encountered matter. Therefore, we have to deal with aspects such as wavelength and polarization of the light and – depending on the size and structure of the matter – with diffraction, refraction and reflection, or absorption and re–emission of (light) energy by an oscillator (matter).

In spectroscopy the terms energy E, wavelength λ, wave–number $\tilde{\nu}$ and frequency ν are often used in the same way since they are all related by proportionality. The wave–number is the reciprocal value of the wavelength

$$\tilde{\nu} = \frac{1}{\lambda} \tag{7.1}$$

which is related to the frequency by the speed of light c

$$\nu = \frac{c}{\lambda} \tag{7.2}$$

and coupled to the energy by Planck's law (h is Planck's constant)

$$E = h\nu \tag{7.3}$$

Usually the energy is expressed in $[cm^{-1}]$, which is the unit of the wave–number, or in $[Hz]$, the unit of the frequency.

There are several possibilities of classifying the various scattering processes. Here the distinction between elastic and inelastic scattering is made. An elastic scattering process refers to an interaction without a permanent exchange of energy between the light and the matter. This restriction of energy exchange does not prohibit a change in direction, but according to Planck's law it does prohibit a change in frequency (wavelength, etc.). An inelastic process, on the other hand, leads to a permanent energy exchange. The energy content of the matter after the interaction is either higher or lower than that in the original state. Due to the conservation of total energy, the energy of the emitted radiation is changed as well, resulting in a change of frequency (wavelength, etc.). This simplified model is shown schematically in Fig. 7.2.

Fig. 7.2. Elastic and inelastic scattering

If a certain volume of particles (aerosols, atoms, molecules, crystals, etc.) is observed with regard to the interaction of an incident light wave, the various scattering processes may take place simultaneously. The light is scattered in all directions with and without a change in frequency. Since part of the incident light is scattered in all directions, the light leaving the volume in the direction of incidence has a lower intensity than the light entering this volume. With a spectrally resolving detector an absorption spectrum can be obtained. Since the non–scattered light is included in the absorption spectrum (except for the small amounts of energy transiting to or from the particles in inelastic scattering processes), the absorption spectrum is complementary to the spectra of scattered light.

A very well known example of an effect caused by scattering is the appearance of the sky, blue during the day–red at dawn and dusk. The scattering particles are the molecules of the air in the atmosphere. Their size is small compared to the wavelength of the visible spectrum and they are randomly oriented. Therefore, Lord Rayleigh's assumption that the intensity of the scattered light is inversely proportional to the fourth power of the wavelength is applicable. The wavelength of red light is around 7000 Å, that of blue light around 4000 Å. Applying the $1/\lambda^4$ dependence for the scattering intensity shows that blue light is scattered about ten times stronger than red light. During the day the sunlight, a continuous spectrum over all visible wavelengths, has to travel a relative short distance through the atmosphere. Since the blue light is scattered much stronger than the red light, the sky appears blue. In the morning and evening the sunlight has to travel a longer distance through the atmosphere before it reaches the ground. The blue light has been scattered so much along the way that mainly the red light remains to be scattered so that the sky appears red.

7.2.2 Elastic Scattering

An overview of the major elastic scattering processes is given in Table 7.2. In this section mainly those processes are described very briefly that are not treated more extensively later in this book. Absorption spectroscopy is treated in the inelastic scattering part, even though the observed wavelength is the same as the incident wavelength. The characteristics of the absorption spectrum, however, mainly result from the effects of inelastic scattering.

Mie and Rayleigh scattering occur in volumes of randomly oriented particles and are used extensively in heat and mass transfer measurements.

Bragg scattering can only be observed in structured arrangements of particles such as crystals and if the distance between adjacent particles is of the same order as the incident wavelength. Therefore Bragg scattering is of less interest as a measuring technique for heat and mass transfer investigations. In crystals the atoms are arranged in equidistant sets of planes. The light emitted from each atom due to the interaction with the incident light interferes and shows a characteristic intensity distribution, very similar to

Table 7.2. Elastic scattering processes

Scattering process	Seeding, scatterer	Frequency shift $\frac{\delta\nu}{\nu}$
Mie–scattering	particles, which are large compared to the wavelength (dust, soot)	0
Rayleigh–scattering	particles, which are small to the wavelength (molecules, microscopic suspended particles)	0
Bragg–scattering	structure of crystals	$\pm(10^{-4} - 10^{-2})$ (Brillouin–scattering)

the effects observed due to diffraction of light at regularly spaced openings (two-dimensional). The strongest intensity is observed if the Bragg condition is fulfilled (Bragg angle). Since the characteristic distance between the atoms in lattice structures is of the order of one Å and the according electromagnetic waves are in the X–ray region, this method is called X–ray diffraction and is very useful in the investigation of the structure of solid–state materials. However, Bragg scattering is utilized in combination with Brillouin scattering (s. below) in an important element of the Laser–Doppler–Velocimeter, the Bragg cell.

7.2.3 Inelastic Scattering

Raman–scattering, absorption and fluorescence techniques are important methods in heat and mass transfer applications that are described in detail in later sections in this book.

Compton–scattering appears when radiation of high frequency interacts with the electron closely bounded in the atom. According to Planck's formula given in Eq. 7.3, high frequency radiation (i.e. X–rays) consists of photons with high energy.

This large amount of energy cannot be absorbed by the electron since the subsequent strong movement of the electron can not be "controlled" by the forces of the nucleus. Therefore the scattering process can be viewed as a collision process between a photon and an electron of the scattering matter. This process can be characterized according to the laws of mechanical collisions. Part of the photon energy is used to move the electron with a certain speed in a certain direction. The photon is also deflected in the opposite direction and contains less energy and therefore a lower frequency than before the collision.

Brillouin–scattering results from statistical density fluctuations due to acoustical vibrations in the scattering medium. These fluctuations travel at the local speed of sound and the frequency of the scattered light is Doppler–

Table 7.3. Inelastic scattering processes

Scattering process	Seeding, scatterer	Frequency shift $\frac{\delta\nu}{\nu}$
Raman–scattering	molecules	$\pm(10^{-5} - 10^{-1})$
Compton–scattering	molecular vibration optical lattice vibration	$\pm(10^{-3} - 10^{-2})$
Brillouin–scattering	acoustic vibration in crystals and liquids	$\pm(10^{-6} - 10^{-5})$
fluorescence	atoms, molecules	$\pm(0 - 10^{-2})$
absorption	atoms, molecules	0

shifted. The relative velocity of the density fluctuation waves depends on the angle of observation, ranging from the speed of sound (observation perpendicular to the plane of the waves) and zero (observation in the plane of the waves). Therefore, the amount of the frequency shift also varies with the angle of observation. The maximum frequency shift is observed in the direction perpendicular to the plane of the waves and is very small compared to other inelastic scattering processes; the maximum relative frequency shift $\frac{\Delta\nu}{\nu_0}$ for liquid and solid media is of the order of 10^{-5}. The highest intensity, on the other hand, is recorded if the Bragg condition is fulfilled (the angle between the direction of the incident light and the direction of the acoustical wave is equal to the Bragg angle). Since there is a frequency shift connected with Brioullin scattering, Bragg–scattering can be even regarded as an inelastic process. On the other hand, there is no permanent exchange of energy connected with the process itself, so that it is treated here as an elastic scattering process.

7.3 Light Scattering Techniques in Heat Transfer

7.3.1 Mie–Scattering

Mie–scattering appears if the scattering particles are large compared to the wavelength. Typical scattering particles for Mie–scattering with visible light are soot or dust particles with diameters between 1 μm and 10 μm. The scattering process is based on a change of the electrical and magnetic properties in the vicinity of the scatterer. Due to the large size of the particles, reflection and diffraction have to be considered beside the actual scattering by the particle. The field of radiation resulting from the interaction between a single particle and an incident light beam can be calculated by the Mie theory with three–dimensional Maxwell equations. The resulting effects on the emitted light compared to the incident light are changes in phase, amplitude and polarization. In addition, the light waves coming from different spots of the particle interfere and a characteristic field of radiation is formed.

Fig. 7.3 shows an example of the polar diagram for the scattering of linearly polarized light by a water droplet (top) and the overall scattering intensity as a function of the diameter of the particle.

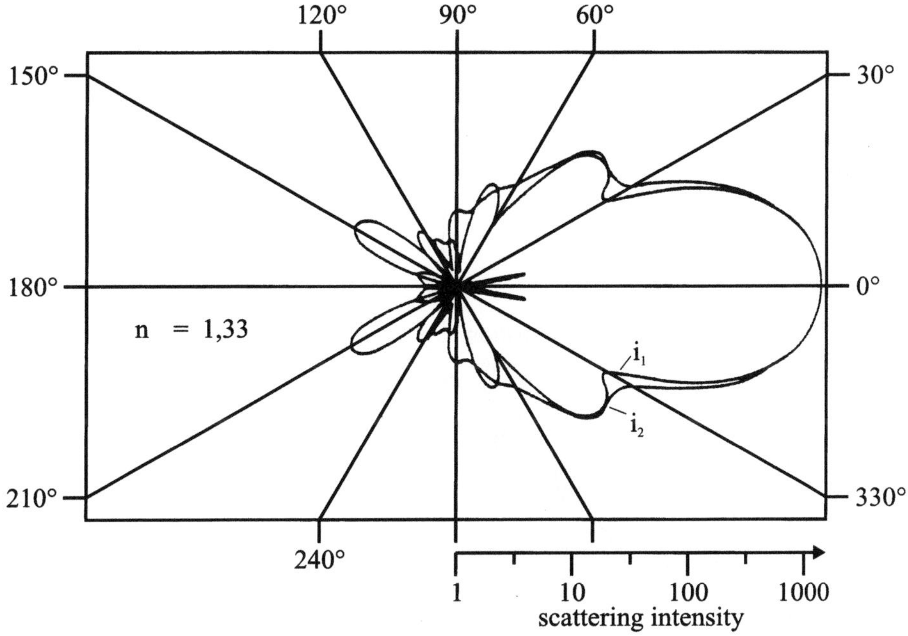

Fig. 7.3. Polar diagram for the Mie–scattering intensity of a water droplet of 1 μm diameter (refractive index n=1.33) illuminated by linearly polarized light. The intensity i is given in dependence of the polarization of the scattered light: $i_1 = q^2 I_\perp, i_2 = q^2 I_\parallel$ [100]

The transition regime in which the particle diameter is slightly above the illumination wavelength ($q > 1$) is usually used in applications of Mie–scattering. Four effects can clearly be seen from the figure:

- the intensity of the light scattered in the forward direction is much larger than in the backward direction
- the overall intensity increases with increasing diameter of he sphere
- the characteristics of the intensity distribution as a function of the angle of observation changes strongly
- the intensity distributions of the two directions of polarization can be different

Generally the particles are assumed to be perfectly spherical; in reality this is rarely the case. The intensity distribution patterns resulting from non–spherical particles are even more complex than those for spheres. However, in practical applications usually only the order of magnitude of the scattering intensity needs to be known. This depends mainly on the incident wavelength, the diameter of the particle and the direction of observation. In most applications the used wavelength and the angle of observation are kept constant while the size and shape of the particles vary to some smaller extent and do not pose a problem to the measurement.

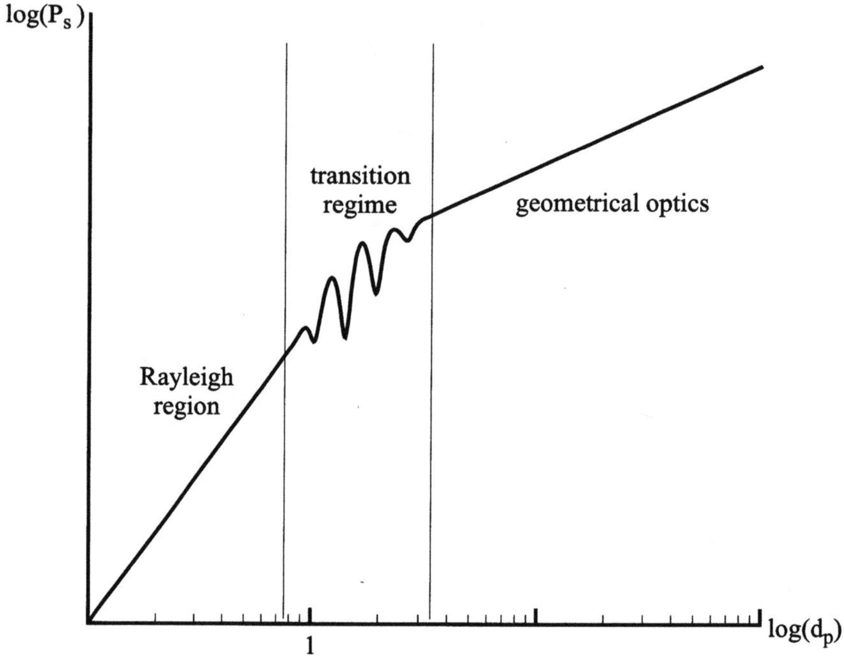

Fig. 7.4. Scattering intensity P_s as a function of normalized particle diameter d_p (ratio of particle diameter and illumination wavelength)

Mie–scattering is used in many applications, mainly in the flow visualization and the determination of particle size. A very common method for the measurement of flow velocities is the Laser–Doppler–Velocimetry (LDV) which is described in greater detail in a separate section of this book.

A newer, more complex method of flow visualization is the Particle Image Velocimetry (PIV). A laser beam is formed into a very thin light sheet illuminating a plane as the measuring volume. For one measurement two consecutive laser pulses are fired within a very short time interval, and the radiation scattered by the particles in the illuminated area is recorded by a two–dimensional camera. A correlation of the particles observed in both pictures is made. Since the time difference between the two exposures is known, the velocity of each particle can be calculated from its distance in the two pictures.

The diameter of scattering particles of sizes between 3 μm and 300 μm can be obtained by Phase–Doppler–Velocimetry (PDV). The scattered light from one particle is observed simultaneously from several directions. The diameter of the particle is proportional to the phase difference, to the wavelength of the incident light, to a geometrical factor for the specific set–up and to the inverse value of the refractive index of the particle [100] (s. Sect. 9).

7.3.2 Rayleigh–Scattering

If light is scattered by particles that are smaller than the wavelength of the light, the process is called Rayleigh–scattering. Unlike Mie–scattering, the size and structure of the particle are not important and the laws of reflection, diffraction and refraction are not applicable in this case. Typical scattering particles are molecules, as those in the example of the color of the sky, or very small particles.

There are two models to describe the interaction of the incident light and the molecule, one is the oscillating dipole and the other the simplified quantum mechanical model. Even though the latter lacks the explanation of some characteristics of Rayleigh–scattering, it is sufficient for the basic understanding of the process. According to this model the molecule first absorbs one photon of the incident radiation, raising it to a virtual state of energy. Since this state is extremely unstable, the molecule immediately returns to its original state, while emitting one photon of light. Since the amounts of absorbed energy is equal to the amount of emitted energy, the emitted light has the same frequency as the incident light (s. Planck's law, Eq. 7.3).

Rayleigh–scattering can effectively be applied to obtain density or temperature measurements in gases and gaseous mixtures. An example of a simple Rayleigh–scattering set–up is shown in Fig. 7.5.

The intensity of the Rayleigh scattered light arriving at the detector is given by the equation:

$$I_r = C_1 \left(\frac{d\sigma}{d\Omega}\right)_{eff} \Omega L N I_0. \tag{7.4}$$

where C_1 is the detection efficiency, $(\frac{d\sigma}{d\Omega})_{eff}$ is the effective differential scattering cross section (or more simply the strength of the scattering process for the observed mixture), Ω is the solid angle of collection, L the length of

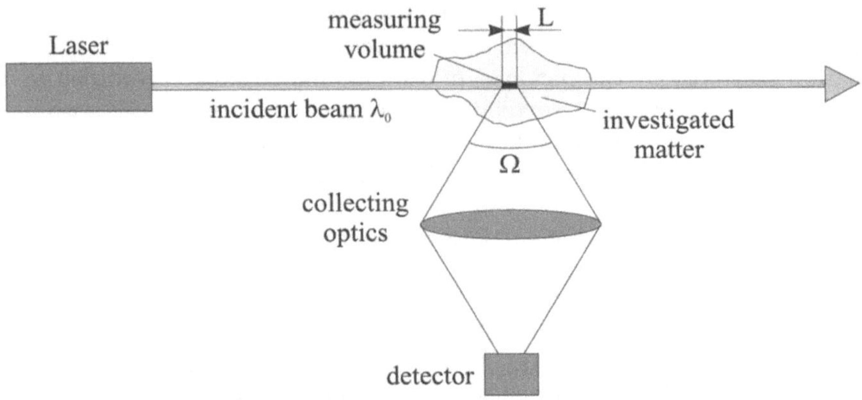

Fig. 7.5. Schematic of a typical set–up for temperature and density measurements with Rayleigh–scattering

the measuring volume, N the number density of molecules, and I_0 the intensity of the laser light in the measuring volume. In the differential scattering cross section the wavelength dependence and the mixture composition are included:

$$\left(\frac{d\sigma}{d\Omega}\right)_{eff} \sim 1/\lambda_0^4 \sum_i x_i \sigma_i \tag{7.5}$$

For a specific set–up and a non–reacting process all factors in Eq. 7.4 are constant, except for the number density. Therefore, the recorded Rayleigh signal is direct proportional to the density of the gas, or, if the pressure is kept constant, inversely proportional to the (absolute) temperature. The system can be easily calibrated under known conditions and online measurements can be carried out. Such measurements where the change in scattered intensity is regarded with respect to time can yield information about the thermodynamic properties of the investigated fluid, such as diffusion coefficients etc. An example for this application of Rayleigh–scattering is photon correlation spectroscopy, discussed in greater detail in a separate section of this book.

7.3.3 Raman–Scattering

The simplified quantum mechanical model that was explained above for Rayleigh–scattering is extended to explain the processes involved both in Raman–scattering and in fluorescence and absorption (s. Sect. 12 and 13). However, due to the inelasticity of these processes, the states of energy of the atoms and molecules have to be regarded in more detail. A diagram of the energetic states of a diatomic molecule is given in Fig. 7.6.

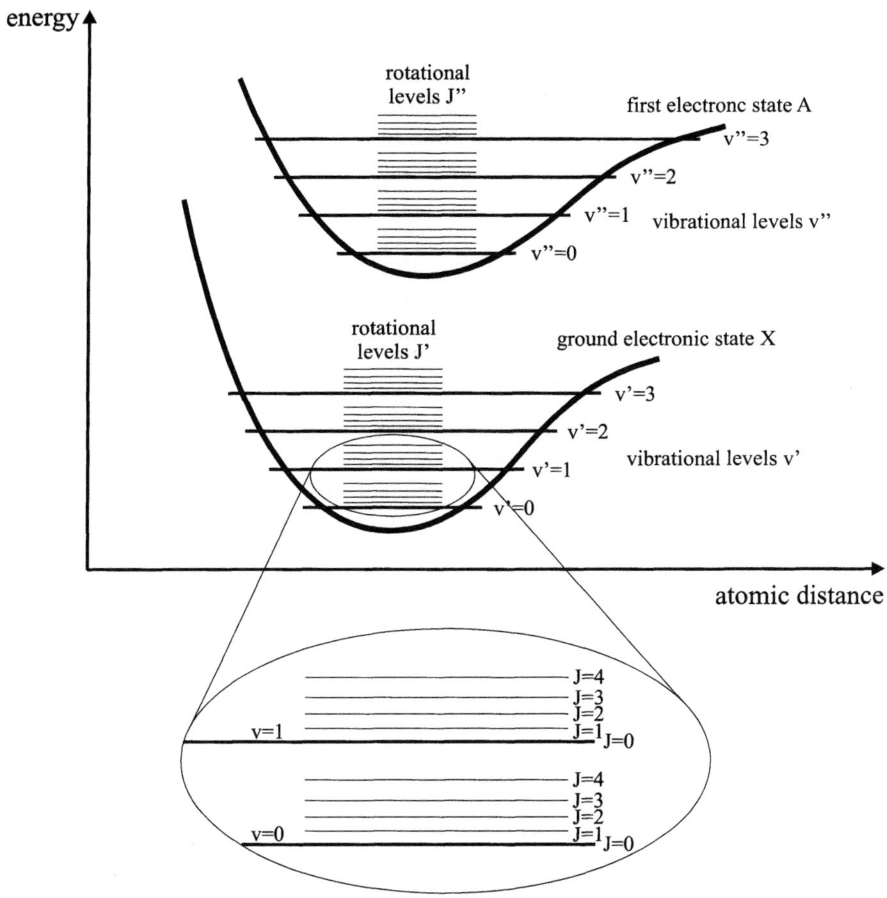

Fig. 7.6. Energetic states of a diatomic molecule. There are three characteristic parameters: the rotational state J, the vibrational state v and the electronic state X, A, B etc. Since in laser spectroscopy only the ground and first electronic state are involved, often the terms are simply denoted by one (lower term) or two (higher term) primes, i.e. (v', J') and (v'', J'')

Bohr's model of the electronic shells in the atom is not exactly valid. However, it does explain many features of electromagnetic radiation absorbed and emitted by atoms. The model says that the protons and neutrons of an atom are concentrated very densely in the center of the atom (nucleus), while the electrons move in paths on spherical shells around this center. The radii of these shells of electronic motion are not random values, they depend on the forces acting between the nucleus and the electrons and, more important, only certain radii are possible. The larger the radius the higher the energy that is contained in the motion of the electron and, therefore, in the atom altogether. The configuration of the electrons with the lowest energy content is called (electronic) ground state; all other configurations are "excited" states. If an electron changes its path of motion, it can only do so in discrete steps corresponding to discrete changes in energy. If this happens in connection with the emission or absorption of light, the energy of the light is equal to the energetic difference of the two involved shells. By applying Planck's law, this energy again corresponds to a certain frequency (or wavelength or wave–number).

Now, if a molecule consisting of two atoms is considered, beside the electronic excitation for each atom there are two more possibilities for energy storage: vibration and rotation. The atoms in the molecule can vibrate against their center of mass; the forces controlling this vibration result from the bonded electrons of the atoms. The third mode of energy storage is the rotation of the molecule around its center of mass. Like the electrons in the atoms, only discrete states are allowed both for rotation and vibration. A detailed and comprehensive discussion of the structure of (atoms and) molecules is given for example by Alonso [86] and, focussing more on the spectroscopically important features, by Ledermann [96]. The differences between adjacent rotational energy states, also referred to as energy levels or terms, is much lower than that between two adjacent vibrational states. Except for some disharmonicities, the energetic difference between adjacent levels, both rotational and vibrational, is equal. The electrons of atoms bonded in a molecule can also be in higher electronic, excited states. The energy difference between a molecule in its ground state and the first electronic state is again much larger than that between vibrational levels. This results in a three stage energy diagram as depicted in Fig. 7.6. Every electronic state contains many vibrational levels, again containing many rotational levels. Since the energy levels are discrete, they are given quantum numbers for indentification. The rotational quantum number is J, the vibrational quantum number v. The electronic levels are not defined by a quantum number, but rather capital letters X, A, B, etc. In the following paragraphs, only the ground state X and first electronic state A are regarded (s. Fig. 7.6).

In Raman–spectroscopy a molecule initially absorbs one photon of the incident light of wavelength λ_0. If visible light is the incident light, this photon contains the energy $E_{\text{abs}} = hc/\lambda_0$. This is more than the molecule can store by rotation or vibration, but often less than the difference between the ground state and first electronic state. Therefore, the photon absorption lifts the molecule up to a highly unstable, virtual level. The molecule immediately drops back to a stable energy level, this time emitting one photon. If this new level is identical with the original level, Rayleigh–scattering is observed (s. above). If the new level is higher or lower than the original level, the scattered light has a different frequency than the incident light. The shift in frequency is referred to as the Raman–shift and is proportional to the energy difference between the two molecular levels involved.

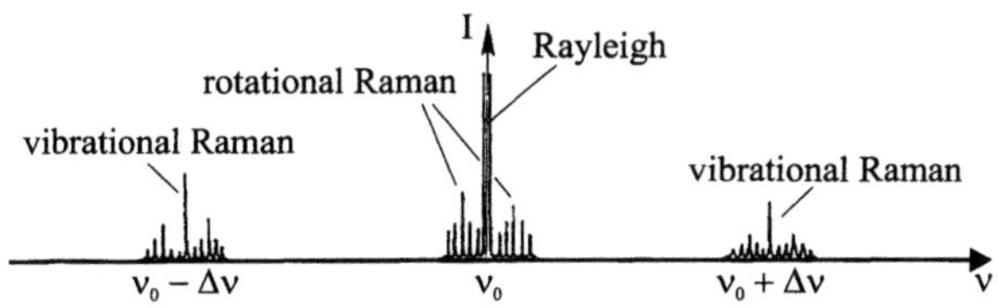

Fig. 7.7. Typical Raman spectrum with rotational and vibrational Raman lines shifted to both sides of the exciting line which is identical with the Rayleigh line

As indicated, the energy difference between rotational levels are much smaller than those of vibrational levels. Therefore, the Raman lines from rotational transitions are much closer to the exciting line than those from vibrational transitions. Fig. 7.7 shows schematically the appearance of a typical Raman spectrum, Fig. 7.8 the energy diagram for Rayleigh, rotational and vibrational Raman–scattering.

Since every species has its own energetic structure, the observed frequency shifts can be related to certain molecules. Raman–scattering Therefore, provides an excellent possibility to detect the concentration of several species in a measuring volume simultaneously. The initial distribution of the molecules of one species among the energy levels in thermodynamical equilibrium is given by Boltzmann statistics. Therefore, the characteristic Raman–spectra of a given species exist as a function of the system temperature, enabling temperature measurements by analyzing the Raman–scattering.

7.3.4 Laser Induced Fluorescence (LIF)

In laser induced fluorescence the molecule of interest also absorbs one photon of the incident laser light. In this case, however, the photon energy is equal to the energetic difference of two energy levels, the original level in the ground electronic state and a corresponding level in the first electronic state (s. Fig. 7.6). Since the energy differences involved are discrete and specific for each species, the frequency of the laser has to be chosen in accordance with the molecule of interest. The states of the upper electronic levels are meta-stable with characteristic lifetimes of about $10^{-5} \sim 10^{-10}$ s. Therefore, after the transition to the upper state the excited molecule drops back to a stable energy level within the ground electronic state as shown in Fig. 7.9.

The emitted radiation again is characteristic for the concentration and temperature of the regarded species. A simultaneous investigation of several species with LIF is generally not possible due to the necessary specific wavelength for excitation.

In order to perform LIF measurements, the structure and distribution of the energy levels of the species of interest along with possible transitions and transition probabilities or transition saturation levels have to be known. As an example Fig. 7.10 shows these levels for NCO and a LIF scan of the $A-X$ system of NCO in a CH_4/N_2O flame [89]. From the figure it is evident

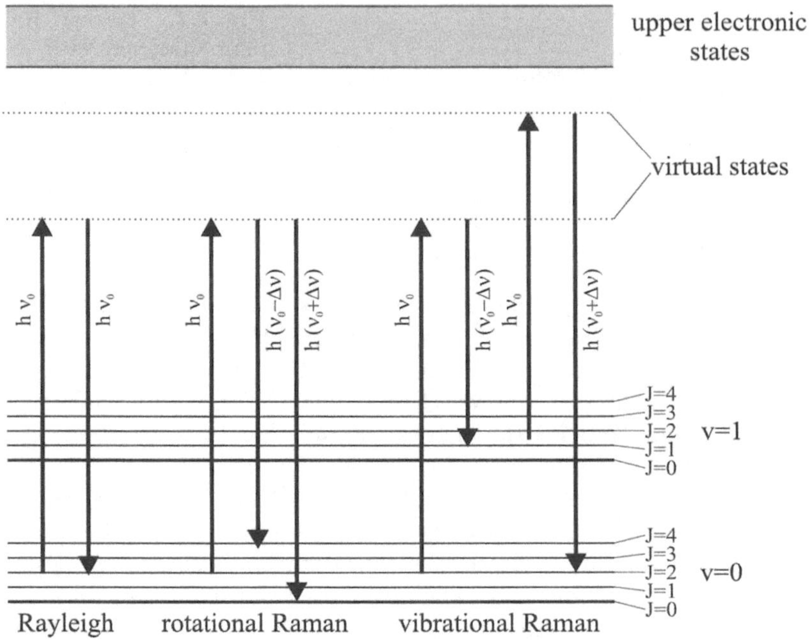

Fig. 7.8. Rayleigh, rotational and vibrational Raman scattering: energy diagrams

that LIF measurements require good background knowledge of the quantum mechanics involved in the processes. The transition of molecules from the upper electronic level to the ground level is not always connected with the emission of light. Collisions with other particles can cause these transitions to occur without light emission; the effect is called collisional quenching. The quenching factors depend on the pressure and temperature, but also on the composition of the investigated system. The non linearity in the LIF signal due to quenching can be avoided by saturating the initial transition. Despite the difficulties involved in these measurements, LIF is a powerful technique for concentration and temperature measurements because its signals are about 5 to 15 orders of magnitude stronger than those of spontaneous Raman– or Rayleigh–scattering.

7.3.5 Absorption

Absorption is not actually a scattering process but rather complementary to all scattering processes. Furthermore it is an important method of determining absolute molecular concentrations and temperatures.

Unlike the set–ups for the other techniques discussed in this chapter, the light detector directly faces the incident laser beam in absorption experiments. The measurement usually represents an integrated measurement over the entire beam path through the measuring volume. Therefore, the spatial

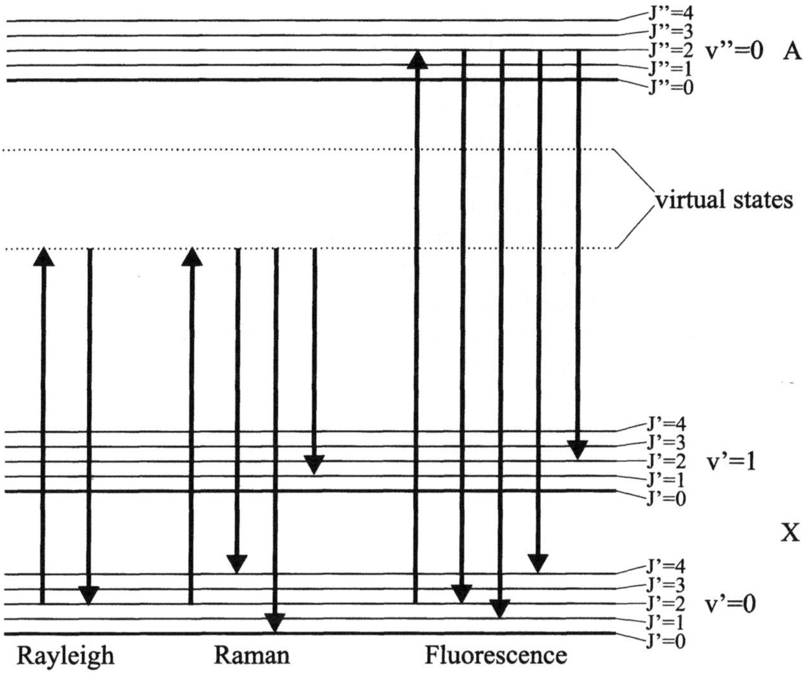

Fig. 7.9. Energy diagrams for Rayleigh, vibrational Raman scattering, and laser induced fluorescence (LIF)

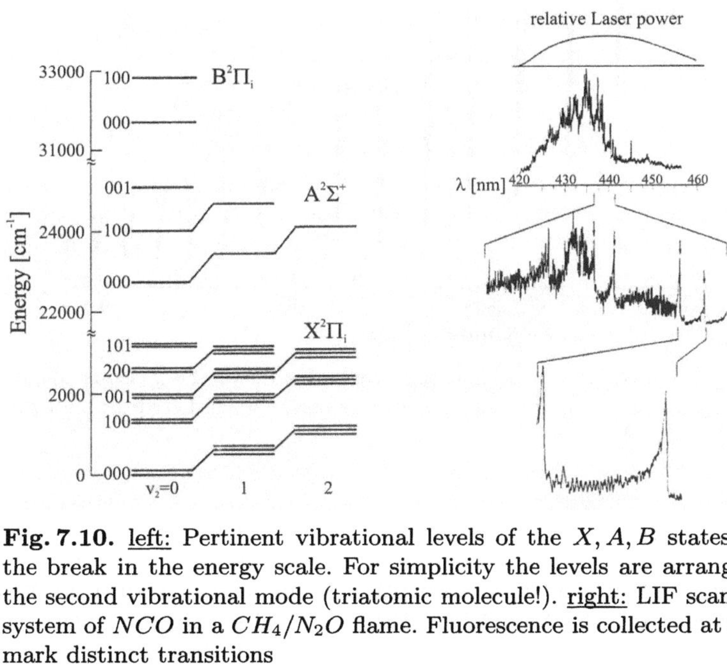

Fig. 7.10. left: Pertinent vibrational levels of the X, A, B states of NCO. Note the break in the energy scale. For simplicity the levels are arranged according to the second vibrational mode (triatomic molecule!). right: LIF scans for the $A - X$ system of NCO in a CH_4/N_2O flame. Fluorescence is collected at 465 nm. Arrows mark distinct transitions

resolution is limited to this length. For absorption measurements, often a tuneable laser is used to scan over a range of frequencies (even though this is not required). The intensity of the light exiting the measuring volume is recorded and plotted against the frequency (or wavelength or wave-number). Whenever the energy of one photon of the laser light is equal to the difference between two energy levels of the species in the measuring volume, the molecules absorb considerable amounts of the laser light and the intensity measured at the exit of the measuring volume decreases. The higher the density of the molecules, the more energy is absorbed; the higher the temperature, the more transitions from higher initial energy levels are recorded. Since the differences between energy levels in the ground electronic state suitable for absorption are small, typical absorption frequencies are in the red or infrared.

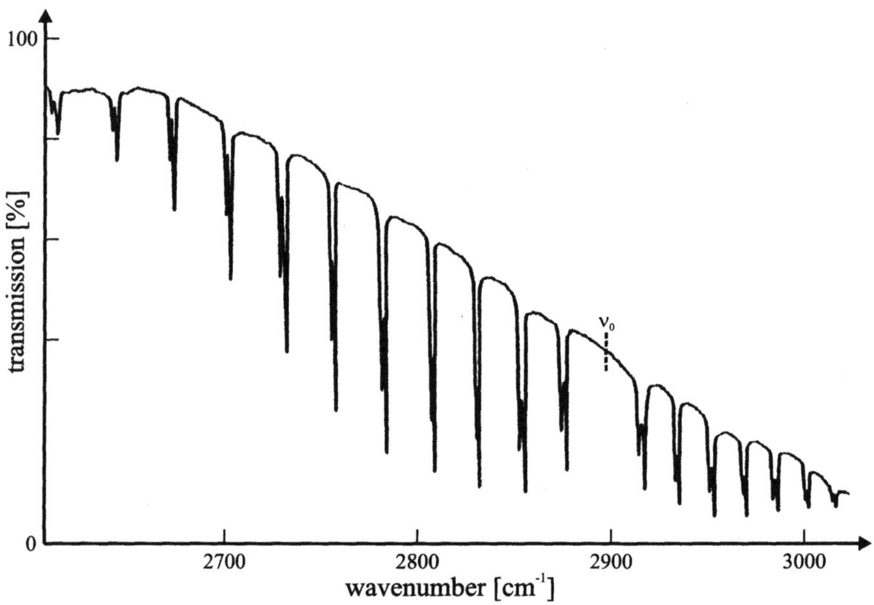

Fig. 7.11. Rotation–vibration absorption spectrum of *HCl* in the gaseous phase. The double lines result from small differences in the energy level distribution of the isotopes $H^{35}Cl$ and $H^{37}Cl$

On the other hand, if electronic absorption is employed, absorption frequencies are typically in the UV (s. Chap. 12). A typical absorption spectrum is shown in Fig. 7.11.

7.4 Concluding Remarks

This section has given a short introduction of the nature of light scattering and the techniques preferably applied in heat transfer investigations. It is only supposed to provide a basic understanding of the effects and the specific applications. For more detailed discussions, especially on the special requirements of experimental tools and set–ups, the other chapters in this book or the literature should be consulted. There are many books and articles on light scattering and its applications of which a few are given in the references.

In many cases different methods can be used for a specific application, each having certain advantages and disadvantages. Selection criteria are:

- signal to noise ratio or measurement accuracy
- spatial resolution
- time resolution
- versatility of the method
- complexity of the experimental set–up
- complexity of the evaluation
- cost of the set–up

Actually the achievable signal/noise ratio is a limiting factor for some applications, especially those using spontaneous Raman techniques. Fig. 7.12 shows the typical scattering cross sections for Mie, Rayleigh, Raman and fluorescence which are proportional to the intensity of the signals of the involved process.

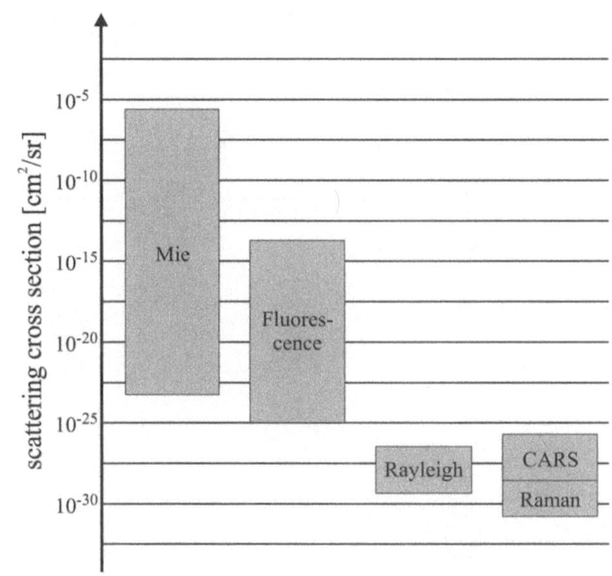

Fig. 7.12. Typical scattering cross sections for Mie, Rayleigh–and Raman–scattering and fluorescence

Sometimes it can be very useful to apply two techniques simultaneously, for example Raman–scattering and Laser–Doppler–Velocimetriy (LDV, based on Mie–scattering). Measurements of a concentration, a temperature or a velocity can be performed by applying one of the presented techniques without interfering with the investigated processes, as opposed to conventional probes. This is especially the case if high velocity (i.e. supersonic) flows or highly reactive systems (i.e. combustion) are to be investigated.

8 Laser–Doppler Velocimetry – Principle and Application to Turbulence Measurements

Andreas Eder, Bodo Durst, and Martin Jordan

8.1 Introduction

The Laser–Doppler Velocimetry (LDV) is an optical measuring method which allows the determination of the velocity of a fluid with a very high temporal resolution. The velocity is measured virtually at one single point, referred to as measurement volume. Steady state as well as transient turbulent flow fields can, therefore, be investigated with a very high spatial accuracy. The measuring principle of LDV is based on the physical effect of Mie scattering (s. Chap. 7). Therefore, the flow has to be seeded with particles capable to follow the flow such that their movement reflects the motion of the flow well.

In comparison to traditional flow measuring devices such as the hot wire anemometer or Prandtl probe, the LDV provides outstanding advantages. Due to the non invasive measuring principle, the flow is not disturbed by a probe. Therefore, flow measurements can be performed for applications not suitable for conventional techniques, e.g., internal combustion engines. Furthermore, it is an absolute measuring technique and, hence, no calibration is required.

Nevertheless, due to the point–oriented nature of the resulting data, the LDV technique should only be applied for flow fields, if local turbulence quantities are intended to be measured. If merely mean velocities are of interest, the application of a Particle Image Velocimeter (PIV, s. Chap. 16) should be taken into consideration which facilitates the planar determination of mean velocities in two or three dimensions simultaneously. On the other hand, PIV requires always optical access to the investigated flow–fields from two directions, whereas LDV measurements can be performed from only one side. Depending on the complexity of the optical set–up, up to three velocity components can be measured simultaneously by means of an LDV system.

Nowadays, LDV systems have established themselves for both, industrial as well as R&D applications. Hence, commercial systems of varying complexity are available which allow the optical set–up to be considered as a "black box" . The intention of this chapter is, therefore, to focus on the evaluation of velocity records, obtained by LDV, and to introduce the measuring technique as well as the optical set–up as far as the general understanding is concerned.

8.2 Principles of LDV

The LDV technique is based on the principle that light, scattered by a particle that is moving relatively to the light source, shows a frequency shift in comparison to the light source which is referred to as the Doppler effect. Although the measuring principle is always the same, many different types of LDV systems have been developed. The most common LDV technique is the dual–beam differential system, where a coherent light beam of the wavelength λ is split equally. The resulting two beams, which are shown in Fig. 8.1, are brought to a focussed crossing at a given angel ϑ which forms the measuring volume. A particle, traversing this volume, scatters light from the two light beams. Hence, this scattered light consists of two components, one corresponding to each beam. Both components obviously have a Doppler shift due to the velocity of the particle. The shift also depends on the direction of the light beam. Since the two beams are not parallel to one another, the two components of the scattered light have a slightly different Doppler shift.

A photo–detector receives the scattered light. The frequency of each of the Doppler shifted components is too high to be followed by the photo–

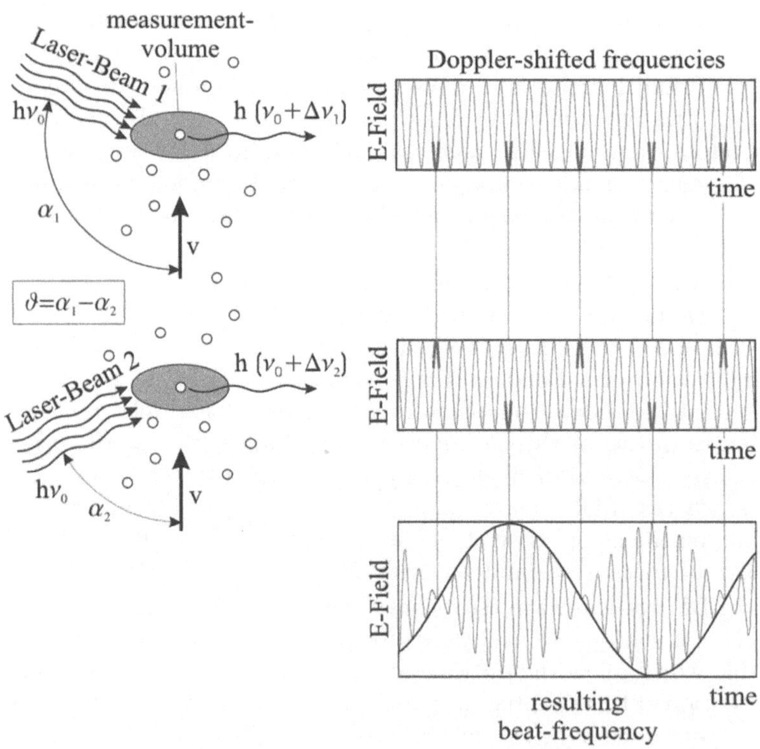

Fig. 8.1. Measuring principle of the Laser–Doppler Velocimetry [101].

detector. However, the superposition of the two Doppler shifted components results in a beat with a frequency that can be resolved by a photo–detector. The velocity of the particle v can now be determined directly from the beat frequency f:

$$v = \frac{f\lambda}{2\sin(\vartheta/2)} \ . \tag{8.1}$$

However, the detected frequency f and thus the calculated velocity v do not depend on the sign of the direction of the velocity. A positive and a negative velocity of the same magnitude result in the same Doppler shift. To overcome this problem, the frequency of one of the beams is shifted by a frequency f_0. The light which is now reflected by a stationary particle interferes with the shift–frequency f_0. Furthermore, a particle moving in positive direction causes Doppler bursts of a higher frequency than the shift, whereas a particle moving in the opposite direction causes a lower frequency than the shift frequency. The resulting velocity of a scattering particle can now be calculated by:

$$v = \frac{(f - f_0)\lambda}{2\sin(\vartheta/2)} \ . \tag{8.2}$$

The dual beam differential system is also referred to as fringe system. The measurement volume can also be regarded as an interference of planar wave fronts of the incident laser light that form a pattern of bright and dark fringes or planes respectively. The fringes are orientated parallel to the optical axis and perpendicular to the plane defined by the two incident beams. A particle passing through this fringe zone scatters only the light of the bright fringes. Therefore, the scattered light consists of intensity oscillations that correspond to the perpendicular velocity component of the particle to the fringe planes. It can be shown (e.g., [102,103]) that both models lead to the same correlation for the calculation of the velocity of the particle (s. Eq. 8.2).

8.3 Optics

A basic set–up of the transmitting optics of an LDV–system is shown in Fig. 8.2. For high–end LDV applications, Ar–ion lasers (three colors: blue 488 nm, green 514.5 nm, and purple 476.5 nm), power up to 10 W) are applied. Furthermore, He–Ne lasers, Nd:YAG laser, as well as laser diodes are often applied. Ar–ion lasers have gained high importance as they can be used for 3D LDV systems, and due to their high power, they can be applied for set–ups with a difficult optical access. It is worth noting that the optics of an LDV–system is specific to a certain type of laser due to the coating of prisms and filters. The laser has to be operated in the TEM_{00} mode, emitting a circular beam with a Gaussian intensity distribution. Further requirements are that the light is monochrome and linearly polarized. Recently, laser–diodes have

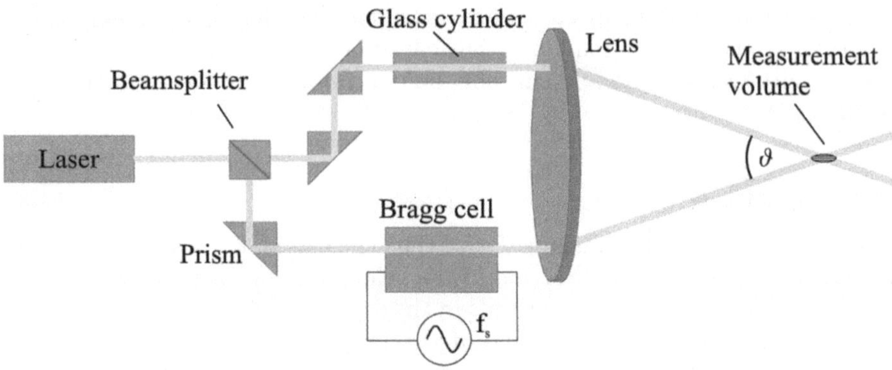

Fig. 8.2. Optical set–up of a dual beam differential or fringe LDV–system.

become technically important due to the tendency of miniaturization of LDV–systems.
In Fig. 8.2, the laser beam is divided by a beam splitter. One beam passes the Bragg cell where the frequency of the laser light is in– or decreased by the shift frequency f_s. The Bragg cell is an acousto–optical modulator and requires a signal generator which is typically operated at 40 MHz. In order to compensate the different optical path of the two beams, the unshifted beam passes a glass cylinder. An achromatic lens focuses the shifted and the unshifted beam so that they intersect with the angle ϑ. From (8.2) it can be derived that ϑ determines the sensitivity as well as the measured velocity–range of the system. The angle ϑ can, e.g., be altered by using a front–lens with a different focal length.

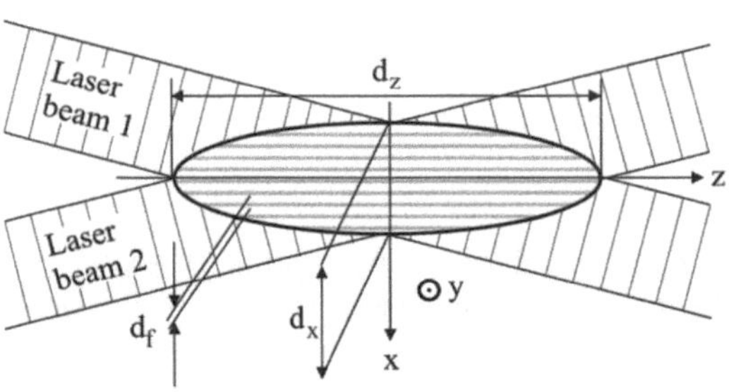

Fig. 8.3. Measurement volume.

The measurement volume is an ellipsoid and has a Gaussian intensity distribution in all three dimensions. The size of the measuring volume can be determined as follows (s. Fig. 8.3):

$$d_x = \frac{d_l}{\cos\frac{\vartheta}{2}}; \quad d_y = d_l; \quad d_z = \frac{d_l}{\sin\frac{\vartheta}{2}} . \tag{8.3}$$

The diameter of the laser beam in the measuring volume d_l is defined as:

$$d_l = \frac{4\lambda F}{\pi d_l'} , \tag{8.4}$$

where d_l' is the beam waist of the unfocused laser beam and F the focal length of the front lens. The fringe separation d_f can be calculated from the wavelength λ and the intersection angel ϑ of the incident beams:

$$d_f = \frac{\lambda}{2\sin\frac{\vartheta}{2}} . \tag{8.5}$$

With (8.5), we can calculate the number of fringes N_f in the measurement volume:

$$N_f = \frac{d_x}{d_f} = \frac{2d_l\tan\frac{\vartheta}{2}}{\lambda} . \tag{8.6}$$

The scattered light can be either detected in forward or backscatter mode. In the backscatter mode the front lens of the LDV optics serves as the focusing lens for the laser beams forming the measurement volume and for the backscattered light and simultaneously. Thus, no additional adjustment of the light collector optics is necessary. Additionally, optical access to the flow being measured is required only from one side. The forward scatter mode is more difficult to handle in case the measuring volume is shifted as the collector optics has to be focussed to the measuring volume separately. Furthermore, the investigated flow must be optically accessible from two sides and the entire system should be free of vibrations. However, the intensity of the scattered light is more than two orders of magnitude higher than that of the backscatter mode for seeding particles with a size commonly used in the order of the wavelength of visible light (s. Fig. 7.3).

8.4 Signal Processing

The scattered light is usually detected by means of a photomultiplier (PM). They distinguish themselves from photodiodes by a remarkably higher amplification and sensitivity. For a detailed description of their way of operation, it is referred to the specialized literature (e.g. [102]). The PM transforms the scattered light into an electronic signal that contains the Doppler frequency but also various spurious components as well. Sources of noise in the LDV signal can originate from:

- Photodetection shot noise,
- secondary electronic noise as well as thermal noise from the preamplifier circuit,
- higher order laser modes (optical noise), and

- light scattered from outside the measurement volume or unwanted reflections.

A typical Doppler burst that originates from a single particle passing through the measurement volume is shown in Fig. 8.4a. The intensity of the Doppler

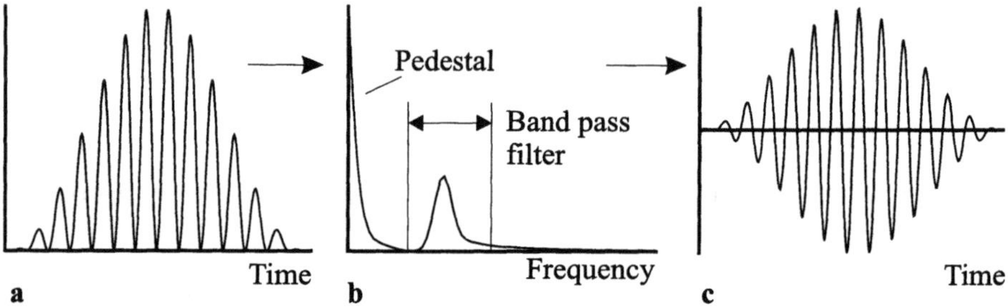

Fig. 8.4. Recorded doppler burst (a) that is transformed into the frequency domain (b) and after band pass filtering re–transformed into the time domain (c).

burst is the square of the superimposed E–Fields shown in Fig. 8.1. Before being processed for Doppler frequency detection the electronic signals are transformed into the frequency domain and band pass filtered in order to eliminate high frequency noise and the low frequent pedestal from the signal (Fig. 8.4 b). The limits of the band pass filter have to be set before the measurement so that all velocities that appear during the measurement are determined. The measurement of high velocity ranges or high velocity jumps requires the band pass filter to be set to a very wide range which can lead to a strong decrease of the signal–to–noise ratio (SNR). A typical burst after filtering is shown in Fig. 8.4 c.

The burst shown in Fig. 8.4 a originates from a single particle in the measurement volume that is smaller than the fringe spacing d_f. Nevertheless, it is possible that more than one particle are in the measurement volume or that particles are larger than the fringe spacing. In these cases, the quality of the unfiltered signal decreases but it is still possible to determine the velocity in a reliable way. This is also due to the progress in the development of new signal processing systems for the detection of the Doppler frequency. Older and less complex systems determine the frequency by measuring the time that elapses for a certain number of zero crossings of a burst, whereas newer systems determine the Doppler frequency by means of an auto–correlation or a fast Fourier transform (FFT) of the signal.

The data are usually sampled from the signal processing system by a PC. Using a 2D or 3D LDV–system, this can be done either arbitrarily for each channel or in a coincidence mode. The coincidence mode facilitates recording only velocities that originate from the same particle. As this mode requires a very accurate alignment of the optics, the data rate is often very poor.

Nevertheless, if Reynolds stress terms (e.g., $u_i'u_j'$) of the flow are intended to be determined, this mode is inevitable.

8.5 Seeding Particles

The properties of the seeding particles are of high importance for the accuracy and the quality of the measurements. Particles whose motion is used to represent that of a fluid continuum should be [103]:

- able to follow the fluctuations of the flow,
- good light scatterers,
- conveniently generated,
- inexpensive,
- non–toxic, non–corrosive, non–abrasive,
- non–volatile, or slow to evaporate,
- chemically inactive, and
- clean.

The type of particles to employ for a particular problem depends on the flow that is to be investigated and on the optical system. Therefore, the importance of the requirements stated above will vary from flow to flow. Besides the density of the particle, a very important property is the particle size in order to follow the flow accurately. Usually, particles with a diameter in the range of the wavelength of visible light are employed . Various sorts of seeding particles can be used depending on the requirements of the flow. For gaseous flows either solids (e.g., SiO_2, MgO, Al_2O_3,TiO_2), droplets of liquids (e.g. water, silicone oil), or smoke are used. In liquid flows, solids such as Al–powder, or latex particles serve as seeding particles. The ability to follow the flow can be calculated by several methods (cf. [103]). In Table 8.1, the maximum acceptable particle diameters are listed for a response of 1 and 10 kHz of the turbulent frequency. These frequencies are typical for the smallest eddies in the flow (s. Sect. 8.6.1) at which significant turbulent energy exists.

Table 8.1. Ability of seeding to follow the flow, [103].

Particle	Fluid	Diameter [μm]	
		1 kHz	10 kHz
Silicone oil	atmospheric air	2.6	0.8
MgO	methane–air flame, 1800 K	2.6	0.8
TiO_2	oxygen plasma, 2800 K	3.2	0.8
PVC	water	16	5.0

8.6 Determination of Characteristic Turbulence–Quantities

The Laser–Doppler Velocimetry is often applied at investigation of heat– and mass transfer phenomena for the determination of statistical turbulence quantities of the fluid–flow. In the following two paragraphs, important turbulence quantities are explained and it is shown, how they can be determined out of a velocity–record of an LDV–system.

8.6.1 Fundamentals of Turbulent Flows

A turbulent flow is characterized by a temporal and spatial irregularity. From a general point of view, every single fluid element follows the mean flow velocity. Nevertheless, the track of an element is completely arbitrary. Turbulent flows consist of many eddies of different size. These eddies change their size continuously and have a finite lifetime. An eddy can transfer its energy to smaller as well as larger eddies. However, more energy is transferred from larger to smaller eddies until the smallest eddies dissipate into an uncorrelated colloidal movement, i.e., via a cascade of continuously smaller eddies, the turbulence energy, produced by the flow, is transformed into internal energy. This is referred to as energy cascade, one of the fundamental characteristics of turbulent flows.

A direct numerical simulation (DNS) of the three–dimensional, vortical and diffusive turbulent motion with governing Navier–Stokes equations will still not be possible in the next future because of the enormous requirement in computing resources. The spatial resolution of the three–dimensional numerical grid for a DNS depends on the Reynolds number Re of the flow and can be estimated to be $O(Re^{9/4})$. Current computer capacities (base 2000) allow computational grids with about 10^7 grid cells, so that Reynolds numbers of $Re \approx 1000$ – which is associated with domain sizes at the order of a few centimeters or less – are close to the limit of what can be dealt with today. The measurement of turbulence quantities is, therefore, very important in order to develop and to improve models (e.g. k–ε, Reynolds Stress, Large–Eddy) for predicting the turbulent flow with CFD–codes in larger scale. These measurements require a very high spatial and temporal resolution which can only be carried out with an LDV–system in a reliable way.

Analogous to the analytical models, statistical techniques are applied when measuring turbulence quantities, where the instantaneous value consists of a mean value and a fluctuation. The flow velocity can, therefore, be expressed by:

$$U(x,y,z,t) = \overline{u}(x,y,z,t) + u'(x,y,z,t) \ . \tag{8.7}$$

In general, two averaging techniques are applied: temporal averaging (*Reynolds averaging*) and mass–weighted averaging (*Favre–averaging*). As the velocity of a flow is, in most cases, recorded from a fixed point, only the Reynolds–averaging shall be regarded in the present context. As depicted in Fig. 8.5, a

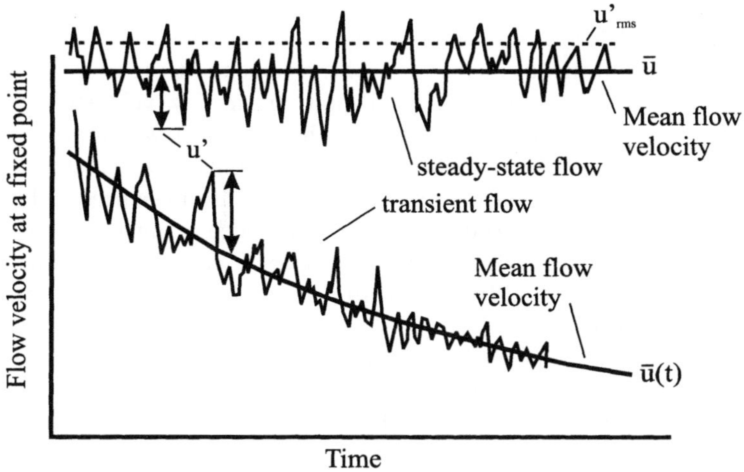

Fig. 8.5. Comparison of a steady–state and a transient turbulent flow.

moving average has to be calculated in order to distinguish between a variable mean value and a fluctuation value. The fluctuation velocity has to meet the requirement $\overline{u'} = 0$.

The characterization of the velocity fluctuations can be performed by means of several quantities. The *turbulence intensity* of every coordinate axis is defined by:

$$\overline{u'^2},\ \overline{v'^2},\ \overline{w'^2}\ . \tag{8.8}$$

In case of $\overline{u'^2} = \overline{v'^2} = \overline{w'^2}$, the mean fluctuations at one point are independent of their directions which is referred to as *isotropic turbulence.* A *homogeneous turbulence* is characterized by $\overline{u'^2} = const_1$, $\overline{v'^2} = const_2$, $\overline{w'^2} = const_3$ in each cross section of the flow.

Measuring results of turbulent fluctuations are often referred to as *root–mean–square (rms)* intensities:

$$\sqrt{\overline{u'^2}},\ \sqrt{\overline{v'^2}},\ \sqrt{\overline{w'^2}}\ . \tag{8.9}$$

The notation u'_{rms} will be used instead of $\sqrt{\overline{u'^2}}$ in the following correlations.

If we relate the "rms intensity" to a mean main flow direction, we can define the *turbulence level* of a flow (also called the *relative turbulence intensity*):

$$Tu = \frac{\sqrt{\overline{u'^2}}}{\overline{u}} \tag{8.10}$$

in case of an isotropic turbulence, or in general:

$$Tu = \frac{\sqrt{\frac{1}{3}(\overline{u'^2} + \overline{v'^2} + \overline{w'^2})}}{\overline{u}} \,. \tag{8.11}$$

It is worth noting that the mean flow velocity in (8.11) does not have to be identical with the instantaneous mean value of the velocity at the point where the fluctuation velocity is measured. An adequate mean flow velocity which is to be inserted in (8.11) depends strongly on the type of flow that is to be investigated. For, e.g., a pipe flow, this can be the undisturbed mean flow velocity at the tube axis, even if off–axis fluctuations are measured behind flow–obstacles such as grids. Furthermore, the turbulence quantity is mostly just an useful measure for (quasi) steady–state flows. Therefore, care has to be taken when using this quantity, especially when comparing turbulence levels of different applications.

According to Kuo [104], the turbulence level is a measure for the force of the eddies in a turbulent flow, i.e., it describes the mean temporal fluctuations of the flow velocity in a fluid. Reynolds [105] states the following reference values for the turbulence level: $Tu \sim 1\%$ for areas behind fine meshed grids, $Tu \sim 10\%$ in a turbulent boundary layer, and $Tu > 10\%$ in a turbulent free jet.

An important parameter which is widely used in numerical simulations is the *turbulent kinetic energy* per unit mass:

$$k = \frac{1}{2}(\overline{u'^2} + \overline{v'^2} + \overline{w'^2}) \,. \tag{8.12}$$

If the relative turbulence intensity defined in (8.11) is expressed by the definition of the turbulent kinetic energy, the turbulence level can be written as:

$$Tu = \frac{\sqrt{\frac{2}{3}k}}{\overline{u}} \,. \tag{8.13}$$

Hence, the turbulence level can be interpreted as a measure for the turbulent kinetic energy with respect to the mean main flow velocity.

Besides the classification of the turbulence in a flow with the quantities shown above, the eddy size distribution is an important parameter for the investigation of turbulent flows.
Within the spectrum of all possible eddy sizes, characteristic length scales are defined. The *macro length scale* describes the so called largest eddies of permanent character in a flow. The order of magnitude of their size is limited by the geometry confining the flow. It will be shown later that only a fraction of the entire turbulent energy in a flow is contained in these eddies. Therefore, their influence can be neglected from a statistical point of view.

Most of the turbulent kinetic energy is contained in eddies with a diameter in the order of the *integral length scale* which are also referred to as energy

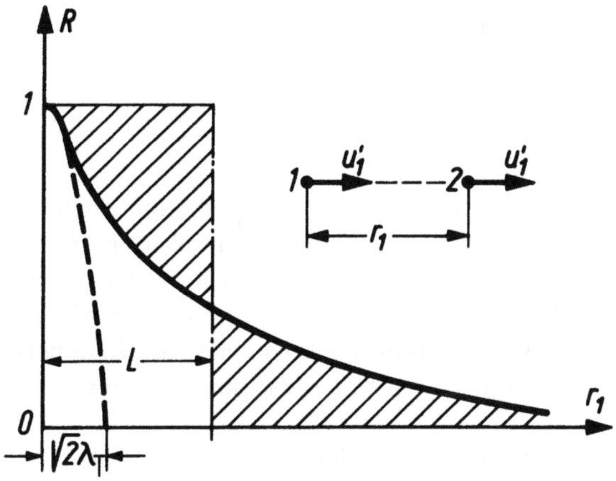

Fig. 8.6. Distribution of the spatial correlation coefficient $\mathcal{R}(r)$ (longitudinal correlation coefficient) [106].

containing eddies. In order to investigate the integral length scale of a flow, the correlation of a fluctuating quantity at two locations in the flow has to be determined. The corresponding correlation coefficient $\mathcal{R}_{ij}$ can, in general, be expressed by:

$$\mathcal{R}_{ij}(\boldsymbol{x}, \boldsymbol{r}) = \frac{\overline{u_i'(\boldsymbol{x}) u_j'(\boldsymbol{x}+\boldsymbol{r})}}{\sqrt{\overline{u_i'^2(\boldsymbol{x})}}\sqrt{\overline{u_j'^2(\boldsymbol{x}+\boldsymbol{r})}}} , \tag{8.14}$$

where $\boldsymbol{x}$ is the vector to the first point at which the velocity is measured, and $|\boldsymbol{r}|$ is the spatial distance between the two points for which the correlation is calculated. In Fig. 8.6, an example of the spatial correlation coefficient as a function of the distance r is shown. It is worth noting that the conditions

$$\lim_{r \to 0} \mathcal{R}(r) = 1 \quad \text{and} \quad \lim_{r \to \infty} \mathcal{R}(r) = 0 \tag{8.15}$$

have to be met. From a physical point of view, the correlation with the finite distance $|\boldsymbol{r}|$ represents – according to Bradshaw [107] – a measure for the force of those eddies which are larger in the direction of $\boldsymbol{r}$ than the amount of $\boldsymbol{r}$, since smaller eddies can not be detected when calculating this correlation. Integrating the correlation $\mathcal{R}(r)$ along r leads to a measure for the so called energy containing eddies (*integral length scale* L):

$$L = \int_0^\infty \mathcal{R}(r) dr . \tag{8.16}$$

Among the integral length scales, the *Eulerian integral length scales* L_l and L_t represent a characteristic quantity for the turbulent eddies. They are determined by calculating the correlation coefficient at two points kept apart by the distance $\boldsymbol{r}$ in the main flow velocity. One has to distinguish between the longitudinal and the transversal integral length–scale:

$$L_l = \int_0^\infty \mathcal{R}_l(r)dr \quad \text{and} \quad L_t = \int_0^\infty \mathcal{R}_t(r)dr \tag{8.17}$$

respectively, where $\mathcal{R}_l(\boldsymbol{r})$ is the longitudinal correlation coefficient with $\overline{\boldsymbol{u}} \parallel \boldsymbol{r}$, and $\mathcal{R}_t(\boldsymbol{r})$ is the transversal correlation coefficient with $\overline{\boldsymbol{u}} \perp \boldsymbol{r}$. L is a characteristic measure for those finite volumes in a fluid that move as an unit. According to Taylor [108], L can be defined as the "average eddy size" in a fluid. The longitudinal and the transversal integral length scale vary in general from each other. Rotta [106] states that in case of isotropic turbulence, the correlation $L_t = 1/2L_l$ is valid.

Based upon the integral length scale, a turbulent Reynolds number can be defined:

$$Re_L = \frac{u'_{rms}L}{\nu} \ . \tag{8.18}$$

Expanding the correlation $\mathcal{R}_t(y)$ in a Taylor's series at $y = 0$ gives a correlation coefficient describing the fluctuations. We can transform this correlation for small values of y (cf. [109]):

$$\mathcal{R}_t(y) \simeq 1 - \frac{y^2}{\lambda_t^2} \ , \tag{8.19}$$

where

$$\frac{1}{\lambda_t^2} = -\frac{1}{2}\left[\frac{\partial^2 \mathcal{R}_t}{\partial y^2}\right]_{y=0} \tag{8.20}$$

and

$$\frac{1}{\lambda_t^2} = \frac{1}{\overline{u'^2}}\overline{\left[\frac{\partial u'}{\partial y}\right]_{y=0}^2} \ . \tag{8.21}$$

According to Hinze [109], $\overline{[\partial u'/\partial y]_{y=0}^2}$ and hence λ_t are a measure for the local fluctuations of u'. Assuming that these fluctuations are caused by the smaller eddies in the flow field, λ_t represents these class of eddies (small eddies in general). In general, the correlation of the longitudinal and the transversal micro length scale can be expressed by:

$$\frac{1}{\lambda_l^2} = -\frac{1}{2}\left[\frac{\partial^2 \mathcal{R}_l}{\partial r^2}\right]_{r=0} \quad \text{and} \quad \frac{1}{\lambda_t^2} = -\frac{1}{2}\left[\frac{\partial^2 \mathcal{R}_t}{\partial r^2}\right]_{r=0} \ . \tag{8.22}$$

Instead of the longitudinal micro length scale, the following definition is very often used:

$$\lambda_T = \frac{1}{\sqrt{2}}\lambda_l = \sqrt{\frac{\overline{u'^2}}{\overline{\left(\frac{\partial u'}{\partial x}\right)^2}}} \ . \tag{8.23}$$

λ_T is referred to as *Taylor microscale*, named after G.I. Taylor. The Taylor microscale neither represents a specific class of eddies, nor, due to its definition, a dissipation length scale. It is very often employed as an approximation of the shear stress s_{xy} ($s_{xy} \sim \overline{u'^2}/\lambda$) [110]. The Taylor microscale is always smaller than the integral length scale. The ratio of these two quantities depends only on the Reynolds number:

$$\frac{\lambda_T}{L} = f(Re) \; . \tag{8.24}$$

Based upon the Taylor microscale, a *turbulent Reynolds number* can be defined:

$$Re_{\lambda_T} = \frac{u'_{rms}\lambda_T}{\nu} \; . \tag{8.25}$$

This quantity represents a criterion for the degree of the turbulence activity of a fluid flow. By applying a correlation that will be introduced below (8.38), it can be shown that this turbulent Reynolds number is proportional to the ratio of time scales of large and small eddies: As in general the eddy life time becomes smaller with an increasing value of λ, (8.38) indicates that the time scale of smaller eddies decreases compared to larger eddies.

$$Re_\lambda \sim \frac{\tau_L}{\tau_\lambda} = \frac{\frac{L}{u'_{rms}}}{\frac{\lambda}{u'_{rms}}} \; . \tag{8.26}$$

τ_L and τ_λ are often referred to as macro or micro eddy lifetime respectively, i.e., this quantity corresponds to the time how long an eddy contains its "identity".

Besides the spatial correlations, time correlations are very important for the analysis of turbulent flows. The *Eulerian time correlation coefficient* can be expressed by:

$$\mathcal{R}_E(\tau) = \frac{\overline{u'(t)u'(t+\tau)}}{\sqrt{\overline{u'^2(t)}}\sqrt{\overline{u'^2(t+\tau)}}} \; . \tag{8.27}$$

In this context, an integral time scale $\mathcal{T}_E$ (analogous to the integral length scale) can be defined. $\mathcal{T}_E$ can be regarded as a reference value for the longest coherent turbulent structures of u' in the flow.

$$\mathcal{T}_E = \int_0^\infty \mathcal{R}_E d\tau \; . \tag{8.28}$$

Analogous to the micro length scale, one can define an *Eulerian micro time scale* τ_E as a measure for the fastest changes in the fluctuations of u':

$$\frac{1}{\tau_E^2} = -\frac{1}{2}\left[\frac{\partial^2 \mathcal{R}_E}{\partial t^2}\right]_{t=0} = \frac{1}{2u'^2_{rms}}\overline{\left[\frac{\partial u'}{\partial t}\right]^2_{t=0}} \; . \tag{8.29}$$

A connection between the spatial and the temporal correlation is described in Taylor's hypothesis. It states that if $\overline{u} \gg \sqrt{\overline{u'^2}}$, the fluctuations at a fixed point of a *homogeneous* turbulent flow with a *constant* mean velocity $\overline{u}$ may behave as if the whole turbulent flow field passes that point with a constant velocity $\overline{u}$. The time series of the fluctuating velocity at that point will then be nearly identical to the instantaneous distribution of the velocity along axis of the main velocity component. This phenomenon is also referred to as *"frozen turbulence"* which helps to visualize this concept. From the equation of motion and taking into account the conditions explained above, Taylor's hypothesis yields to:

$$\overline{\left(\frac{\partial u'}{\partial x}\right)^2} = \frac{1}{\overline{u}^2}\overline{\left(\frac{\partial u'}{\partial t}\right)^2} . \tag{8.30}$$

Because of $x = \overline{u}t$, one can define:

$$L_l = \overline{u}T_E \text{ , and} \tag{8.31}$$

$$\lambda_l = \overline{u}\tau_E . \tag{8.32}$$

Therefore, one can evaluate under certain conditions important length scales by measuring the mean flow velocity and calculating the time–correlation coefficient.

The micro length scales introduced above are not the smallest length scales in a turbulent flow. Kolomogorov [111] has shown that the smallest length scale, that is also referred to as dissipation length scale, solely depends on the dissipation–rate ε as well as the kinematic viscosity ν. Tennekes et al. [110] developed the following ideas for the dissipative regions: he considered that the turbulence fine structure is composed of randomly distributed tubes with diameters of the order of the Kolmogorov microscale l_s and a spacing of the order of the Taylor microscale λ_T, cf. Fig. 8.7.

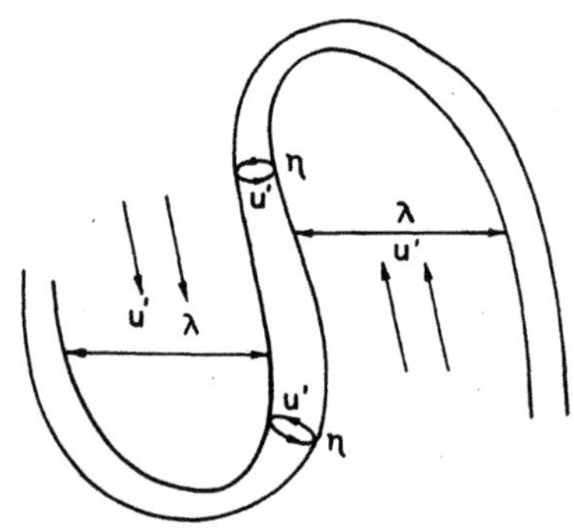

Fig. 8.7. Tennekes's model of dissipative regions in a turbulent flow, after Chomiak [112].

The smallest length scale (Kolmogorov micro scale) and the corresponding velocity scale are defined as:

$$l_s = \left(\frac{\nu^3}{\varepsilon}\right)^{\frac{1}{4}} \quad \text{and} \quad u_s = (\varepsilon\nu)^{\frac{1}{4}} \ . \tag{8.33}$$

With (8.33), a Kolmogorov time scale can be defined:

$$\tau_s = \frac{l_s}{u_s} = \sqrt{\frac{\nu}{\varepsilon}} \ . \tag{8.34}$$

A derivation of the dissipation rate ε is given, e.g., by Hinze [109]. The amount of turbulent kinetic energy per unit mass for large eddies is proportional to u'^2. This can be explained by the idea that the instantaneous energy transfer rate due to dissipation of larger eddies to smaller ones by viscous effects is proportional to u'/L. Hence, energy in the order of $u'^2 \cdot u'/L$ is transferred to smaller eddies. Therefore, the energy transfer rate (dissipation rate) yields to:

$$\varepsilon \sim \frac{u'^3}{L} \ . \tag{8.35}$$

In case of isotropic turbulence, the dissipation rate ε is equivalent to the decrease of turbulent kinetic energy k:

$$\varepsilon = -\frac{d\left(\frac{1}{2}\overline{k^2}\right)}{dt} = -\frac{3}{2}\frac{d\overline{u'^2}}{dt} = 15\nu\overline{\left(\frac{\partial u}{\partial x}\right)^2} \ . \tag{8.36}$$

The definition of the Taylor micro scale inserted in (8.36) yields to [109]:

$$\varepsilon = 15\nu\frac{\overline{u'^2}}{\lambda_T^2} \ . \tag{8.37}$$

It is worth noting that the small scale turbulence can be regarded as isotropic for turbulent Reynolds numbers $Re_\lambda > 100$.

The following relations between the length scales introduced above can be derived: Equating (8.35) and (8.37), we get an relation between the integral length scale and the Taylor length scale:

$$\frac{\lambda_T^2}{L} = \frac{15}{A}\frac{\nu}{u'_{rms}} \ , \tag{8.38}$$

Combining (8.33) and (8.35), we get

$$\frac{l_s^4}{L} = \frac{1}{A}\frac{\nu^3}{u'^3_{rms}} \ , \tag{8.39}$$

and the combination of (8.33) and (8.37) leads to

$$\frac{\lambda_T}{l_s^2} = \sqrt{15}\frac{u'_{rms}}{\nu} \ . \tag{8.40}$$

The coefficient A is in a range between 0 and 1. Abdel–Gayed et al. [113] found a value of 40.4 for $15/A$ in case of isotropic turbulence, whereas Dryden [114] suggests a value of 48.64. A comparison of these lengths in dependence of the turbulence intensity is shown in Fig. 8.8.

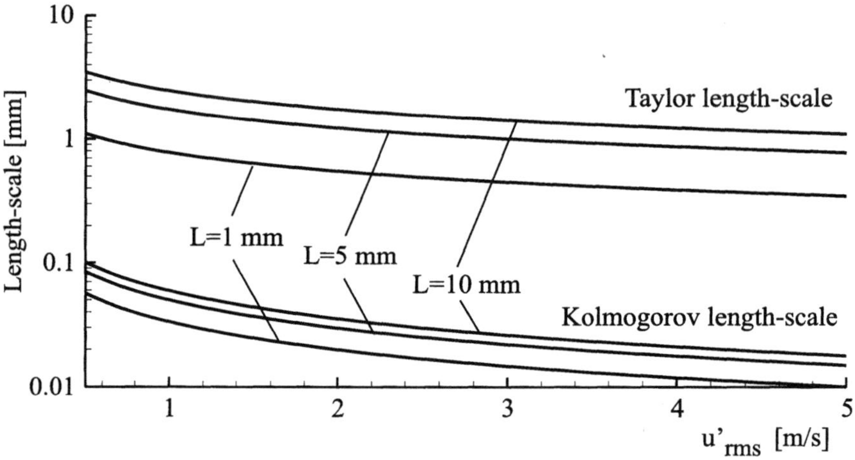

Fig. 8.8. Taylor and Kolmogorov length scale of an isotropic air–flow (293 K) as a function of the integral length scale and the fluctuation velocity.

As shown above, correlations – either spatial or temporal – facilitate to determine specific turbulence characteristics such as three–dimensional structures or instantaneous turbulent flow situation. A further method for the analysis of turbulent flows is represented by the spectral distribution of the turbulence frequencies that are contained in the statistical–temporal velocity records obtained by an LDV.

Turbulent motion is represented in general by a superposition of recurring eddies of different time and length scales. For the spectral analysis in particular, adequate spectra are:

- *Frequency spectra*, applied to temporal correlations,
- *Wavenumber spectra*, applied to spatial correlations.

The frequency spectrum is limited at the lower end by the frequency ω_L – corresponding to the time scale of the largest eddies – and at the upper end by ω_U, where the latter is theoretically the frequency of the smallest eddies. As it is not possible to determine this frequency by means of experiments, ω_U is often used as the upper noise limit of the measuring system.

The *power spectral density* can be derived from the time correlation [105]:

$$\mathcal{R}(\tau) = \int_0^\infty \phi(\omega) \cos \omega\tau d\omega \tag{8.41}$$

and

$$\phi(\omega) = \frac{2}{\pi} \int_0^\infty \mathcal{R}(\tau) \cos \omega\tau d\tau \ , \tag{8.42}$$

where ϕ is the normalized spectral density ($\int_0^\infty \phi(\omega)d\omega = 1$, dimension: [time]).

Analogous to the frequency spectra, we can define a wavenumber spectrum for the spatial correlations $\mathcal{R}(r)$. The wavenumber k corresponds to the reciprocal value of the length scale:

$$k(L, \lambda, \nu) \sim 1/(L, \lambda, \nu) \ , \tag{8.43}$$

and is of the dimension rad/m. Following (8.41) and (8.42), the wavenumber spectrum can be defined as:

$$\mathcal{R}(r) = \int_0^\infty \phi(k) \cos krdr \tag{8.44}$$

and

$$\phi(k) = \frac{2}{\pi} \int_0^\infty \mathcal{R}(r) \cos krdr \ . \tag{8.45}$$

By considering some mathematical restrictions (cf. [105]) and with a few dozen lines of algebra, one can define these equations via a complex Fourier transformation:

$$\mathcal{R}(r) = \int_{-\infty}^\infty \phi k e^{ikr} dk \tag{8.46}$$

and

$$\phi(k) == \frac{2}{\pi} \int_{-\infty}^\infty \mathcal{R}(r) e^{ikr} dr \ . \tag{8.47}$$

It is not possible to measure wavenumber spectra directly. Indirect possibilities are provided by both, the Fourier analysis described as well as by applying Taylor's hypothesis upon the frequency spectra. Therefore, the frequency ω can be expressed as a function of the wave number k, and the correlation distance r as a function of the time shift τ:

$$k = \frac{\omega}{\overline{u}} \quad \text{and} \quad r = \overline{u}\tau \ . \tag{8.48}$$

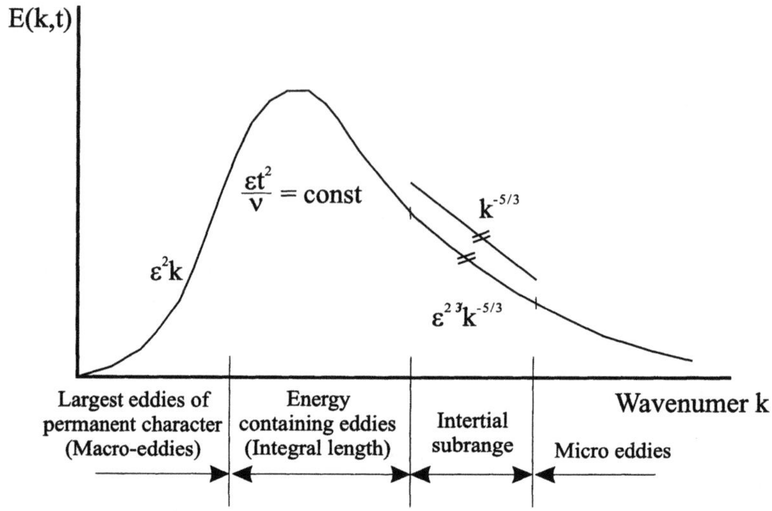

Fig. 8.9. Three dimensional wavenumber spectrum.

Applied to (8.41) and (8.42) one can define:

$$\mathcal{R}\left(\frac{r}{\overline{u}}\right) = \overline{u} \int_0^\infty \phi k\overline{u} \cos krdk \tag{8.49}$$

or

$$\phi(k\overline{u}) = \frac{2}{\pi\overline{u}} \int_0^\infty \mathcal{R}\left(\frac{r}{\overline{u}}\right) \cos krdr \tag{8.50}$$

respectively. A comparison with (8.41) and (8.42) leads to:

$$\phi(k) = \overline{u}\phi(\omega) \text{ with } \omega = \mathrm{k}\overline{\mathrm{u}} \,. \tag{8.51}$$

In Fig. 8.9, a typical three dimensional wave number spectrum is shown. In the region of high turbulent Reynolds numbers (corresponding to a small eddy size), an area of wavenumbers exists in which viscosity effects can be neglected in contrast to the energy transfer by "inert " effects. This area is referred to as *inertial subrange*, the spectrum is only a function of the dissipation ε:

$$E(k,t) \sim const. \times \varepsilon(t)^{\frac{2}{3}} k^{-\frac{5}{3}} \,. \tag{8.52}$$

8.6.2 Measurement of Turbulence–Quantities

The first step in each analysis of velocity records obtained by an LDV system is to prove whether the records contain any spurious data which are caused by noise introduced by the photo–multipliers and the processing electronics. Recent commercial LDV systems have implemented an algorithm that checks the data, whereas older models need to perform this check as a first post–processing task. The elimination of spurious data can be performed by many possibilities: one common technique is to define a a rough moving average and a running standard deviation and to eliminate all data outside this data–bandwidth.

The rms turbulence intensity of a flow is derived by:

$$u'_{rms} = \sqrt{\frac{1}{n_E}\sum_{n=1}^{n_E}(U_n - \overline{u}_n)^2} \ , \tag{8.53}$$

where n_E is the number of measured points within the time–interval for which u'_{rms} is to be determined. The most important task, which is also the task that requires the sure instinct of the analyst, is the separation of the obtained data into a running mean $\overline{u}_n$ and the fluctuation velocity as shown in Fig. 8.5. The corresponding mean value of the nth measured quantity is defined as:

$$\overline{u}_n = \frac{1}{n_M}\sum_{i=n-n_M/2}^{i=n+n_M/2} U_i \ . \tag{8.54}$$

The application of (8.54) contains the problem of choosing the right number of measured points n_M for calculating a moving average. Using too few points, the moving average can contain information of the fluctuations and, on the other hand, using too many points could lead to a situation in which global changes of the mean velocity could be neglected which would lead to an enormous over–estimation of the turbulence intensities. Durst [115] developed the following procedure in order to find the right number of data points in case of using a 2D or 3D LDV system: The measured turbulence quantities obtained for each coordinate are taken to be uncorrelated. Correlated portions in the velocity signal originate from global changes in the flow. A measure for the correlation of two signals is represented by the correlation coefficient r:

$$r = \frac{\sigma_{xy}}{\sigma_x \sigma_y} \ , \tag{8.55}$$

where σ_x and σ_y are the standard deviations. If two signals are completely uncorrelated, r yields to zero, with an increasing degree of correlation r yields to 1.

If we apply this to the fluctuations of the velocity, it is possible to check if the fluctuations still contain a correlated portion after the velocity records have been separated into a mean and fluctuating value. Hence, one can define:

$$\sigma_x = \sqrt{\frac{1}{n_E}\sum_{n=1}^{n_E}(u'_n - \overline{u'_n})^2} \ , \tag{8.56}$$

$$\sigma_y = \sqrt{\frac{1}{n_E} \sum_{n=1}^{n_E} (v'_n - \overline{v'_n})^2} \ , \text{ and} \tag{8.57}$$

$$\sigma_{xy} = \frac{1}{n_E} \sum_{n=1}^{n_E} (u'_n - \overline{u'_n})(v'_n - \overline{v'_n}) \ . \tag{8.58}$$

An example of r as a function of used points for the calculation of the mean velocity is shown in Fig. 8.10. The following points have to be considered

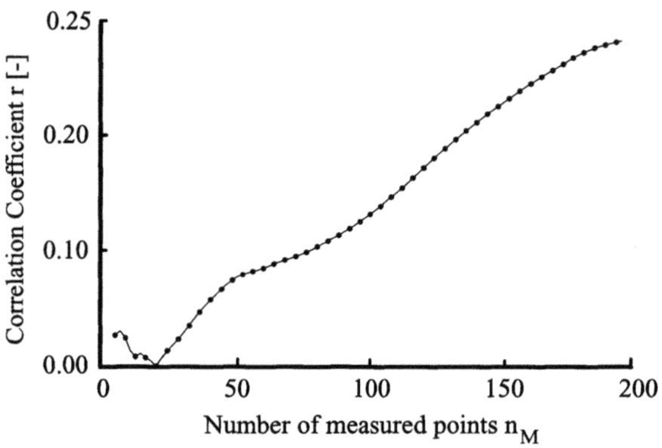

Fig. 8.10. Determination of an optimum number of measured points for the calculation of a transient mean flow velocity [115].

when applying this technique:

- The data–rate of the recorded signals has to be as high as possible (already more than 10.000 kHz at least are required for slow flows (e.g., < 10 m/s) in order to detect high frequency turbulence.
- The investigation of transient flows can require a section by section definition of n_E for the determination of n_M. In this case, the time–interval for n_E has to be greater than the integral time scale $\mathcal{T}_\mathcal{E}$, cf. (8.28). The problem of this procedure is that the evaluation of $\mathcal{T}_\mathcal{E}$ requires the separation of the velocity records into a mean– and fluctuating value. In order to overcome this problem, $\mathcal{T}_\mathcal{E}$ can estimated by means of (8.32), assuming an integral length scale L of about 10% of the confining geometry and an average value for $\overline{u}$ within this interval. In case the determination of the integral length scale (s. below) differs in order of magnitude of this first approximation, one has to start from the beginning and to determine an adequate value of n_E in an iterative way.

The quality of the recorded velocity signal can be determined by calculating the power spectral density (PSD). The PSD–function Φ can be derived, as described above, from the time–correlation coefficient $\mathcal{R}$ – which is also referred to as auto correlation function ACF – and applying Taylor's hypothesis.

In practice, the ACF is determined for an equally sampled velocity record by:

$$\mathcal{R}_k = \mathcal{R}(\tau) = \mathcal{R}(k/f_a) \sum_{j=0}^{J-K} u_j u_{j+k}, \quad k = 0, ..., K-1 , \tag{8.59}$$

where J is the total number of velocity samples in the evaluation interval and τ is the time lag for which the ACF is computed, whereas f_a is the sampling frequency of the velocity record. Applying the Fourier transform yields to the PSD in complex form:

$$\Phi = \frac{1}{f_a} \sum_{k=0}^{2K-2} \mathcal{R}_k e^{-2\pi ijk/(2K-1)}, \quad j = 0, ..., 2K-2 . \tag{8.60}$$

A typical trace of a PSD evaluated from a measured velocity recored is shown in Fig. 8.11.

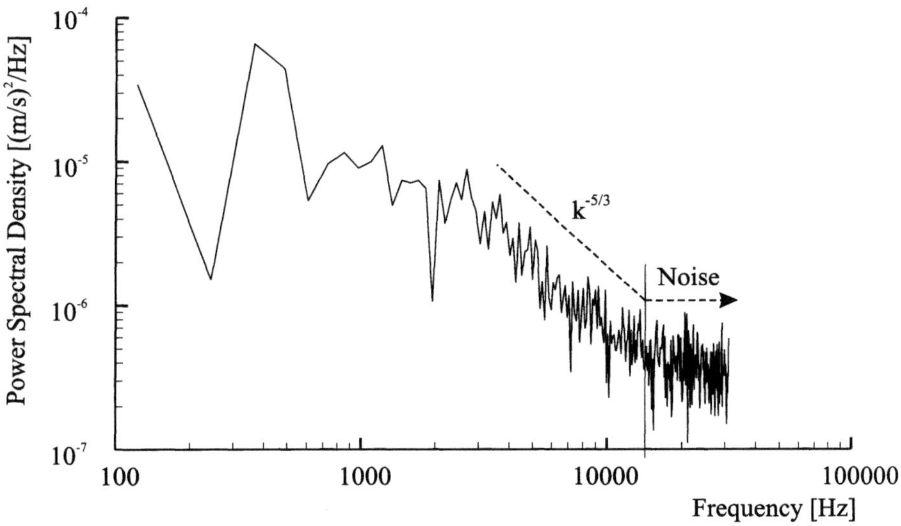

Fig. 8.11. Power Spectral Density distribution of a measured LDV record [115].

The PSD can be used to assess the SNR of the recorded velocity. In the PSD–distribution the signal noise is constant. We know from Fig. 8.9 that within the regime of the inertial subrange, the PSD has a gradient which is proportional to $k^{-5/3}$. The eddies of the dimension of the Kolmogorov microscale are too small to be detected by means of an LDV–system. Therefore, we can define an upper end frequency for the velocity signal. This information can be used to filter the velocity signal with a low pass filter in order to eliminate noise from the signal (s. Fig. 8.11).

After this step, we have now a data set for the evaluation of turbulence quantities such as the turbulence intensities (s. above). Furthermore, the ACF can now be calculated again in order to determine a suitable time–interval for the evaluation of these quantities. The integral time scale is the integral of the ACF (s. 8.28). The ACF can be approximated by the function:

$$\mathcal{R}_{approx} = (1 + a\tau + b\tau^3)e^{c\tau} , \tag{8.61}$$

where a,b, and c are fitting constants. The upper integration limit is the first zero–crossing of the ACF. The time interval for the determination of turbulence quantities has to be larger than this integral time scale. If the requirements for Taylor's hypothesis are met, one can evaluate the integral length scale by multiplying the integral timescale with the mean velocity according to (8.31). If the requirements for Taylor's hypothesis are not met, the length scale can only be determined by means of a simultaneous measurement of the turbulent flow at two different positions. Hence, two LDV–systems are required in order to calculate the spatial correlation coefficient.

9 Phase Doppler Anemometry (PDA)

Bernd Ofner

9.1 Introduction

Phase Doppler Anemometry (PDA), also known as Particle Dynamics Analysis (PDA), Phase Doppler Particle Analysis (PDPA), Phase Doppler Difference Method (PDDM) and Phase Doppler Interferometry (PDI), is a laser optical instrument for simultaneous measurement of the velocity and size of spherical particles. Additionally it can be used for concentration and mass flux measurements.

PDA enables measurements of high accuracy and a very high spatial resolution. Since it is based on an absolute physical effect PDA does not require any calibration. The principle of velocity measurement is the same as that used in Laser Doppler Velocimetry (s. Chap. 8). The velocity is determined by analyzing the frequency shift of light which is scattered by a particle flowing through the interference volume of at least two intersecting laser beams. For particle size measurements PDA makes use of the fact that scattered light which is received from two different directions of observation has a phase difference, which is proportional to the diameter of the particle. Since this relation is valid for both, transparent and non–transparent particles PDA can be used for different applications, such as for the investigation of droplets in water–, fuel– or painting–sprays, of bubbles in liquids and of solid spheres like powders, glasses or metal balls.

9.2 General considerations for the application of PDA

Application of PDA requires the consideration of the following items:

- The particles (droplets, bubbles, solids) have to be spherical; otherwise the diameter and consequently the mass flux cannot be determined correctly. Since the surface tension supports the formation of spheres, this requirement is usually fulfilled for small droplets and bubbles.
- Generally, the size of the particle can be in the range of some micro– to several millimeters. However, the measurable size range depends on several factors including the optical configuration (e.g. laser power, focal length, beam separation and location of detectors), the particle material, the photo detectors as well as the signal processor.

- The velocity of the particles generally can be analyzed up to 500 m/s, depending on the signal processor and on the measurement configuration. But since some optical parameters, such as the focal length or the beam separation control the measurable diameter and velocity range in opposite ways, either the maximum velocity or the diameter range will be limited.
- If transparent particles are analyzed, refraction is commonly used as the predominant scattering mode. In this case, the phase–diameter relationship is dependent on the refractive indices of the particle and the surrounding medium which therefore have to be known. In addition to that, the transparent particles have to be homogeneous. Any present impurities have to be much smaller than the wavelength of the laser light and their quantity has to be low.
- The particle concentration in the two phase flow should be moderate. A high number of particles increases the risk of measurement uncertainties due to multiple scattering effects and due to low signal to noise ratio.
- An optical access to the measurement area, usually from two directions, has to be available. The medium which surrounds the particle, e.g. liquid or gas, has to be transparent for the wavelength of the laser beams. The windows (if necessary) should be plane without curvature. A planar window reduces the probability of the laser beams being refracted unequally and ensures the intersection necessary to form the measurement volume.

Before deciding on PDA, the costs for such a measurement system should be taken into consideration. The prices for PDA–systems (in year 2000) vary between \$80.000 for 1D–systems and up to \$250.000 for 3D–systems. Additionally, a precise traversing system is usually necessary for an exact positioning of the measurement volume in the two phase flow due to the small diameter of the measurement volume.

9.3 Principles of PDA

The principles of PDA were introduced by Durst and Zaré (1975). Several publications followed, including technical descriptions as well as applications of PDA to two phase flows. The most known are those of Bachalo and Houser (1984), Bauckhage and Flögel (1984) and Saffmann et. al. (1984). A detailed review of the history and the early developments of PDA was given by Hirleman (1996). Tropea (1999) described the more recent developments and gave an outlook to novel variations and improvements of the Phase Doppler Technique.

9.3.1 Light–scattering by particles

The Phase Doppler measurement technique is based on the mechanism of light scattering by particles. The scattering mechanism consists of reflection, refraction, diffraction (and absorption), (s. Chap. 7). A complete and exact mathematical description of light scattering by small, spherical particles was given by Mie (1908) in his famous theory, which is based on the solution of Maxwell's equations. A detailed review, consisting of two parts, about light scattering by moving particles was published by Raszillier and Durst (1992). Further information can be found in publications by Van de Hulst (1981), Negus (1982), Bohren (1983), Naqwi (1991), Gouesbet (1991), Albrecht (1994, 1995) and Borys (1998).

For PDA, reflection as well as first order and second order refraction are the most important scattering modes. Refracted light of a higher order is not commonly used since the intensity is too low. In case of refraction, the relative refractive index $m = n_p/n_m$, which is the ratio of the particle's refractive index n_p and the refractive index of the surrounding medium n_m, is very important. The basic principles of reflection and refraction of light by spherical particles are shown in Fig. 9.1.

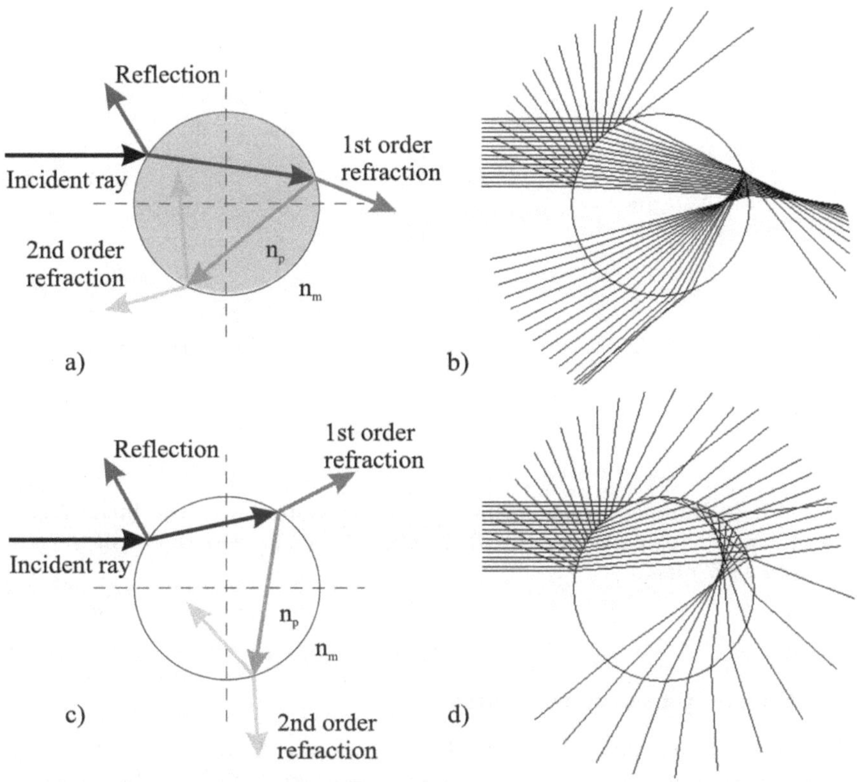

Fig. 9.1. Light scattering by droplets ($n_p > n_m$) and bubbles ($n_p < n_m$); general definitions and calculations with raytracing–software for a water droplet in air (**a, b**) and for a air bubble in water (**c, d**)

The incident ray gets partially reflected as well as once (1st order) and twice (2nd order) refracted. The intensity ratio of the reflected and refracted part of the scattered light, which depends upon the relative refractive index m, the polarization and the angle of the incident ray, can be calculated by the Fresnel coefficients. The scattering angle is defined by Snell's law. Both calculations are valid for droplets ($m > 1$) as well as for bubbles ($m < 1$).

9.3.2 Optical parameters of a Phase Doppler Measurement System

The important optical parameters of a Phase Doppler measurement configuration are the beam intersection angle 2Θ, the scattering angle φ and the elevation angle Ψ, as shown in Fig. 9.2. Additionally, the polarization of the laser beams (parallel or perpendicular to the YZ–plane), the focal length of the detector unit, and the size and shape of the detector aperture have a significant influence on the measurement.

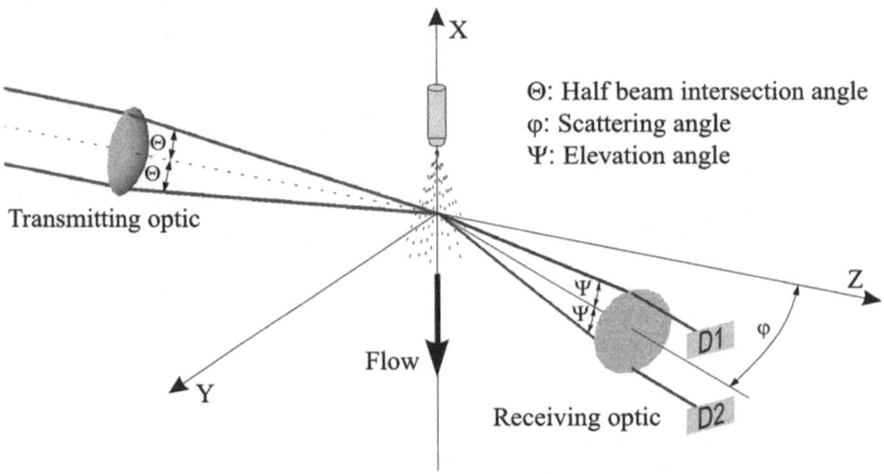

Fig. 9.2. Phase Doppler measurement set–up; 1D–system with two incident laser beams and two photo–detectors ($D_{1/2}$), located above and underneath of the scattering plane (YZ–plane)

9.3.3 Phase–diameter relationship

Each particle which flows through the measurement volume scatters light from the two incident laser beams (intersection angle 2Θ) to different directions, as shown in Fig. 9.3. The scattered light waves of the two beams interfere and create a signal which is called *Doppler burst* (s. Chap. 8, Fig. 8.4). The *Doppler burst* can be measured by means of a photo–detector.

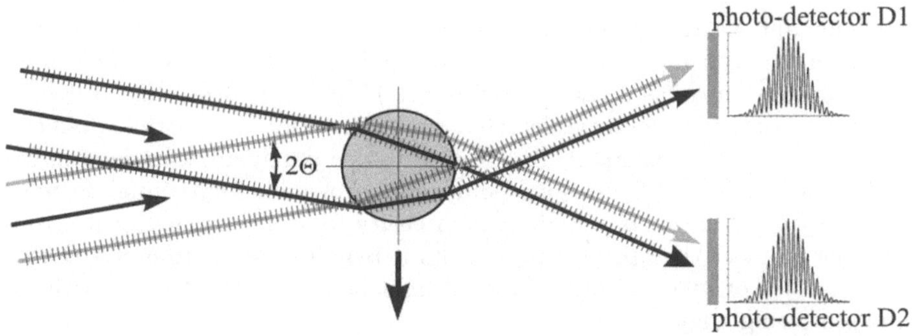

Fig. 9.3. Light scattering by particles flowing through the measurement volume and interference of the two light waves to a *Doppler burst* at the photo detectors

The optical path length of the scattered light is dependent on the position of the photo–detectors (D1, D2). Since both photo–detectors are located at different positions they receive the same *Doppler burst* with a measurable phase difference but with the same frequency. Durst and Zaré (1975) have found that this phase difference has a linear dependence on the diameter of the particle. An increase of the particle diameter causes an increase of the absolute phase difference and vice versa. This relationship is valid for reflection as well as for refraction. The mathematical equations which describe the phase–diameter relationship for reflection and for first order refraction are shown in (9.1) and (9.2). An exact equation for second order refraction is not available. A solution can be found only by means of iterative calculations.

Reflection:

$$\Phi = \frac{2\pi d_p}{\lambda} \frac{sin\Theta \, sin\Psi}{\sqrt{2(1-q)}}; \qquad sin\Theta \, sin\Psi << q \tag{9.1}$$

1st order refraction:

$$\Phi = \frac{-2\pi d_p}{\lambda} \frac{m \, sin\Theta \, sin\Psi}{\sqrt{2(1+q)(1+m^2 - m\sqrt{2(1+q)})}}; \qquad m > 1 \tag{9.2}$$

where $\quad q = cos\Theta \, cos\Psi \, cos\varphi$

Equation (9.1) shows that in case of reflection the phase–diameter relationship is independent of the relative refractive index m. This fact is essential for measurements in which the refractive index n_p of the particles is not known.

Since the phase difference Φ can be measured, and the wavelength λ, angles Θ, Ψ, φ and relative refractive index m are usually known, the particle diameter d_p can be calculated by using either (9.1) or (9.2). Therefore one predominant scattering mode is necessary in order to ensure a linear phase–diameter relationship, because reflection and refraction cause a phaseshift with opposite signs, as it can be seen in (9.1) and (9.2). Otherwise, if light of different scattering modes is detected simultaneously, the particle diameter calculations will be incorrect. In addition to one dominant scattering mode, the measurement configuration should also provide a high signal to noise ratio (SNR) by means of a high light intensity to avoid measurement errors. In order to meet the above requirements it is indispensible to consider that the intensity of reflected and refracted light depends upon the polarization and scattering angle φ.

Intensity charts in the form of log scale polar plots, as shown in Fig. 9.4, are a helpful means for illustration of the intensity ratio of the different scattering modes as a function of these parameters. Commercial software is available for calculations of such intensity charts.

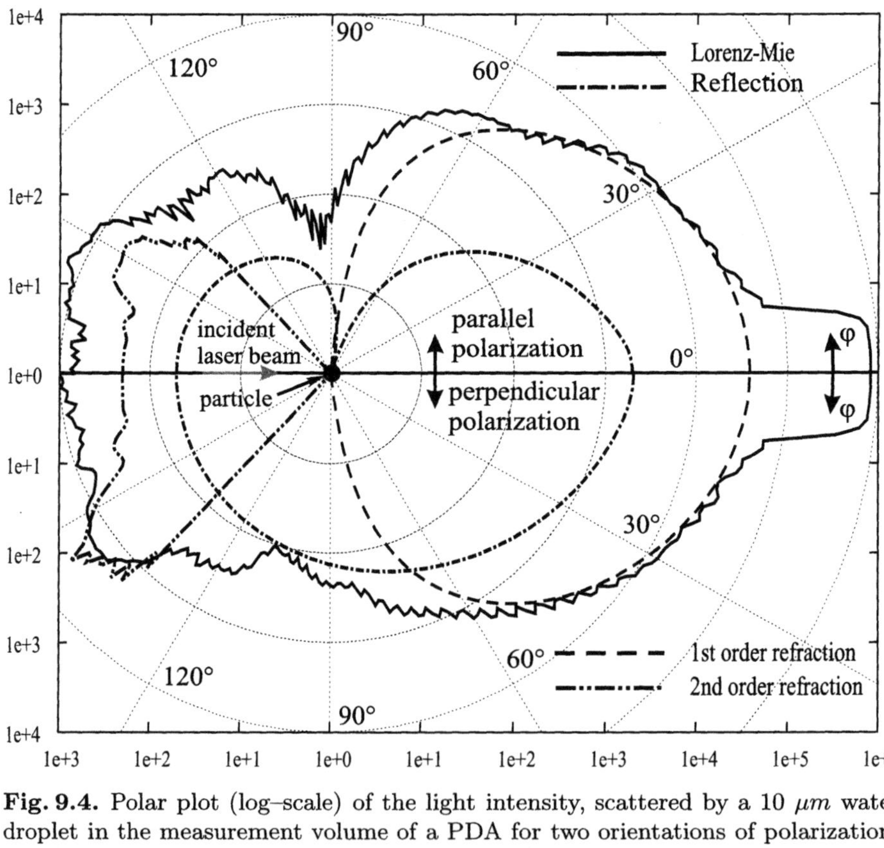

Fig. 9.4. Polar plot (log–scale) of the light intensity, scattered by a 10 μm water droplet in the measurement volume of a PDA for two orientations of polarization: parallel (upper half) and perpendicular (lower half) to the scattering plane (YZ–plane); scattering modes: reflection, first and second order refraction as well as total intensity of scattered light (Lorenz–Mie)

Figure 9.4 shows a log–scale intensity distribution of light which is scattered by a 10 μm water droplet in air. The full line represents the total light intensity calculated with the Lorenz–Mie theory. The dotted lines indicate the light intensity distribution based upon either reflection or first order and second order refraction. At small scattering angles φ, the intensity of the first order refracted light is much higher than that of the reflected light, independent of the polarization. If the light of the incident laser beams is polarized parallel to the scattering plane (upper half of Fig. 9.4), the intensity of the reflected light becomes zero at a scattering angle of about 74°. This angle is called "Brewster angle". At this angle, the intensity of the refracted light is very close to the total light intensity (Lorenz–Mie). This indicates that light is scattered mainly by one dominating scattering mode (refraction) which is necessary for a linear phase–diameter relationship. Therefore the Brewster angle should be used in PDA measurements. In case of perpendicular polarization (lower half of Fig. 9.4), the intensity of the refracted light becomes zero at 82.4° which makes reflection the dominating scattering mode. At angles greater than 90°, light is mainly scattered by reflection and 2nd order refraction.

If one dominating scattering mode is ensured, the phase–diameter relationship can be assumed as linear and looks similar to Fig. 9.5 a. The biggest measurable particle diameter depends upon the sensitivity of the phaseshift to the particle diameter or on the gradient of the curve in the phaseshift vs. diameter diagram, respectively. Since the phase difference can be analyzed only between 0° and 360°, particles which exceed this diameter limit might lead to measurement errors. A large particle which causes a phaseshift of 420° would be determined as a small particle with a phaseshift of 60° as it is described in Fig. 9.5 a. In order to minimize those measurement errors due to the 2π–ambiguity, a third detector (D3), located close to one of the other two detectors (e.g. D1), is commonly used, compare Fig. 9.5 b. This configuration results in a lower phaseshift of the signals between D1 and D3 and, consequently, in a lower gradient in the phase–diameter diagram, as shown in Fig. 9.5 c. Therefore, the phaseshift measured with D1/ D3 determines the location of the particle diameter in the measureable sections (2π–jumps) of $D1/D2$. By applying this method, a high resolution and accuracy can be combined with a large diameter range.

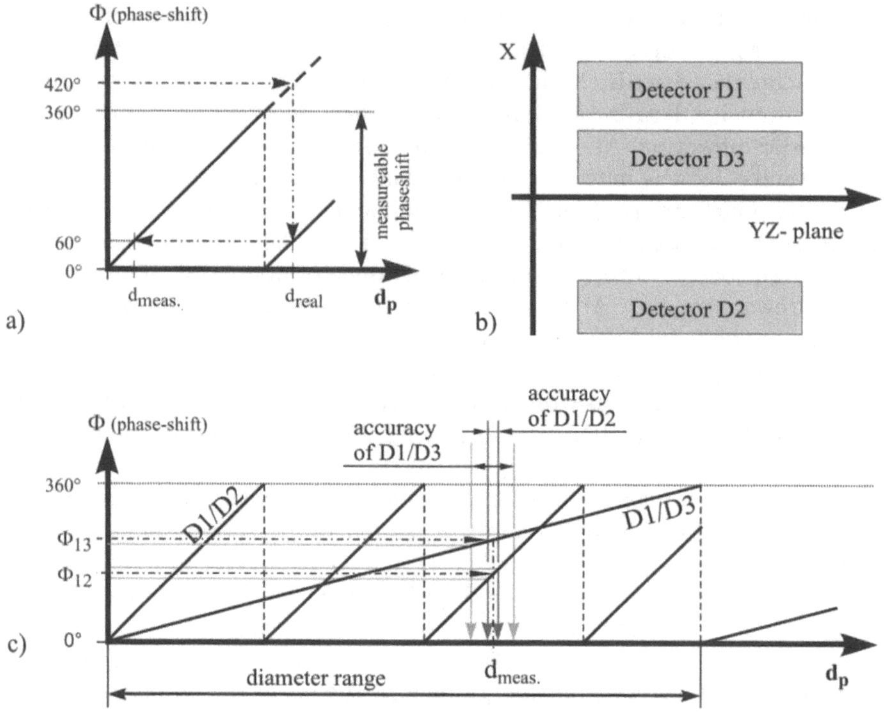

Fig. 9.5. Phase–Diameter relationship, 2π–ambiguity and extension of the diameter range

9.4 Measurement accuracy

Although the Phase Doppler technique is based on an absolute measurement principle without the need for calibration, it cannot provide a 100 % measurement accuracy. Therefore it is important to minimize all the possibilities for measurement errors.

Generally, the measurement accuracy increases with the quality of the signals which have to be analyzed. This requires a high signal–to–noise ratio (SNR) which can be provided by a high light intensity as well as an accurate intersection of the laser beams to form the measurement volume. Measurement conditions under which the light gets partially absorbed or under which a high particle density decreases the intensity of the scattered light usually cause increased noise and low signal intensity.

The basic requirement for a correct diameter measurement, however, is a linear phase–diameter relation. This can be affected by several phenomena: non–homogeneities and surface distortions (non–sphericity) of the particles as well as multiple scattering effects all of which change the phaseshift and should therefore be avoided. Moreover, the optical configuration, as shown in Fig. 9.2, can be optimized for a linear phase–diameter relationship, compare (9.1) and (9.2). The scattering angle and polarization have to ensure one dominating scattering mode (s. Fig. 9.4).

Data acquisition in high particle concentration conditions requires an accurate setting of the voltage for the photo detectors. With a low voltage setting only the strong signals from large particles will be detected whereas the weak signals from small particles can not be distinguished from the noise. Increasing the high voltage also increases the noise and may cause oversaturation of the detectors when signals of large droplets are received.

Another possible measurement error is caused by the Gaussian intensity profile of the laser beams. This fact, however, is only problematic for large transparent particles which exceed 30% of the width of the measurement volume. The *Gaussian beam intensity effect* is described in Fig. 9.6. It can clearly be seen that the dominance of either reflected or refracted light is dependant upon the position of the particle in the laser beam. This might lead to multiple scattering mode effects. In this case the phaseshift to diameter conversion will be inaccurate, because the diameter calculation can be based only on one single scattering mode (s. Chap. 9.3.3).

In addition to the *Gaussian beam intensity effect*, the so called *Slit–effect* is also a well known problem. The *Slit–effect* is caused by the limited area (slit) of the measurement volume, which is visible to the receiving optic. A description of the *Slit–effect* is shown in Fig. 9.7. Particles which are partially outside of the slit can scatter light by reflection or by refraction, depending on their position in the laser beam. This results in measurement uncertainties and non–linearity in the phase–diameter relationship, similar to the *Gaussian beam intensity effect.*

Inaccurate measurements can also be caused by very small ($< 5\ \mu$m) particles which show strong oscillations in the phase–diameter relation. In order to reduce or avoid these measurement uncertainties, the manufacturers of PDA–Systems have designed different optical and electronic modifications for their systems (s. Tropea, 1999).

9.5 Applications of PDA

In recent years applications for PDA techniques have developed rapidly. As a result, PDA measurement techniques have become extremely popular in a wide variety of scientific fields. Due to the importance of sprays in many technical, chemical and medical fields, investigations of the size, velocity, concentration and mass flux of droplets in water, fuel or painting sprays are the conventional applications for PDA. Additionally, PDA can be used for velocity and size measurements of bubbles in liquids, e.g. in the field of (chemical) process engineering. Furthermore, the Phase Doppler technique is

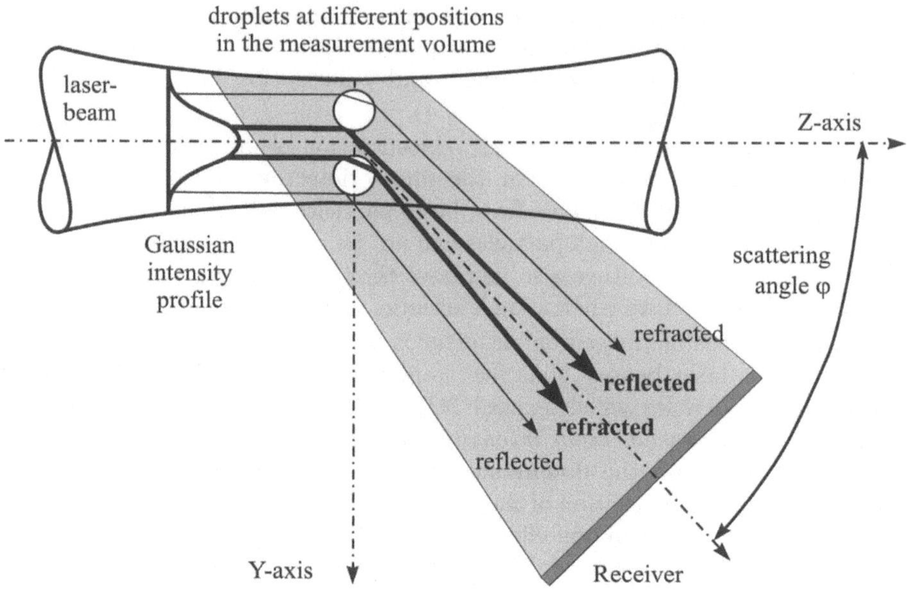

Fig. 9.6. Gaussian beam intensity effect: depending on the position of the particles in the measurement volume, either refracted or reflected light is dominant; this leads to non–linearity in the phase–diameter relationship of large transparent particles which result in measurement errors

used for sizing spherical solids, such as powders, industrial dust, glass and metal balls. Such measurements can also be automated for quality control.

Investigation of High Pressure Diesel Fuel Injection Systems as an Example for the Application of PDA

Motivation

The task of fuel injection systems is the atomization of liquid fuel. The aim is the generation of a spray with small droplets and a high kinetic energy, i.e. a high droplet velocity, in order to improve the mixture formation of the fuel with the air, evaporation of the droplets and combustion. PDA is an important laser–optical instrument for the evaluation of the performance of different injection systems and nozzle types.

Measurement conditions and choice of the optical configuration

Due to the high droplet concentration in Diesel fuel sprays, a high laser power is required to ensure the laser beams penetration into the spray center. Moreover, the measurement volume has to be as small as possible to avoid the movement of several droplets through the measuring volume at the same time. A small measurement volume decreases measurement errors due to the

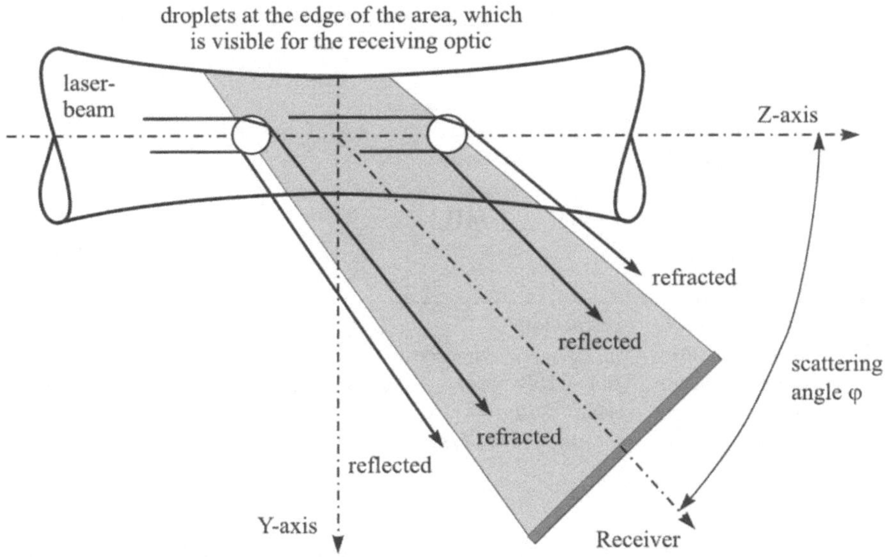

Fig. 9.7. Slit effect: droplets which are partially outside the visible area of the receiving optic, will be detected either by reflected or refracted light, depending on their position in the measurement volume. This causes measurement errors due to non–linearity in the phase–diameter relationship

slit and the trajectory effect [116]. Furthermore, a small measurement volume increases the intensity of the scattered light and therefore increases the signal to noise ratio (SNR). The mean droplet diameter of the investigated sprays was smaller than 20 μm and conversely the droplet velocity in the center of the spray reached more than 300 m/s. Therefore a compromise in the choice of the measurement set–up was necessary. In order to obtain a small measurement volume and a high spatial resolution in diameter, the focal length of the transmitting probe has be short and the beam separation has to be large. The high velocity of the droplets, however, requires a large fringe spacing which can only be realized by a narrow crossing angle of the laser beams, i.e. a large focal length and a small beam separation. The best results were achieved with a configuration as shown in Table 9.1. Comparative measurements by means of a high speed camera have proved that this configuration of the PDA enabled the measurement of droplet sizes and velocities even in the very dense region of the spray tip center.

Results of the PDA measurements

PDA measurements of the droplet sizes and the droplet velocities were carried out under room temperature conditions and ambient pressure (0.1 MPa). The measurement configuration can be seen in Fig. 9.8. In each spray 50 spots were analyzed, distributed over a plane at distances of 20 mm from

Table 9.1. Configuration of PDA–system for measurements in a high pressure Diesel spray

Measuring volume		
diameter:	47	μm
length:	1.96	mm
fringe spacing:	10.6	μm
laser power:	500	mW

Optical configuration		
scattering mode:	refraction	
scattering angle:	27	$degrees$
diameter range:	0.4 – 156	μm
velocity range:	-106 – 372	$\frac{m}{s}$
bandwidth:	45	MHz

the nozzle. In order to get a representative statistic, 20.000 droplets were measured at each location.

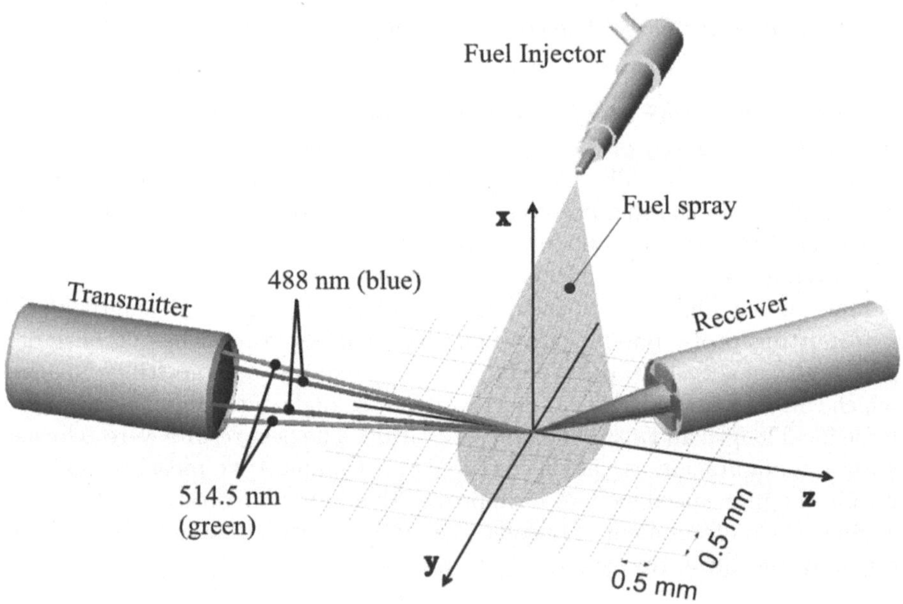

Fig. 9.8. Configuration for measurements in Diesel sprays (2D–PDA–system)

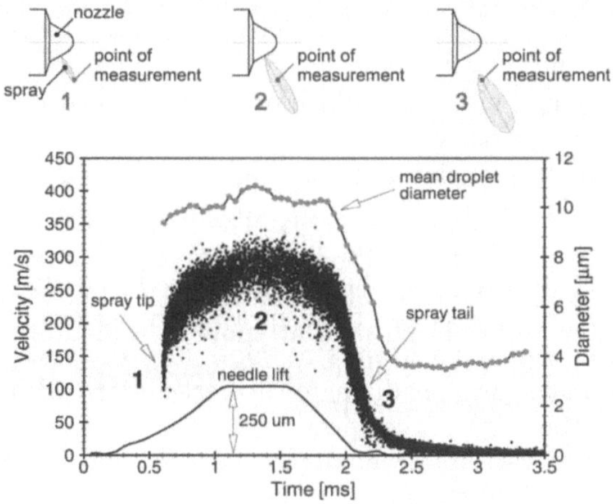

Fig. 9.9. Time series of the droplet velocity and the mean droplet size (D_{10}) in the center of a high pressure Diesel spray at a distance of 20 mm from the nozzle (20.000 validated droplets)

The results of the PDA experiments show a characteristic and reproducible structure in the time series of the droplet size and the droplet velocity in the center of the spray (s. Fig. 9.9).

The center of the spray tip covers a wide range of velocities from 80 m/s up to more than 300 m/s. The highest droplet velocities appear right behind the spray tip and remain at a constant level as long as the needle lift is constant. At the end of the injection the droplet velocity decreases rapidly.

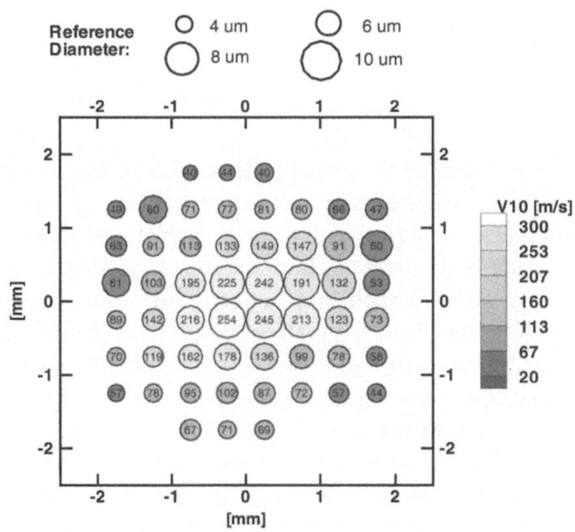

Fig. 9.10. Distribution of the mean droplet diameter (D_{10}) and the mean droplet velocity in a plane perpendicular to the spray axis

At ambient pressures 0.1 MPa the profile of the mean droplet diameter (D_{10}) is similar to the profile of the velocity, as shown in Fig.9.9. The smallest droplets have the lowest velocity and vice versa which proves that the droplet size is correlated with the droplet velocity.

The distribution of the mean droplet size and the mean velocity in a plane perpendicular to the spray axis is shown in Fig. 9.10. The mean values were calculated in a time interval with constant needle lift. The highest velocities and biggest droplets appear in the center of the spray. The velocity distribution has a high gradient towards the edge of the spray with peak values of up to 260 m/s in the center and only 20 m/s at spots in a distance of 1.5 mm from the center. The mean droplet diameter decreases from 10 μm in the center down to 4 μm at the edge of the spray.

Acknowledgement

The author wishes to thank Dantec Measurement Technology for their kind support. It is further acknowledged that the presented measurements were financed by the Bavarian Research Foundation (Bayerische Forschungsstiftung) and the automotive companies AUDI, BMW and MAN.

10 Dynamic Light Scattering

Boris Kruppa, Martin Pitschmann, and Johannes Straub

10.1 Introduction

As the name implies, Dynamic Light Scattering analyzes the temporal behavior of light scattered by a sample fluid, enabling a number of properties of the sample to be determined. The theoretical foundations of this method were laid at the beginning of the century, but the high resolution required to detect and measure the small frequency shifts and narrow spectral linewidths of the scattered light imposed stringent demands on both the light source and the detection system.

Only with the development of the laser in the 1960s and the concurrent widespread use of photomultipliers as detection elements did measurements become feasible. The continuous improvement of these components and the introduction of the digital correlator as a means of measuring spectral linewidths has improved the accuracy and simplified the operation of the method. Today, dynamic light scattering has ever more applications in the fields of physics, chemistry and biology, such as the measurement of diffusion and diffusion–related properties or of molecular characteristics and structural changes in fluids.

10.2 Overview

The primary process of dynamic light scattering is known as Rayleigh scattering. As discussed in the introductory chapter on scattering processes, Rayleigh scattering deals with particles which are small compared to the wavelength of the incident light such that their interaction does not appreciably change the phase or amplitude of the light wave. The molecules of the fluid itself are typical scatterers, but microscopic suspended particles also fulfil the Rayleigh criterion. The effect was first observed by Tyndall [160], who studied the scattering of light on aerosols and was able to explain the color of the sky and the polarization of scattered light. Rayleigh [157], [158] cast these observations into theoretical form and showed that the scattered light intensity was proportional to the square of the particle diameter and inversely proportional to the fourth power of the wavelength of light.

Classical electromagnetic wave theory can be used to describe the scattering process involved (Fig. 10.1). In the presence of an electric field $\boldsymbol{E}_0$, the individual molecules are polarized and the positive and negative charges are distorted, aligning themselves in the direction of the applied field. A dipole moment $\boldsymbol{P}$ parallel to the direction of the field is induced, creating an oscillating dipole which radiates electromagnetic energy $\boldsymbol{E}_S$ in all directions. The magnitude of the resultant radiation field $\boldsymbol{E}_S$ is dependent on the molecular polarizability α as $\boldsymbol{E}_S = \alpha \cdot \boldsymbol{E}_0$. This polarized scattered light is of the same frequency as the incident light, a characteristic of Rayleigh scattering.

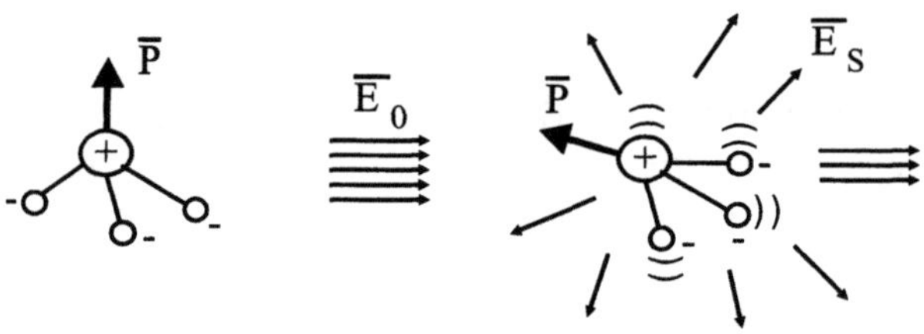

Fig. 10.1. Rayleigh scattering mechanism

In classical Rayleigh theory, the molecules are treated as independent scatterers. Thus the resultant intensity of scattered light is simply the sum of the individual contributions of the molecules. While this theory can predict the scattered intensities of dilute gases quite accurately, it fails in the fluid region, the predicted intensities being more than an order of magnitude larger than actually measured. The explanation lies in the fact that the individual scattering contributions in fluids and dense gases interact, making an exact estimate of the overall intensity very difficult. This problem was solved in the early 1900s by Smoluchowski [159] and Einstein [150], who developed the fluctuation theory of light scattering. The concept can be demonstrated by a simple thought experiment.

If we assume the molecules of a substance to be motionless and ordered in a strict lattice structure, an incident monochromatic light beam of frequency ω_o will produce no resultant scattering intensity I_s due to destructive interference (Fig. 10.2 a). By observing a scattering center at any point P in space, we will always find another which is half a wavelength apart such that both contributions cancel. The light beam is only visible in the forward direction.

If the molecules are next assumed to be randomly oriented, as they would be in any typical fluid, but still motionless, some scattered intensity of constant value results. This intensity is however less than that predicted by the Rayleigh theory since there is still a great amount of destructive interference. The frequency spectrum of the scattered light $S(\omega)$ reveals a sharp line at the incident light frequency ω_0 if Rayleigh scattering is the only process involved (Fig. 10.2 b).

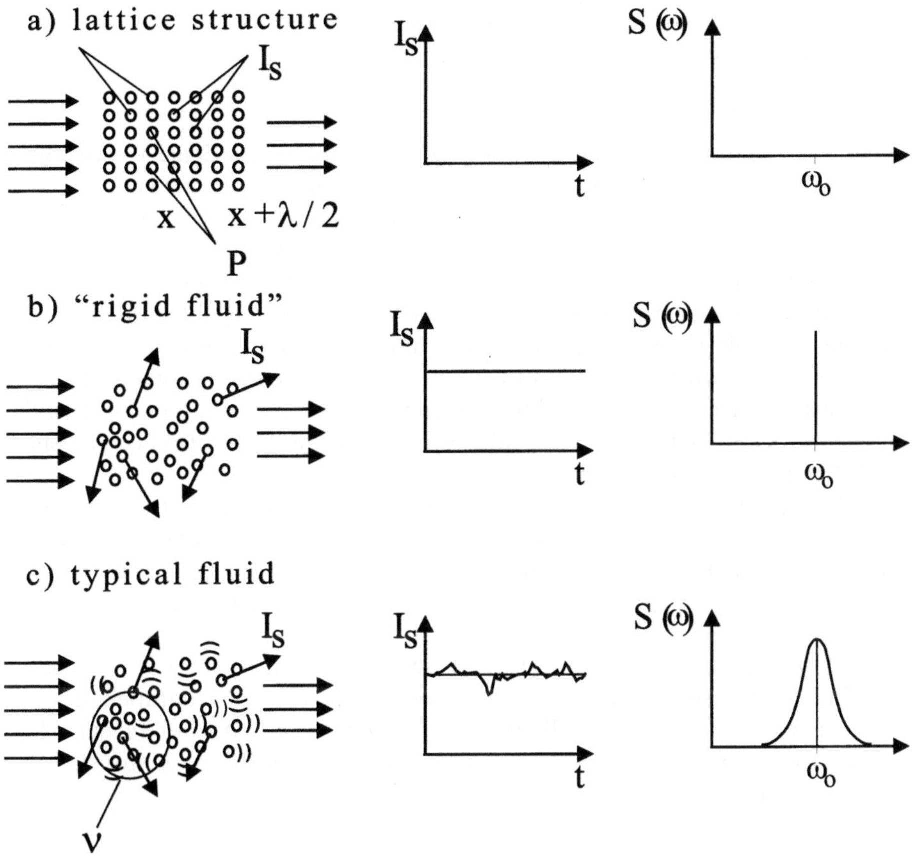

Fig. 10.2. Demonstration of the fluctuation theory of light scattering. **a**) lattice structure. **b**) "rigid fluid ". **c**) typical fluid

In fluids however, the molecules are constantly in random motion, known as thermal or Brownian Motion. Light scattered from these molecules will fluctuate in intensity. The resulting spectrum is still centered around the incident frequency but is now broadened somewhat due to the characteristics of the molecular motion (Fig. 10.2 c).

If we now observe a volume element ν in Fig. 10.2 c which is small enough to be considered microscopic but large enough to contain a sufficient number of molecules, the thermal motion will cause the number density to fluctuate in time, even for fluids in a macroscopic state of equilibrium.

By considering the large number of molecules that are in the element ν, we may overlook its particle nature and treat the fluid as a continuum. The paths and positions of the molecules need not be taken into account and macroscopic thermodynamic properties such as density can be used to describe the fluctuations. Thus we measure small density fluctuations around macroscopic equilibrium values in fluids

$$\varrho(t) = \varrho_0 + \Delta\varrho(t) \tag{10.1}$$

where ϱ_0 denotes the constant macroscopic equilibrium density and $\Delta\varrho(t)$ the fluctuating component. Other thermodynamic properties such as pressure, entropy and temperature fluctuate as well. These fluctuations are in essence dissipative, that is, decaying towards 'macroscopic equilibrium'. The forces governing these processes are characteristic properties of the fluid. The concept of a continuum and small fluctuations around equilibrium values allows us to apply the hydrodynamic equations of motion (laws of conservation of mass, momentum and energy) to investigate the temporal behavior of the fluctuations in fluids. This yields transport properties as thermal diffusivity **a**, the diffusion coefficient D for mixtures, the sound absorption coefficient β and the speed of sound C_S.

Finally, Onsagers [155] law of reciprocal relations states that the laws governing fluid behavior on a microscopic scale are identical to those governing macroscopic thermodynamic behavior, so that by investigating the temporal behavior of microscopic fluctuations we can extract the transport properties of the fluid. The fluid may also be seeded with suspended particles small enough to fulfil the Rayleigh criterion. Depending on their mass, the particles follow the thermal motion of the fluid molecules in a dampened form. A temporal investigation of light scattered off these particles reveals their respective diffusion coefficients through the fluid, and finally, the size and mass distribution of the particles themselves.

10.3 Light Scattering Theory

The application of dynamic light scattering has been extended to several fields of science. Depending on the experimental characteristics and on what property is to be measured, the assumptions and theoretical approaches differ, making a general treatment of light scattering theory far too large a task for the scope of this contribution. For this reason we restrict ourselves to translational diffusion, treating an important application: the measurement of thermal diffusivity of pure fluids.

Even with these restrictions, the theoretical considerations involving light scattering are rather extensive, which is why we emphasize only the important aspects here. Our objective is to supply an overview. A detailed derivation can be found in standard literature such as Berne & Pecora [145], Chu [147], Cummins & Pike [149] and Pecora [156].

10.3.1 Scattering Geometry and Assumptions

Although light scattering theory can be described by a molecular approach, eg. Fixmann [151] and Zwanzig [161], fluctuation theory is more commonly used because it treats the fluid as a continuum, allowing the use of macroscopic thermodynamics. The results are consistent with molecular theory for the applications treated here.

Fig. 10.3 depicts a model of the geometry used in light scattering theory. Usually the laser beam is focused into the fluid and the scattered light is observed at a far point P at an angle Θ by means of a detector.

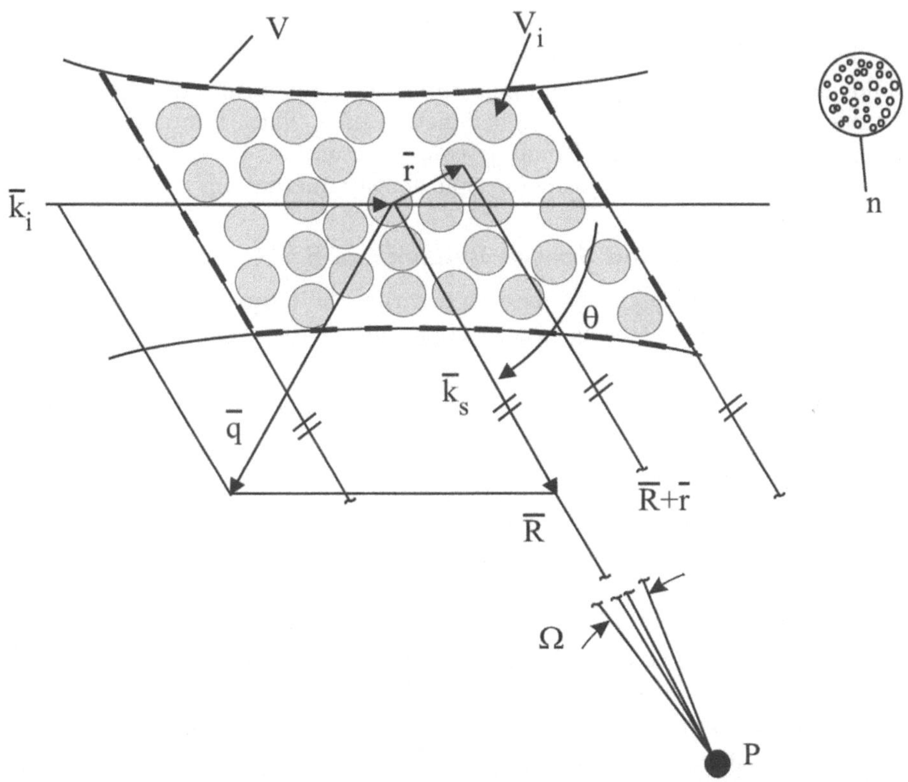

Fig. 10.3. Light scattering geometry

The scattering volume V is defined by the optics of the light source (by how precisely the laser beam can be focused down to a point), and by that of the detection system (characterized by the solid angle Ω). V consists of a large number of imaginary microscopic volumes ν which, in analogy to Fig. 10.2 c, contain the actual molecules that produce the scattering. The fluid is assumed to be isotropic. ν is still small compared to the incident wavelength ($\nu \ll \lambda^3$), so that the incident electric field is homogenous over the volume. Since ν is treated as a continuum (the molecules in fluids and dense gases being packed closely together), the molecular polarizability α responsible for Rayleigh scattering is replaced by the corresponding macroscopic property, the dielectic constant ε. The thermal motion of the molecules causes the dielectric constant of ν to fluctuate as $\varepsilon(t) = \varepsilon_0 + \Delta\varepsilon(t)$, with ε_0 being the mean macroscopic dielectric constant and $\Delta\varepsilon(t)$ signifying the fluctuation of ε. $\Delta\varepsilon$ is assumed to be scalar, meaning that we do not consider depolarized light scattering. The resultant scattered intensity at the far point P is the sum of all scattering contributions of the microscopic volumes ν, each of which is treated as an independent scatterer, i.e. the scattering contributions of each element ν are uncorrelated.

Generally, the fluctuations in the dielectric constant $\Delta\varepsilon(r,t)$ are a function of the temperature T and density ϱ of the fluid. However, for gases and fluids in a moderate range of temperatures, the temperature dependence can be neglected and $\Delta\varepsilon(r,t)$ is only a function of ϱ. Thus, fluctuations in the dielectric constant, which ultimately cause fluctuations in the scattered light intensity, directly reflect the microscopic density fluctuations within the fluid.

10.3.2 Temporal and Spatial Behavior of Scattered Light

Classical wave theory can be used to describe the temporal and spatial behavior of Rayleigh scattering. The electromagnetic radiation obeys Maxwell's equations:

$$\nabla \cdot \boldsymbol{D} = 0 \tag{10.2a}$$

$$\nabla \cdot \boldsymbol{B} = 0 \tag{10.2b}$$

$$\nabla \times \boldsymbol{E} = -\frac{\partial \boldsymbol{B}}{\partial t} \tag{10.2c}$$

$$\nabla \times \boldsymbol{H} = \frac{\partial \boldsymbol{D}}{\partial t} \tag{10.2d}$$

whereby we assume no isolated charges and currents (the volume charge density ρ and free current density σ are zero). $\boldsymbol{D}$ denotes the electric displacement and $\boldsymbol{B}$ the magnetic field. Since the detector, responding to light intensity I, is usually a square law detector ($I = |\boldsymbol{E}|^2$), we are primarily interested in variations of the electric field. The incident electromagnetic wave is written in its complex form:

$$\boldsymbol{E}(\boldsymbol{r},t) = \boldsymbol{E_0} \cdot e^{i(\boldsymbol{kr}-\omega t)} \tag{10.3}$$

$\boldsymbol{E_0}$ is a complex amplitude, $\boldsymbol{k}$ the wave vector with $|\boldsymbol{k}| = \frac{2\pi}{\lambda}$ and ω the angular frequency of the wave. The scattering geometry is that of Fig. 10.3 with the incident wave $\boldsymbol{k_i}$ polarized perpendicularly to the scattering plane. The resultant scattered field $\boldsymbol{E_s}$ at a far point P is the superposition of all fields radiated from volume elements $d^3r(\nu)$ at a position $\boldsymbol{r}$ with respect to the scattering center (random walk in three dimensions). $\boldsymbol{E_S}$ can, after a rather lengthy calculation, be expressed as

$$\boldsymbol{E_S}(\boldsymbol{R},t) = \frac{-\boldsymbol{E_0}\boldsymbol{k_s}^2}{4\pi \boldsymbol{R}\varepsilon_0} \cdot e^{i(\boldsymbol{k_s R}-\omega_i t)} \int_V e^{i\boldsymbol{qr}} \Delta\varepsilon(\boldsymbol{r},t) d^3r \tag{10.4}$$

The scattering vector $\boldsymbol{q}$, as shown in Fig. 10.4, defines the scattering geometry. It results from the difference between the incident and scattered wave vectors $\boldsymbol{q} = \boldsymbol{k_i} - \boldsymbol{k_s}$. The product $\boldsymbol{q} \cdot \boldsymbol{r} = (\boldsymbol{k_i} - \boldsymbol{k_s}) \cdot \boldsymbol{r}$ is the phase delay that results from the path difference $(a-b)$ between light scattered from a volume element ν at $\boldsymbol{r}$ relative to the one at the origin (Fig. 10.3 and 10.5).

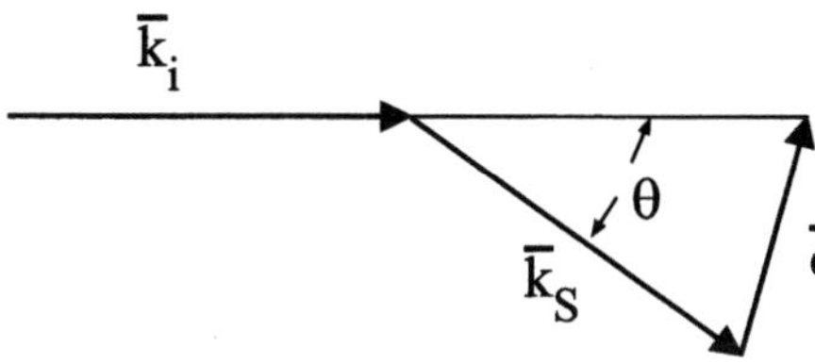

Fig. 10.4. Definition of the scattering vector

If we assume elastic light scattering, then $|\boldsymbol{k_i}| = |\boldsymbol{k_s}|$ and the value of $\boldsymbol{q}$ (s. Fig. 10.4) becomes:

$$|\boldsymbol{q}| = \frac{4\pi n}{\lambda_L} \cdot \sin\left(\frac{\Theta}{2}\right) \tag{10.5}$$

where λ_L is the wavelength of light in vacuum and n the refractive index of the fluid. The scattered field $\boldsymbol{E_S}$ therefore consists of a monochromatic carrier wave $\int_V e^{i\boldsymbol{qr}} \cdot \Delta\varepsilon(\boldsymbol{r},t)d^3r$ which constitutes the fluctuating signal. The integral is actually the spatial Fourier transformation of the dielectric constant fluctuations $\Delta\varepsilon(\boldsymbol{r},t)$ into $\boldsymbol{q}$–space (the component of the fluctuations in $\boldsymbol{q}$–direction):

$$\Delta\varepsilon(\boldsymbol{q},t) = \int_{\mathbf{V}} e^{i\boldsymbol{qr}} \cdot \Delta\varepsilon(\boldsymbol{r},t)d^3r \tag{10.6}$$

Equation (10.4) can therefore be written as:

$$\boldsymbol{E_s}(\boldsymbol{R},t) = \frac{-\boldsymbol{E_0}\boldsymbol{k_s}^2}{4\pi\boldsymbol{R}\varepsilon_0} \cdot e^{i(\boldsymbol{k_s}\boldsymbol{R}-\omega_i t)} \cdot \Delta\varepsilon(\boldsymbol{q},t) \tag{10.7}$$

Since $\Delta\varepsilon$ is only a function of the density ϱ, the information of the fluctuating dielectric constant, and of the density fluctuations $\Delta\varrho(\boldsymbol{q},t)$ is thus contained in the temporal fluctuations of the scattered light signal $\boldsymbol{E_s}(\boldsymbol{R},t)$.

10.3.3 Correlation Functions

The aforementioned microscopic fluctuations represent a stochastic process resulting from the random motion of the molecules or particles. The signal possesses typical characteristics of a noise pattern. To obtain the information

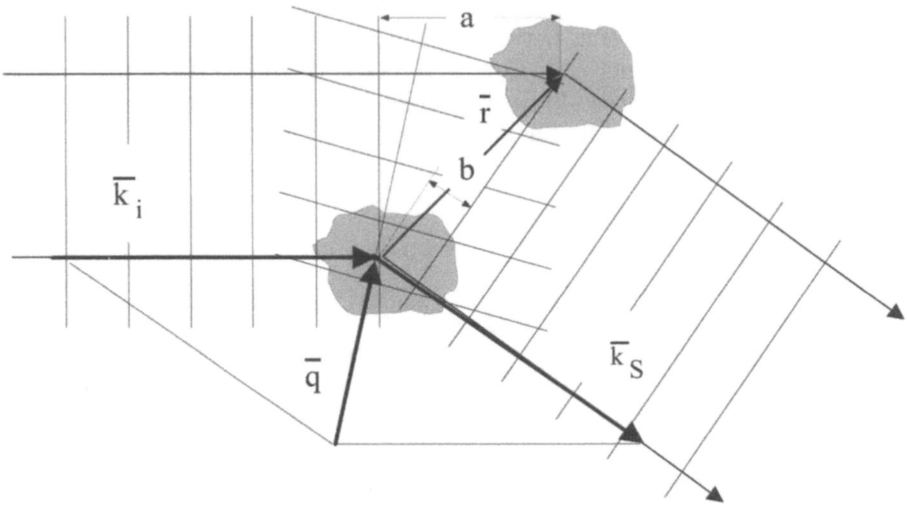

Fig. 10.5. Section of the scattering geometry

on transport properties present in the fluctuating density signal we can analyze its temporal behavior by digital correlation. Another possibility is to decompose the signal into its frequency components by Fourier transformation with a spectrum analyzer, thus obtaining its power spectrum.

Most experiments today are performed with digital correlators since these instruments are fast and introduce no noise of their own, thus possessing an efficiency of 100%. Modern spectrum analyzers can obtain the spectral information with comparable speed. However, they invariably introduce a bandwidth error into measurements. We therefore concentrate on the concept of time correlation functions and digital correlators in this section. A typical fluctuating signal, in this case the fluctuating part of the microscopic density $\Delta\varrho(t)$ in fluids, is shown in Fig. 10.6.

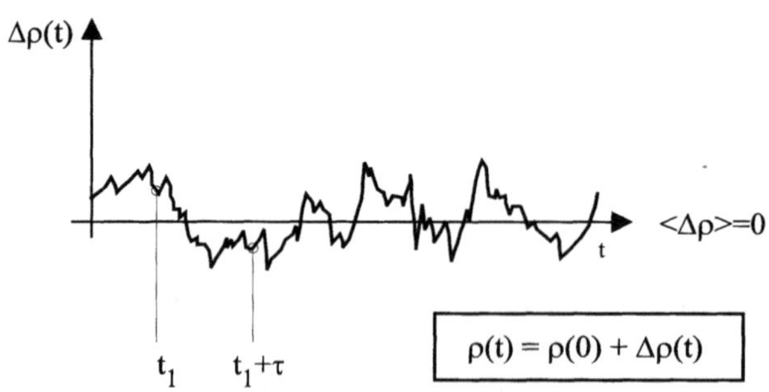

Fig. 10.6. Fluctuating density signal

Provided the system is in macroscopic equilibrium ($\varrho_0 = \text{const}$), the signal will fluctuate around a mean value $< \Delta\varrho >$ given by:

$$< \Delta\varrho > \;=\; \lim_{T\to\infty} \frac{1}{T} \int_0^T \Delta\varrho(t)\mathrm{d}t \tag{10.8}$$

with T as integration time. In this case $< \Delta\varrho > = 0$. Generally, at a certain time t_1, $\Delta\varrho$ will differ from the same signal a variable time τ later, $\Delta\varrho(t_1) \neq \Delta\varrho(t_1 + \tau)$. If τ is chosen to be small ($\tau \to 0$), then the signal will not have changed appreciably during τ and the correlation between both terms is large. If τ is increased, the deviation is more likely to increase, and if this comparison is repeated often the correlation is lost. A measure of this interdependence can be expressed in a correlation function $G(\tau)$ which is defined as:

$$G(\tau) \;=\; < \Delta\varrho(t_1)\cdot\Delta\varrho(t_1+\tau) > \;=\; \lim_{T\to\infty} \frac{1}{T} \int_0^T \Delta\varrho(t_1)\cdot\Delta\varrho(t_1+\tau)\mathrm{d}t \tag{10.9}$$

The brackets <> denote the temporal averaging. If this process is repeated ($T \to \infty$), the correlation function becomes independent of the starting time t.

$$G(\tau) \;=\; < \Delta\varrho(0) \cdot \Delta\varrho(\tau) > \tag{10.10}$$

$G(\tau)$ is a measure of the mutual relationship in a signal. If the signal is completely random, then the correlation function will remain flat (i.e. uncorrelated), even after long times ($T \to \infty$). For nonperiodic signals such as those which are the case here, $G(\tau)$ is a decaying function for τ varying from 0 to infinity (Fig. 10.7).

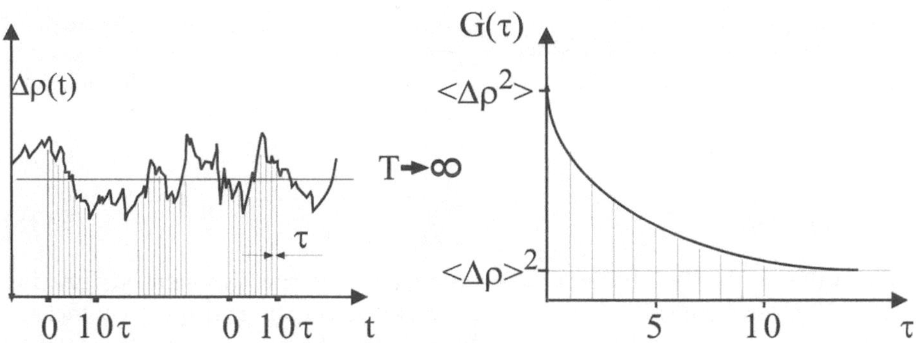

Fig. 10.7. Typical correlation function of a non–periodic signal

If we correlate the signal over a long period ($T \rightarrow \infty$) for $\tau \rightarrow 0$ (i.e. almost no time lag between the two signals to be correlated), then the value of the correlation function is simply the square of the signal:

$$G(\tau)_{\tau\rightarrow 0} \quad = \quad <\Delta\varrho(0)\cdot\Delta\varrho(0)> \quad = \quad <\Delta\varrho(0)^2> \qquad (10.11)$$

For long delay times $\tau \rightarrow \infty$, the signal contributions are totally uncorrelated, and their respective averages are the mean value of the fluctuation:

$$\tau \rightarrow \infty <\Delta\varrho(0)> \quad = \quad <\Delta\varrho(\tau)> \quad = \quad <\Delta\varrho> \qquad (10.12)$$

Thus, the correlation function drops to the square of the mean signal:

$$G(\tau)_{\tau\rightarrow\infty} \quad = \quad <\Delta\varrho(0)\cdot\Delta\varrho(\tau)> \quad = \quad <\Delta\varrho><\Delta\varrho> \quad = \quad <\Delta\varrho>^2 \qquad (10.13)$$

Usually, the correlation function is required in its normalized form $g(\tau)$, as shown in Fig. 10.8, which is simply $G(\tau)$ divided by its initial value $G(0)$:

$$g(\tau) \quad = \quad \frac{G(\tau)}{G(0)} \quad = \quad \frac{<\Delta\varrho(0)\cdot\Delta\varrho(\tau)>}{<\Delta\varrho(0)^2>} \qquad \text{with } 0 < g(\tau) < 1 \qquad (10.14)$$

The exact form of the correlation function depends on the physical process under study. As will be shown in the next section, $g(\tau)$ is usually a decaying exponential function for dissipative fluctuations about equilibrium:

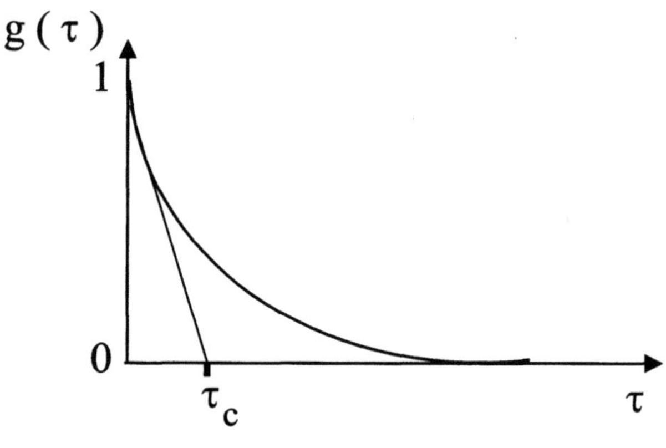

Fig. 10.8. Normalized correlation function with characteristic decay time τ_c

$$g(\tau) = \exp(-\tau/\tau_c) \qquad (10.15)$$

The value τ_c is the characteristic decay time of the exponential function (Fig. 10.8) and contains the actual information on the transport coefficient. $g(\tau)$ represents the statistical mean behavior of the microscopic density fluctuations.

In practice, digital correlators are used to measure $g(\tau)$. These instruments, which are basically fast CPU's, possess a large number of sampling channels or stores (I) which represent distinct points of the correlation function. The delay time τ is split into discrete time intervals Δt and counted by a running parameter j with $j = 1...I$ (number of channels):

$$\tau = j\Delta t(j = 1...I). \tag{10.16}$$

Δt represents the fundamental sample time of the correlator. Thus, each successive channel represents an increasing time lag τ in the correlation function. The actual correlation is achieved by means of shift registers (one for each channel). The momentary value of the signal $i(t)$ arriving at the correlator is multiplied by the delayed signal distribution in each shift register $i(t + j\Delta t)$ before it is added to the contents of the individual sampling channels. Fig. 10.7 schematically demonstrates idealized measurements with a 10–channel correlator. The correct setting of the sample time Δt is very important when correlating, as this defines the time window within which the fluctuations decay.

Another method of extracting information from the density fluctuations is to Fourier transform $G(\tau)$ them into frequency components ($\tau \rightarrow \omega$):

$$S_{\Delta\varrho}(\omega) = \frac{1}{2\pi} \int_{-\infty}^{\infty} < \Delta\varrho(0) \cdot \Delta\varrho(\tau) > e^{i\omega t} \mathrm{d}t, \tag{10.17}$$

obtaining the power spectral density of the signal. This transformation is known as the Wiener–Khintchine theorem and is shown in Fig. 10.9.

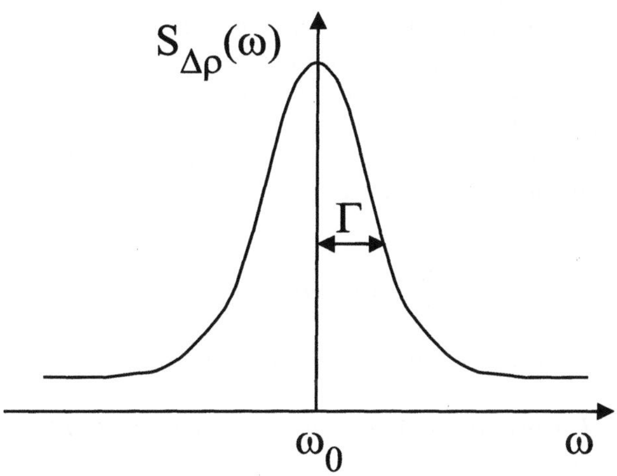

Fig. 10.9. Typical power spectrum

The form of this curve, arising from dissipative fluctuations (i.e. an exponential decaying correlation function), is a Lorentzian distribution centered at the incident laser frequency ω_0 with a linewidth Γ. Γ is a direct measure of the dissipative process involved. Since the power spectrum and the corresponding correlation function are Fourier transform pairs, Γ is directly related to the characteristic decay time:

$$\Gamma = \frac{1}{\tau_c}. \tag{10.18}$$

In practice, spectrum analyzers are used to obtain the integrated power density of the signal at a set frequency ω. By tuning the filter through a frequency range, the power spectral density is measured.

10.3.4 Hydrodynamic Fluctuation Theory

Up to this point, we have shown that the fluctuating electric field (or intensity of scattered light) contains information on density fluctuations in a fluid arising from the collective motion of the molecules. Temporal investigation of the signal by means of correlation, or frequency decomposition by spectrum analysis, reveals the characteristic behavior of the fluctuations. What is still required is the connection between the microscopic fluctuations and macroscopic transport phenomena. In this derivation, we are considering fluctuations arising from the collective motion of large numbers of molecules (the volume elements ν), allowing us to treat the fluid as a continuum.

Molecular interactions such as collision times and inter–particle distances are thus avoided, and only relatively long–lived and long–ranged fluctuations are considered. Macroscopic laws, such as the hydrodynamic laws of motion, can then be used to describe the temporal behavior of the fluctuations. The basis of this theory is the regressional hypothesis of Onsager [155] which states that the spontaneous fluctuations among the individual microscopic volume elements ν 'regress' back to equilibrium according to the same equations which describe the macroscopic relaxation process.

The laws of hydrodynamics are the laws of conservation of mass, momentum and energy stated here in their general form:

$$\frac{\partial \varrho}{\partial t} + \mathrm{div}(\varrho \boldsymbol{u}) = 0 \quad \text{(mass)} \tag{10.19a}$$

$$\frac{\partial (\varrho \boldsymbol{u})}{\mathrm{dt}} + \mathrm{div}(\varrho \boldsymbol{u}\boldsymbol{u}) + \mathrm{grad} P + \mathrm{div}(\boldsymbol{\tau}) = 0 \quad \text{(momentum)} \tag{10.19b}$$

$$\varrho \frac{\partial e}{\partial t} + \mathrm{div}(\varrho \boldsymbol{u} e) - \lambda \nabla^2 T + \mathrm{div}(p \boldsymbol{u}) + \mathrm{div}[\boldsymbol{\tau} \cdot \boldsymbol{u}] = 0 \quad \text{(energy)} \tag{10.19c}$$

where $\boldsymbol{u}$ is a velocity vector, $\boldsymbol{\tau}$ the stress tensor, e is the sum of the internal and kinetic energies and λ the thermal conductivity.

The fluctuations of these properties are determined for instance by substituting $\varrho_0 + \Delta\varrho$ for the density ϱ or $\boldsymbol{u}_0 + \Delta \boldsymbol{u}$ for the velocity $\boldsymbol{u}$. By assuming that these fluctuations are small around equilibrium values, the above equations can be linearized, resulting in 5 equations (conservation of momentum is a vector equation) with 7 variables. Two of the scalar variables can be eliminated by applying local equilibrium thermodynamic equations of state. Usually density and temperature are chosen as the independent variables. The resulting linearized hydrodynamic equations are solved by Fourier–Laplace analysis [154], the fluctuations such as density $\Delta\varrho(\boldsymbol{r},t)$ being space– ($\boldsymbol{r} \rightarrow \boldsymbol{q}$) and time–transformed ($t \rightarrow s$), obtaining $\Delta\varrho(\boldsymbol{q},s)$.

The correlation function of these fluctuations describes the light scattering spectrum. For density fluctuations, the solution is obtained in the form of a correlation function and by Laplace inverting ($s \rightarrow t$) as:

$$\begin{aligned} g(\tau) &= \frac{< \Delta\varrho(\boldsymbol{q},0) \cdot \Delta\varrho(\boldsymbol{q},\tau) >}{< \Delta\varrho(\boldsymbol{q},0) \cdot \Delta\varrho(\boldsymbol{q},0) >} \\ &= \left[\frac{c_p - c_v}{c_v}\right] \cdot \exp(-aq^2\tau) + \frac{c_v}{c_p} \cdot \exp(-\beta q^2\tau) - \cos(C_s q\tau) \end{aligned} \quad (10.20)$$

c_p and c_v denote the specific isobaric and isochoric heat capacities, respectively. a is the thermal diffusivity, β the sound absorption coefficient and C_s the local speed of sound. The spectral density can be obtained by a Fourier transform ($\tau \rightarrow \omega$) of (10.20), according to (10.17). The result is presented schematically in Fig. 10.10.

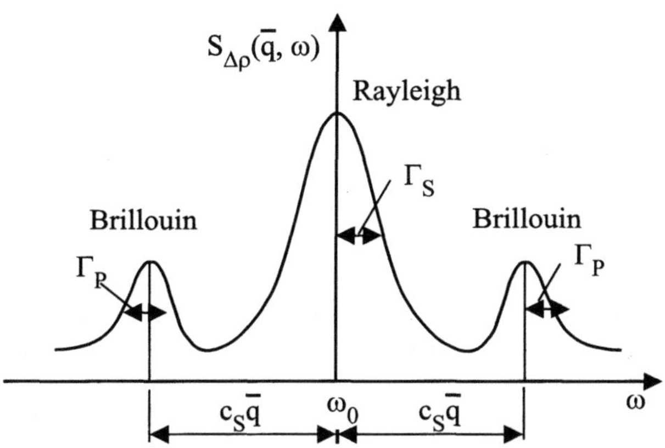

Fig. 10.10. Spectrum of light scattered off density fluctuations

The spectrum of light scattered from density fluctuations is the sum of three Lorentzian distributions. The largest contribution results from the central or Rayleigh line, which peaks at the incident light frequency. The broadening of this spectrum, as expressed by its half–width at half–maximum, is a direct measure for the rate of diffusion, which for pure fluids is the thermal diffusity a (the half–width is in general denoted by Γ as in Fig. 10.9). The

Rayleigh contribution to the spectrum is often interpreted as local entropy fluctuations at constant pressure.

The next two terms represent a doublet, also known as the Mandelshtam–Brillouin doublet. This contribution can be interpreted as local adiabatic pressure fluctuations which give rise to sound waves travelling through the fluid. Light scattered off these waves is slightly frequency shifted in both directions in analogy to the Doppler effect; the amount of frequency shift is dependent on the local speed of sound and the scattering vector. This scattering process is known as Brillouin scattering and has been briefly described in the general scattering section of this book. As with the Rayleigh spectrum, the Brillouin spectrum also possesses a dissipative element (the sound waves decay), a measure of which is the sound absorption coefficient β.

Thus the thermal diffusivity a, diffusion coefficient D (for mixtures), the speed of sound C_s and the sound absorption coefficient β can be obtained by investigating the spectrum of light scattered off local density fluctuations in fluids. In this contribution we only deal with the Rayleigh spectrum investigating diffusion and diffusion related properties.

10.4 Experimental Methods

There are two methods which can be employed in dynamic light scattering experiments; those which measure the intensity of the scattered light directly are known as the single beam, self–beat or homodyne method, while those which superimpose a second reference beam or a local oscillator with the scattered light are called dual beam or heterodyne methods. The major components used in both set–ups are essentially identical; when measuring in heterodyne mode, some additional optical components (e.g. optical fibers, couplers, mirrors, or beam splitters) are used to realize the local oscillator.

10.4.1 Homodyne Method

Figure 10.11 shows a typical experimental set–up of a light scattering apparatus employing the homodyne or heterodyne detection method. We will first treat the major components before dealing with the characteristics of the light detection.

Frequency stabilized (etalon) argon–ion lasers are often used as light sources in light scattering experiments due to their high stability and output power. When increasing the intensity of the incident light beam the signal–to–noise ratio S/N of the photomultiplier is improved. However, for two reasons the light intensity should be kept at the required minimum. First, unwanted convection in the fluid should be avoided. Secondly, indefinable stray light from the cell and the glass window surfaces should be minimized. As shown below, the correlation function is influenced by this stray light, which complicates the data evaluation.

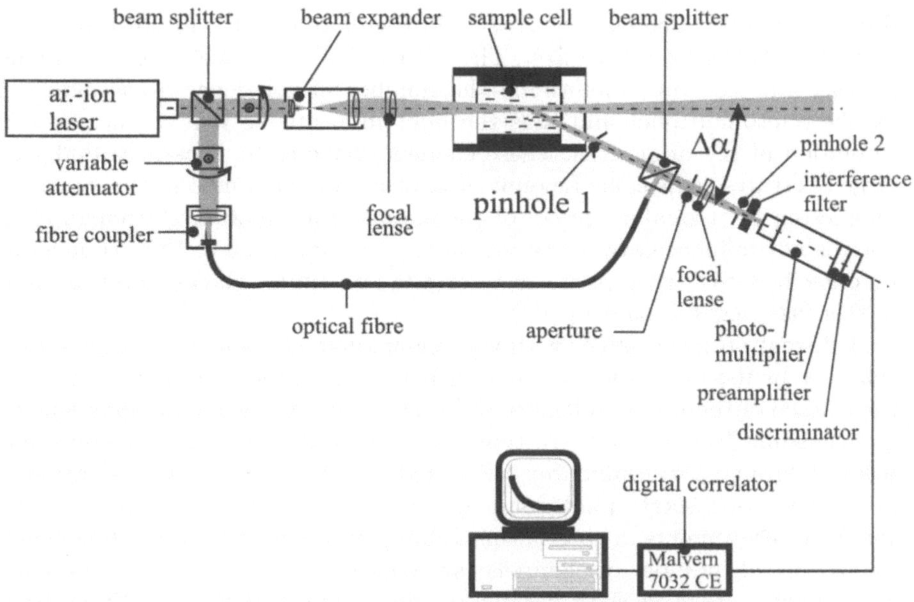

Fig. 10.11. Experimental set–up in light scattering experiments

The required laser power lies between 1 and 300 mW depending on the scattering ability of the fluid. For experiments with large scattering intensities, such as those involving light scattered off particles or those investigating the critical region of fluids, lower light levels of a few milliwatts suffice, making helium–neon lasers applicable.

As the theory of light scattering assumes Gaussian statistics of the incident light, the laser beam is usually directed through a beam expander before it is focused into the scattering volume. The beam is polarized with the polarization vector perpendicular to the scattering plane. The intensity of the scattered light is then observed at the variable angle Θ by means of a photomultiplier tube. Recent advances made with avalanche photodiodes which possess advantages such as a higher quantum efficiency and compact dimensions make these instruments useful for detection [153].

In order to reduce self correlation effects (after–pulse and dead time effects) which appear at short delay times τ, two photomultipliers can be used in cross correlation. The disadvantage is the reduction of the signal due to beam splitting.

An important factor which determines the signal–to–noise ratio S/N is the amount of power scattered into a single coherence area A_{COH}. Thus the detection capability of the light scattering apparatus A_{COH} defines an area over which signal contributions arising from different parts of the scattering volume V are correlated at the photomultiplier detector surface (allowing its treatment as an ideal point–detector), which is inversely related to the dimensions of V. The laser beam should therefore be focused down to small diameters, typically $d < 0.1\ mm$. The elements of the optical system (pinhole 2, lens, and aperture) determine the number of coherence areas detected by the photomultiplier and thus the light intensity, as well as the angular resolution of the apparatus. These elements have to be chosen carefully as large apertures reduce the amount of information contained in the signal as manifested by a smaller spread of the exponential function. Although S/N itself is not influenced by these measures, the evaluation of the exponential function becomes more prone to error and run times increase substantially in this case (e.g. Cummins [149]).

Different optical systems with varying angular, on– and cross–axis resolutions can be used to project the light generated in the scattering volume onto the surface of the photomultiplier [146]. Dual pinhole, single imaging lens or dual imaging lens systems are typically employed. After amplification and passing through a discriminator, the signal $i(t)$ is fed into a digital correlator in the form of discrete photocount events within a certain sampling interval Δt. Improvements in correlation techniques and correlator features (such as size, speed, number of channels) are continuous. In particular, the ability of correlators to work in parallel (or more) modes with logarithmic time scales allows a simultaneous determination of various diffusive processes with different time scales. Particle sizing is a typical application. The obtained correlation function $g(\tau)$ is then analyzed, usually by a nonlinear least–squares fit, resulting in a direct determination of diffusivity.

The derivation of light scattering theory in Sect. 10.3.2 to 10.3.4 has shown that the correlation function of the density fluctuations is an exponential function and that the characteristic decay time τ_c is related to diffusivity $\tau_c = 1/aq^2$. Furthermore, the electric field and the density fluctuations are directly related as:

$$\Delta\varrho(\boldsymbol{R},t) \sim \Delta\epsilon(\boldsymbol{R},t) \sim \boldsymbol{E_s}(\boldsymbol{R},t). \tag{10.21}$$

Therefore, the normalized correlation function of the scattered electric field directly measures diffusivity:

$$g_1(\tau) = \frac{< |\boldsymbol{E_s}(0)| \cdot |\boldsymbol{E_s}(\tau)| >}{< |\boldsymbol{E_s}(0)|^2 >} = \exp(-aq^2\tau). \tag{10.22}$$

Photomultipliers are square–law detectors and only respond to intensity fluctuations:

$$I_S(t) = \boldsymbol{E_S}(t) \cdot \boldsymbol{E_S}^*(t), \tag{10.23}$$

the "*" denotes the complex conjugate value of the field. Thus it is not the so–called first–order correlation function $g_1(\tau)$ that is measured in homodyne experiments, but the second–order correlation function $g_2(\tau)$:

$$g_2(\tau) = \frac{< |\boldsymbol{E_S}(0)|^2 \cdot |\boldsymbol{E_S}(\tau)|^2 >}{< |\boldsymbol{E_S}(0)|^2 >^2} = \frac{< I_S(0) \cdot I_S(\tau) >}{I_S^2}, \tag{10.24}$$

with $< |\boldsymbol{E_S}(0)|^2 >^2 = I_S^2$.

The informational content of $g_1(\tau)$ is not lost in $g_2(\tau)$, provided that the scattered light has a Gaussian distribution. This requires that the total scattered intensity arises from a large number of statistically independent subregions (as assumed in Sect. 10.3.1). The statistical properties of the higher moments are then only a function of the first order correlation function:

$$g_2(\tau) = 1 + g_1(\tau)^2. \tag{10.25}$$

The correlation function now takes the form:

$$g_2(\tau) = 1 + [\exp(-aq^2\tau)]^2 = 1 + \exp(-2aq^2\tau) \tag{10.26}$$

or, if not normalized:

$$G_2(\tau) = I_S^2 + I_S^2 \exp(-2aq^2\tau). \tag{10.27}$$

In homodyne experiments we therefore measure the square of the signal $\boldsymbol{E_S}(\boldsymbol{R}, t)$ and the resulting exponential decay is multiplied by a factor of 2. The homodyne method is generally employed when the scattered light intensities are high and no other stray sources of light are present. This condition can be achieved by means of an additional pinhole (pinhole 1) which effectively screens out sources such as the flare of the cell windows.

10.4.2 Heterodyne Method

At low fluctuation intensities, signal enhancement is achieved by superimposing a second coherent beam or local oscillator of constant intensity with the scattered light beam. This can be done in numerous ways so as to include light scattered from the interior window surface. This is achieved by shifting the scattering volume closer to the window surface. Another effective method is to split the laser beam before the scattering volume, superimposing the local oscillator and scattered light beam at the detector surface. This set–up has the advantage that the intensity of each light beam can be adjusted easily by a variable attenuator.

To minimize time consuming alignment when using mirrors to bypass the sample cell, a polarization preserving optical fiber can be used instead, as shown in Fig. 10.11. The electric field arriving at the detector is then the sum of the scattered field $\boldsymbol{E_S}$ and the local oscillator $\boldsymbol{E_{LO}}$. Again, since photomultipliers respond to intensity, the second–order correlation function (not normalized) becomes:

$$G_2(\tau) = <| \boldsymbol{E_S}(0) + \boldsymbol{E_{LO}}(0) |^2 \cdot | \boldsymbol{E_S}(\tau) + \boldsymbol{E_{LO}}(\tau) |^2> . \tag{10.28}$$

Multiplying out this expression results in 16 terms. By assuming $|\boldsymbol{E_{LO}}|$ to be constant, that is $|\boldsymbol{E_{LO}}(0)| = |\boldsymbol{E_{LO}}(\tau)|$, and $\boldsymbol{E_{LO}}$ and $\boldsymbol{E_S}$ to be statistically independent, Eq. (10.28) is reduced to three terms:

$$G_2(\tau) = < I_{LO}^2 > + < 2I_{LO}(\boldsymbol{E_S}(0)\boldsymbol{E_S^*}(\tau)) > + < I_S(0)I_S(\tau) > . \tag{10.29}$$

With Eq. (10.22) and (10.27) the resulting correlation function can be written as:

$$G_2(\tau) = (I_{LO} + I_S)^2 + 2I_{LO}I_S\exp(-aq^2\tau) + I_S^2\exp(-2aq^2\tau) \quad (10.30)$$

which is the addition of two exponential functions, a heterodyne term $2I_{LO}I_S \cdot \exp(-aq^2\tau)$ and a homodyne term $I_S^2 \cdot \exp(-2aq^2\tau)$, differing in their decay rates by a factor of two. Since the signal detected at the photomultiplier is always a superposition of light scattered off other objects (cell walls, windows, dust etc.), Eq. (10.30) represents the general form of the correlation function for both methods. The simultaneous evaluation of two exponential functions is always difficult and prone to error. Thus, the homodyne technique assumes that the intensity of stray light I_{LO} is negligible ($I_S \gg I_{LO}$) reducing Eq. (10.30) to Eq. (10.27), while the heterodyne method assumes ($I_{LO} \gg I_S$) and reduces Eq. (10.30) to:

$$G_2(\tau) = (I_S + I_{LO})^2 + 2I_{LO}I_S\exp(-aq^2\tau). \quad (10.31)$$

This condition can be assured by varying the intensity of I_{LO} with respect to I_S.

Heterodyning can be a very effective method for measuring diffusivities and diffusion coefficients at distances further away from the critical point, especially in the liquid region. In those regions where intensity fluctuations are very weak the statistical accuracy may be improved by a factor of 4. Heterodyning, however, requires coherent mixing of the scattered light and reference beam signals. The sensitivity is comparable to that of a Michelson interferometer.

Intensity fluctuations of the local oscillator I_0 and external vibrations are sources of error which are difficult to avoid completely. Due to the weak signal I_S, run times vary between one to three hours.

10.5 Measurement of Thermal Diffusivity

As already mentioned, dynamic light scattering has numerous applications in physics, chemistry and biology. A detailed overview is given by Pecora [156]. Since even a brief treatment of these applications would exceed the scope of this contribution, we will only describe an application from the field of physics, the measurement of thermal diffusivity a of pure fluids.

The behavior of microscopic density fluctuations can be described, as shown, by the hydrodynamic equations of motion. The resulting spectrum consists of entropy fluctuations at constant pressure (Rayleigh spectrum) and pressure fluctuations at constant entropy (Brillouin spectrum). The fluctuations in entropy represent a purely dissipative mode of heat transfer. Since entropy is primarily a function of temperature T, the decay of these fluctuations is expressed by the fourier heat law:

$$\frac{\partial}{\partial t}\Delta T(\boldsymbol{r},t) = a\,\nabla^2\left(\Delta T(\boldsymbol{r},t)\right) \tag{10.32}$$

which is basically the law of conservation of energy (10.19a–c) assuming $p =$ const. The solution of this equation, after performing the necessary Fourier transformation into $\boldsymbol{q}$–space and integrating the differential equation, is the exponential function:

$$\Delta(\boldsymbol{q},t) = \Delta T(\boldsymbol{q},0)\exp(-a\,|\boldsymbol{q}|^2\,t). \tag{10.33}$$

The decay of temperature gradients in macroscopic as in microscopic systems follows a simple exponential behavior. The measurement of the characteristic decay time τ_c by means of correlation directly yields the thermal diffusivity:

$$\tau_c = \frac{1}{a\,|\boldsymbol{q}|^2}. \tag{10.34}$$

Substituting for the scattering vector $\boldsymbol{q}$, we obtain:

$$a = \frac{1}{\tau_c[\frac{4\pi n}{\lambda_L}\sin(\frac{\Theta}{2})]^2} \tag{10.35}$$

which is the equation determining thermal diffusivity. τ_c is obtained from a regression analysis of the correlation function. The scattering angle Θ has to be determined along with the refractive index n of the fluid, although the error made by assuming $n = \text{const} = 1$ is below 1% for $\Theta < 10°$ [8]. The incident wavelength of light λ_L is usually kept constant.

Fig. 10.12 shows thermal diffusivity measurements of the alternative refrigerant R152a along the coexistence curve and several sub– and supercritical isotherms plotted against the reduced density ϱ/ϱ_c. ϱ_c denotes the critical density of the fluid. Measurements were made using both methods described earlier. In the extended critical region characterized by $0.3 < \varrho/\varrho_c < 1.8$, where scattering intensities are relatively large, the homodyne method was employed. The measurements lying outside this region were obtained by the heterodyne method. The accuracy of these measurements is dependent upon the investigated region of state and lies between 0.5% and 2.0%. In the gas region at lower densities (typically $\varrho/\varrho_c < 0.2$), this method is subject to increasing errors due to low density fluctuations and resulting low scattered light intensities and therefore cannot be applied.

When compared with other conventional methods of measuring thermal diffusivity or thermal conductivity, such as the instationary hot wire technique or the parallel plates method, dynamic light scattering offers three major advantages.

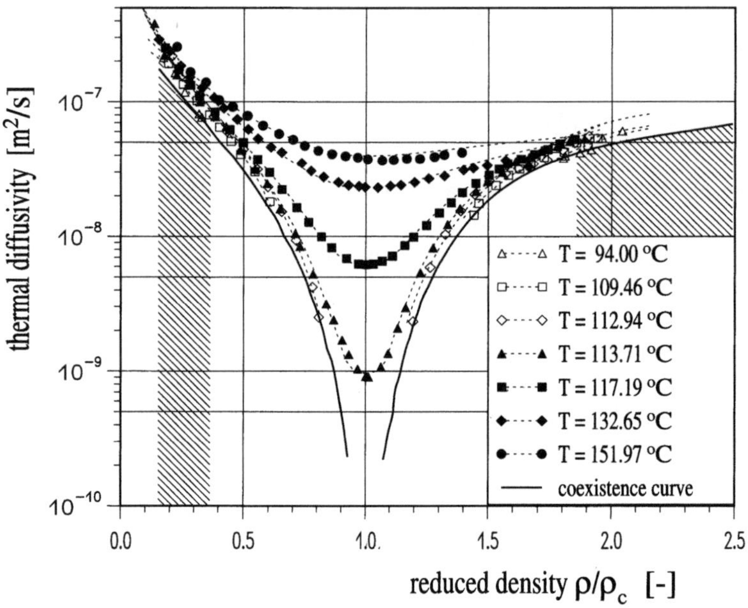

Fig. 10.12. Thermal diffusivity measurements of the refrigerant R152a. The shaded regions represent measurements conducted with the heterodyne technique

1. The investigated substance is in a state of thermodynamic equilibrium with no temperature gradients being present in the fluid.
2. The measurement is absolute. No lengthy calibrations are necessary, since only the decay time and not the spread or the position of the exponential function is required.
3. The time required for one measurement point is relatively short, even for heterodyne experiments.

11 Raman Scattering

Gernoth Strube

11.1 Introduction

Raman spectroscopy is a non–intrusive method for measuring species concentrations, concentration ratios and temperatures of molecules. While Smekal had predicted the Raman effect in 1923 based on theoretical considerations, it was the Indian scientist C.V. Raman who first observed the effect now carrying his name since 1928. Today, after only 60 years, the development of measuring methods and suitable systems based on the Raman effect has resulted in a wide selection of measuring systems which are available for various applications of Raman spectroscopy. The main reason for this rapid development is the tremendous progress in laser technology and electronics.

The fact that Raman spectroscopy is mostly applied to molecules results from the characteristic behavior of molecules with regard to their possible energetic states. Every molecule shows specific, discrete rotational and vibrational energy levels. Therefore, the frequency shift between the scattered and the incident light corresponds to an observed molecule. This feature of Raman scattering enables the simultaneous investigation of different species in a single measuring volume. If the Raman lines of multi–atomic molecules interfere with one another, there are usually other transitions which can be investigated.

Raman spectroscopy is applicable to any transparent medium regardless of its physical state. In heat and mass transfer problems the media are usually either liquid or gaseous, while chemical and biological applications often deal with solid samples. Since the scattering cross sections of liquids are several orders of magnitude higher than those of molecules in the gaseous phase, this has to be considered for the selection of the system components, which will be discussed later.

Based on the Raman effect, several different methods for the measurement of temperature and species concentrations have been developed, some of them relying on special molecular resonance effects. However, a detailed discussion of each of the various methods with their advantages and disadvantages is not possible in the framework of this book. The most widely used methods are Spontaneous Raman Scattering (SRS) and the Coherent Anti–Stokes Raman Scattering (CARS). CARS provides higher intensity light signals which is especially useful for gaseous applications. On the other hand, the set–up and the evaluation of the results are much more complicated than those of SRS.

11.2 Theoretical Basics of Raman Spectroscopy

An understanding of Rayleigh and Raman scattering is best provided by a simplified model of energy exchange during the interaction of a molecule and light hitting the molecule. Each molecule exhibits discrete energy levels within the ground electronic state (higher electronic states are not regarded in Raman scattering). The energy of a molecule is stored partly as rotational and partly as vibrational energy. This yields a specification of the energy state of a molecule by one rotational J and one vibrational quantum number v. The energetic difference between adjacent rotational levels is much smaller than that between adjacent energy levels, therefore each vibrational level contains many rotational levels. For a diatomic molecule the distribution of possible energy levels with their respective quantum numbers (v,J) are schematically shown in Fig. 11.1.

The energy content of a specific term $E(v, J)$, divided by $h{\cdot}c$ ($h =$ Planck's constant; $c =$ speed of light) and expressed in [cm^{-1}], is [169]:

$$\begin{aligned} E(v,J) &= \omega_e(v+1/2) - \omega_e x_e(v+1/2)^2 + \omega_e y_e(v+1/2)^3 + \cdots \\ &+ [B_e - \alpha_e(v+1/2)]J(J+1) - [D_e + \beta_e(v+1/2)]J^2(J+1)^2 \\ &+ 2D_e/3{\omega_e}^2(12{B_e}^2 - \alpha_e\omega_e)J^3(J+1)^3 + \cdots \end{aligned} \tag{11.1}$$

Except for the quantum numbers v and J all quantities in (11.1) are molecular constants. ω_e, B_e and D_e are characterized by the mass of the involved atoms and their bond structure. $\omega_e x_e$ and $\omega_e y_e$ in the vibrational terms represent deviations from an ideal harmonic oscillator, α_e and β_e represent rotation–vibration coupling effects. A more detailed, comprehensive discussion of the structure of molecules can be found for example in [162,166]. However, in order to perform Raman scattering experiments a detailed knowledge of the involved quantum mechanics is not absolutely necessary; usually the values for the expected Raman shifts can be obtained from tables found in the literature.

Now the events occurring in the scattering process shall be discussed. In the beginning the molecule is assumed to be in its energetic ground state and the incoming light to be monochromatic of wavelength λ_0 (or wave–number $\tilde{\nu}_0$ ($\tilde{\nu}_0 = 1/\lambda_0$)). The molecule absorbs one photon of the incident radiation with an energy content of $E = h \cdot c \cdot \tilde{\nu}_0$. This energy absorption lifts the molecule to a virtual state above the stable states of the ground electronic state, but below the first electronic state. Within a very short period of time ($\sim 10^{-14}$ s; [169]) the molecule returns from this virtual state to a stable energetic state within the ground electronic state, this time emitting

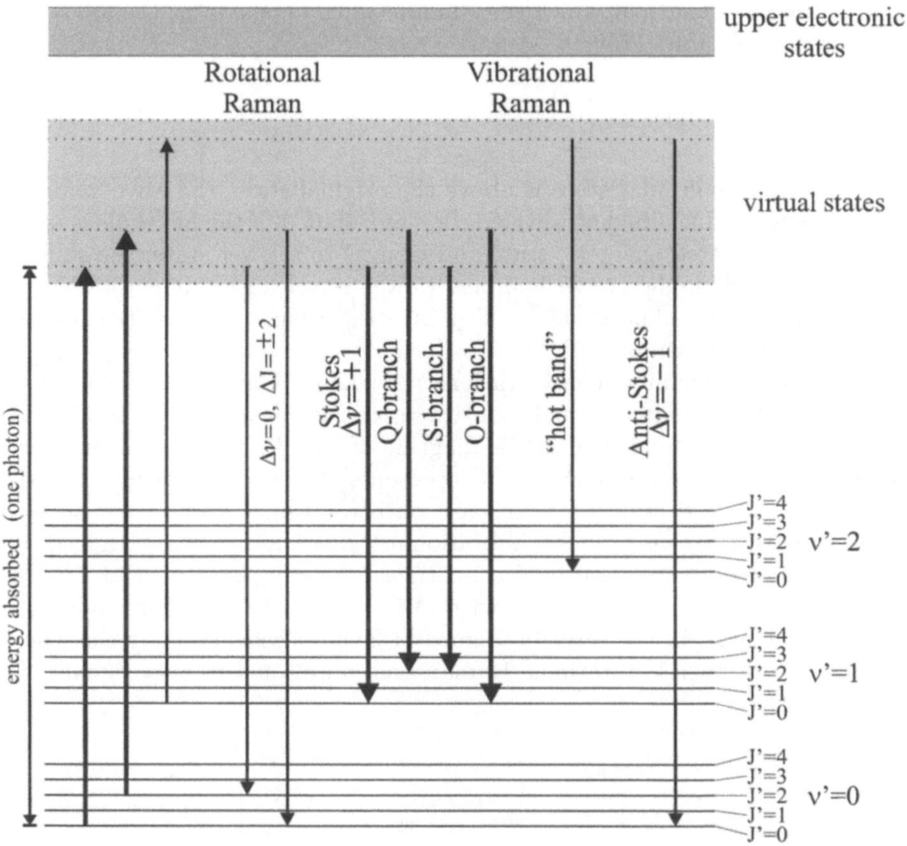

Fig. 11.1. Simplified energy model for a diatomic molecule and possible transitions

one photon of light. If the original and the new energy state are identical, the amount of energy emitted is equal to the amount of energy originally absorbed.

Due to the proportionality of energy and wave–number this equality implies that the wave–number or wavelength of the incident and emitted radiation are equal as well. If the final energy state of the molecule is above the original state a small portion of the absorbed energy remains in the molecule which means that the emitted photon contains less energy than the absorbed photon. This leads to a wave–number (or wavelength) shift $\Delta\tilde{\nu}_R$ ($\Delta\lambda_R$) of the emitted light compared to the incident light:

$$\Delta\tilde{\nu}_R = E(v, J) - E(v + \Delta v, J + \Delta J) \tag{11.2}$$

For the wavelength at which the Raman signal appears λ_R, the following formula is applicable [170]:

$$\lambda_R = \left(\frac{1}{\lambda_0} - 1.0003\Delta\tilde{\nu}_R\right)^{-1} \tag{11.3}$$

If the final molecular state is above the original state, the wave–number shifts to lower values (higher wavelength; "red–shift"), if the molecule is originally in an elevated energetic state and returns to a lower state, the shift is directed towards higher wave–numbers (lower wavelength; "blue–shift"). In spectroscopy a red–shift is called "Stokes"–transition, a blue–shift is called "Anti–Stokes"–transition.

Quantum mechanical considerations only allow a number of molecular transitions. For a diatomic molecule these transitions are $\Delta v = 0, \pm 1$ and $\Delta J = 0, \pm 2$ and are included in the diagram of Fig. 11.1. A transition characterized by $\Delta v = 0$ and $\Delta J = 0$ represents Rayleigh scattering, transitions with $\Delta\tilde{\nu} = 0$ and $\Delta J = \pm 2$ are pure rotational Raman transitions (Stokes or Anti–Stokes branch). Vibrational transitions are characterized by $\Delta v = \pm 1$, where pure vibrational transitions ($\Delta J = 0$) are called Q–branch and vibration–rotation transitions with $\Delta J = +2$ or $\Delta J = -2$ are called S–branch or O–branch, respectively. Figure 11.2 shows schematically the characteristics of Rayleigh, rotational Raman and vibrational Raman signals.

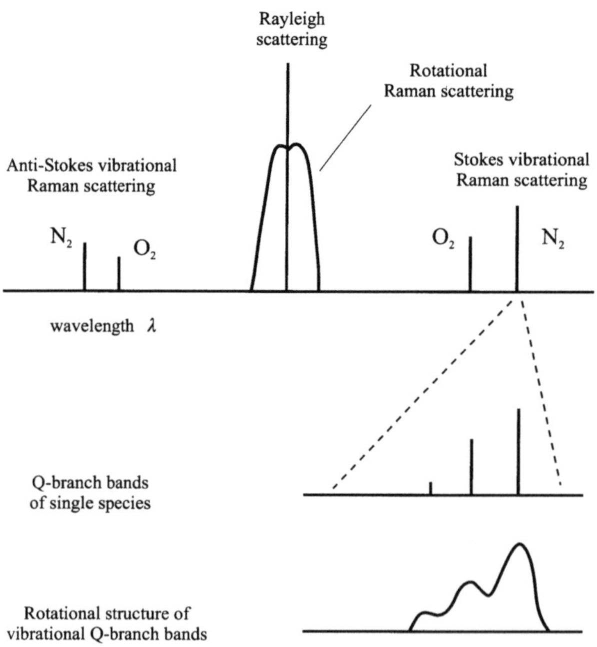

Fig. 11.2. Schematic of Rayleigh, rotational Raman and vibrational Raman scattering of O_2 and N_2. The shown details are the splitting of the vibrational Stokes Raman signals into a series of fundamental vibrational bands (ground state band, first upper state ('hot') band, second upper state band,..) and the spreading of this fundamental band series due to rotational structure

11.2.1 Concentration Measurements

Since the difference between the energy levels of a molecular species is characteristic for the molecule depending on the atoms and the acting forces in the molecule, the wave–number shifts which are encountered in Raman scattering are also characteristic for the considered molecule. Usually the vibrational Raman signals can easily be resolved spectrally due to the relatively large difference in wave–number shift.

Figure 11.3 shows the spectral position of the first Stokes vibrational Q–branch ($v = 0 \rightarrow v = 1$; $\Delta J = 0$) of technically important gases when an excimer–laser (XeCl) with a wavelength of 308 *nm* is used for excitation. Longer laser wavelengths lead to bigger spectral distances between the signals, at the same time, however, the intensity of the Raman signals decreases. If the Raman lines of two species are very close together (i.e. CO_2 and N_2O at 320.7 *nm*), there are usually other transitions suitable for spectroscopic purposes (CO_2 at 321.8 *nm* and N_2O at 330.7 *nm*).

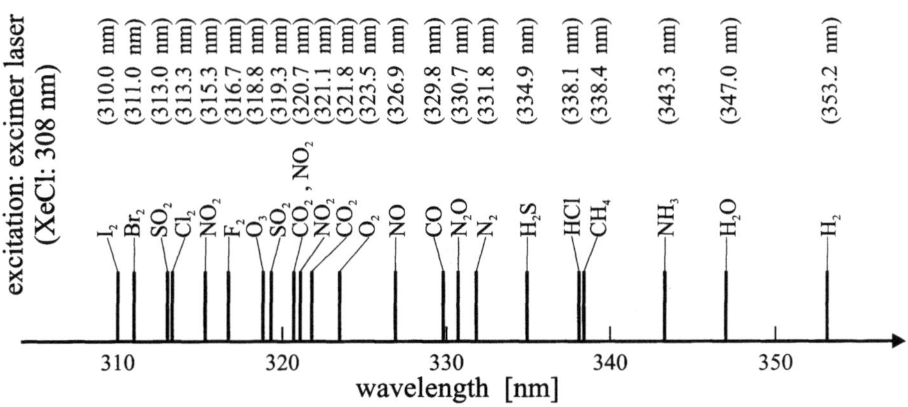

Fig. 11.3. Wavelength of the Stokes Q–branch for technically important gases when excited with an excimer laser (XeCl) at 308 *nm*

The intensity I_R of the observed Raman signal is proportional to the differential cross section $d\sigma/d\Omega$ of the molecule, the solid angle of observation Ω_E, the observed length of the measuring volume l, the number density of the molecules in the initial energy level (v, J) $N_{v,J}$, and the irradiance of the laser beam I_0. The proportionality factor C_1 represents the efficiency of the detection system, yielding the following equation:

$$I_R = C_1(\frac{\mathrm{d}\sigma}{\mathrm{d}\Omega}) \cdot \Omega \cdot l \cdot N_{v,J} \cdot I_0 \tag{11.4}$$

The differential cross section $d\sigma/d\Omega$ depends on the strength of the interaction between the incident radiation and the considered molecule in the energetic state before the transition. Beside yielding values for the calculation of the absolute signal intensity of a specific transition the differential cross section is important for the derivation of the formulae used to determine concentrations and temperatures from the obtained Raman spectra. The differential cross section is:

$$\frac{\mathrm{d}\sigma}{\mathrm{d}\Omega} = C_2 \cdot \tilde{\nu}_R^4 \cdot \gamma_v \cdot A(b_{J+\Delta J,J}, \alpha) \tag{11.5}$$

$\tilde{\nu}_R$ is the wave–number of the scattered radiation ($\tilde{\nu}_R = \tilde{\nu}_0 \pm \Delta\tilde{\nu}_R$), γ_v is a factor for the regarded vibrational transition (Stokes: $\gamma_v = v+1$; Anti–Stokes: $\gamma_v = v$) and the factor A describes the transition probability depending on the regarded rotational transition and polarization considerations for the applied set–up. Here, $b_{J+\Delta J,J}$ denotes the Placzek coefficient, α the molecular polarizability tensor. A detailed discussion can be found in [171].

Generally intensity ratios of two specific signals are used for the measurements; therefore, the values of the proportionality factors used in (11.4) and (11.5) are not important as long as the same experimental set–up is used for each measurement.

The number of molecules $N_{v,J}$ in the regarded energy level (v, J) can be obtained from thermodynamical statistics according to the Boltzmann distribution function as:

$$N_{v,J} = N \cdot g_i(J) \cdot (2J+1) \cdot \frac{\exp[-hcE(v,J)/kT]}{Q_v \cdot Q_J} \tag{11.6}$$

where N is the number density of molecules of the regarded species. The terms $g_i(J)$ and $(2J+1)$ result from degeneracies of the rotational state J [171]. Q_v and Q_J are the rotational and vibrational partition functions

$$Q_{v,J} = Q_v \cdot Q_J = \sum_v \sum_J g_i(J) \cdot (2J+1) \cdot \exp[-hcE(v,J)/kT] \tag{11.7}$$

with h as Planck's constant, c the speed of light, k Boltzmann's constant, $E(v, J)$ the energy of the state with the quantum numbers v and J and T as the absolute temperature.

Equations (11.4) to (11.7) yield

$$I_R = C_3 \cdot \tilde{\nu}_R^4 \cdot \gamma_v \cdot A(b_{J+\Delta J,J}, \alpha) \cdot I_0 \cdot N \cdot \frac{g_i(J)(2J+1)\exp[-hcE(v,J)/kT]}{\sum_v \sum_J g_i(J) \cdot (2J+1) \cdot \exp[-hcE(v,J)/kT]} \tag{11.8}$$

Equation (11.8) shows that for a given experimental set–up and constant temperature the intensity of a Raman line is directly proportional to the number of molecules in the measuring volume, i.e. species concentration. Therefore, a detailed knowledge of all factors in the above equations is not required. It is sufficient to calibrate the set–up at a known temperature and pressure for each species that is expected in the investigated system. If lower accuracy of the measurements is sufficient for the purpose of the investigations, a simple calibration using nitrogen as a reference may be made. Then the Raman signals from the actual species have to be multiplied by calibration factors accounting for the differing Raman cross sections of nitrogen and the species. These factors have been measured for many species at various laser wavelengths and are available in the literature.

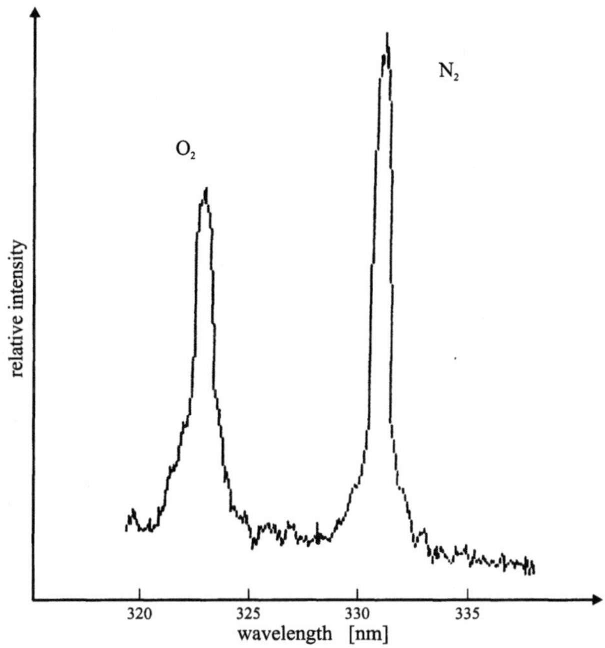

Fig. 11.4. Stokes vibrational spectrum (unresolved rotational lines) of air at ambient temperature and pressure (raw data); $\lambda_0 = 308\ nm$. The differential cross section for oxygen and nitrogen are approximately the same if the peaks are corrected for the quantum efficiency of the detection systems and scaled according to the λ^4 relation

Figure 11.4 shows vibrational spectra of air at ambient conditions with the unresolved rotational lines. The ratio of the intensity of the oxygen and nitrogen Raman signals (area underneath peak) is equivalent to their concentration ratio in the gas after corrections for the deviations in quantum efficiency of the detection system and scaling according to the λ^4 relation are made. This means that the cross sections for oxygen and nitrogen are the same. This is not the case for nitrogen and hydrogen, as can be seen in Fig. 11.5. This figure shows the Stokes vibrational Raman spectra of nitrogen and hydrogen at ambient temperature and pressures of 1, 2 and 3 *bar*. The differential cross section of hydrogen is about four times higher than that of nitrogen. The spectra in Fig. 11.5 also show the linearity of the Raman effect. Since the measuring volume and the temperature are held constant during the experiments, the number of molecules in this volume is proportional to the pressure and again proportional to the Raman signal. If the pressure of the investigated medium is constant, the vibrational Raman spectrum of a non reacting species yields temperature information due to the inverse proportionality between temperature and density provided by the equation of the ideal gas. However, at higher temperatures, where a significant number of molecules may initially be in upper state bands, the $(v + 1)$ factor in the transition intensity equations leads to errors if it is not accounted for.

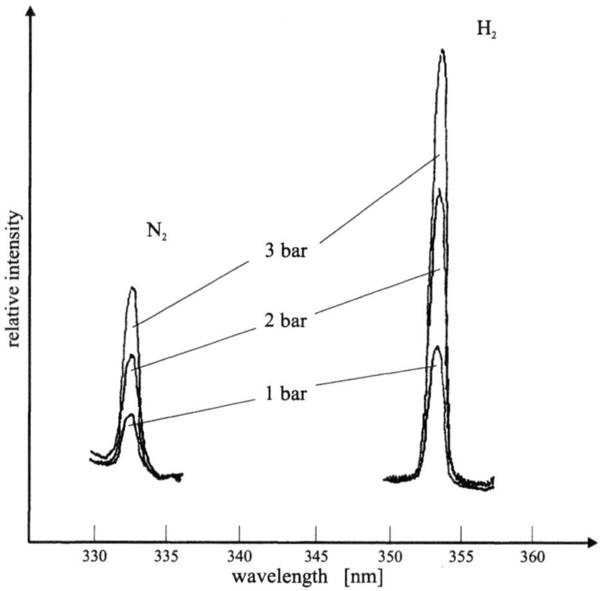

Fig. 11.5. Stokes vibrational spectra of nitrogen and hydrogen at ambient temperature and various initial pressures; $\lambda_0 = 308$ *nm*. The linearity of the spontaneous Raman effect can be clearly seen: the intensity of the Raman signal (area underneath the peak) rises constantly with the pressure. Also, the differential cross section of hydrogen is approximately four times higher than that of nitrogen

11.2.2 Temperature Measurement

General Considerations

The measurement of temperature with Raman scattering is possible due to the temperature dependence of the population distribution of the energy levels of the molecules. This population distribution is described by the Boltzmann equation and is included in (11.6). Figure 11.6 shows the population of the vibrational energy levels at various temperatures, Fig. 11.7 the population of the rotational levels for oxygen, nitrogen, and hydrogen. Equation (11.6) can be applied to evaluate the temperature of a system by comparing the intensities of different Raman transition signals. If only vibrational transitions are used, the vibrational temperature is obtained, if rotational transitions are used the rotational temperature is obtained. For systems in thermodynamic equilibrium the rotational and the vibrational temperature are equal. Only in high–velocity flow fields (supersonic nozzles etc.) or in highly reactive systems (combustion processes) the vibrational temperature with a relaxation time of 10^{-6} s may be "frozen" while the rotational temperature with a relaxation time of 10^{-8} s is not [170]. The differential scattering cross–section of rotational transitions are at least one order of magnitude larger than those of vibrational transitions, yielding a much higher signal/noise ratio. Therefore, thermometry by rotational Raman scattering usually gives more accurate results than thermometry by vibrational Raman scattering. On the other hand, the rotational lines of the different species tend to interfere with each other, especially in systems with many different species. There is no general rule of preference to either of the methods.

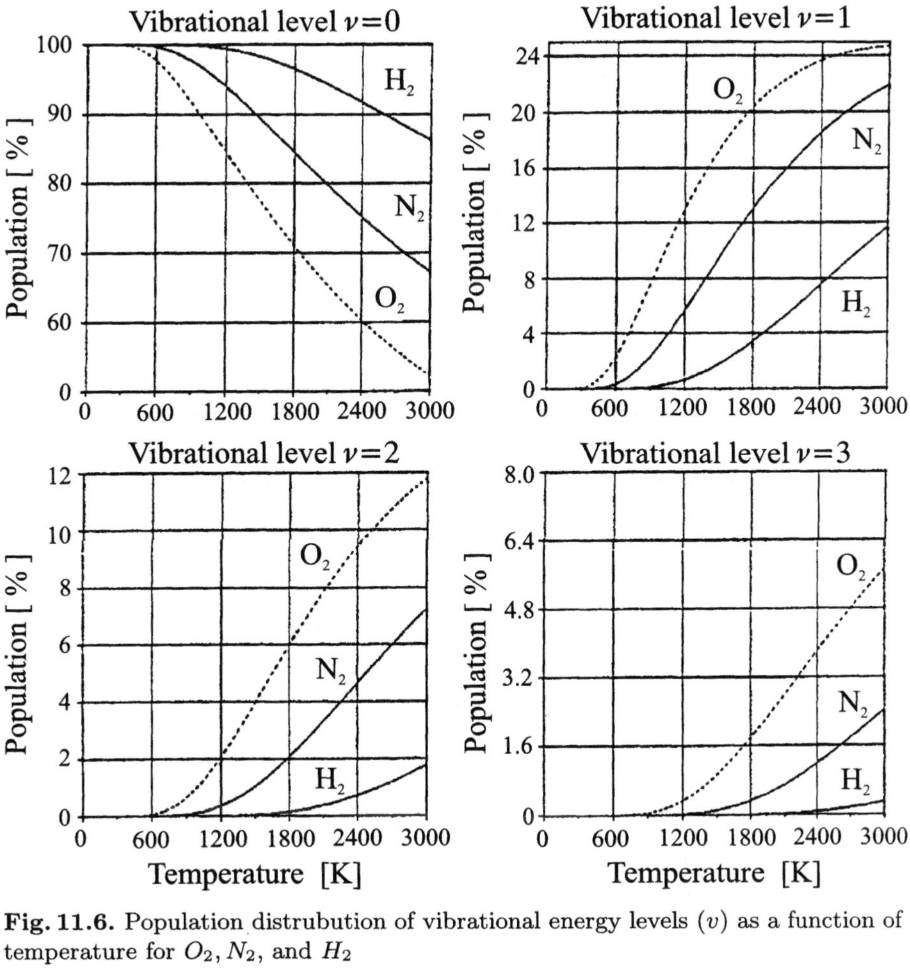

Fig. 11.6. Population distrubution of vibrational energy levels (v) as a function of temperature for O_2, N_2, and H_2

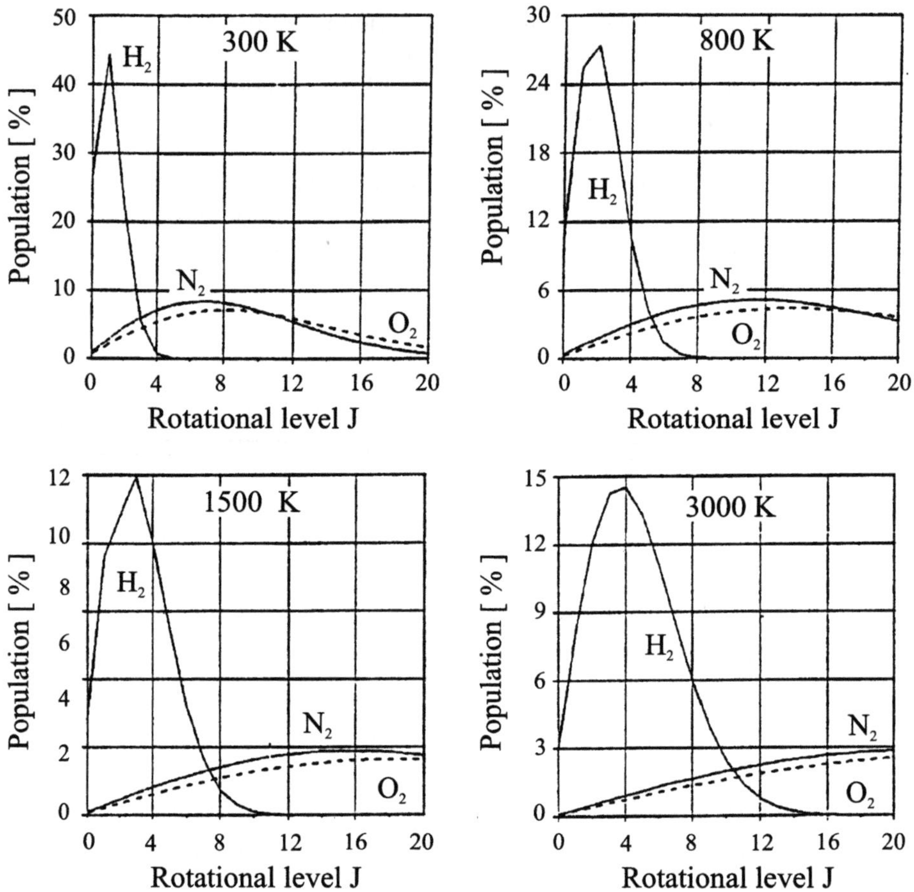

Fig. 11.7. Population distribution of the rotational levels (J) at $T = 300, 700, 1000$ and 3000 K for O_2, N_2, and H_2

Thermometry by Rotational Raman Spectroscopy

If a pure rotational Stokes transition within the ground vibrational state is considered ($\Delta v = 0; \Delta J = 2$), the application of (11.1) and (11.8) gives the intensity of the Raman signal of the initial rotational level J as:

$$I_{R,S}(J) = C_4 \cdot (\tilde{\nu}_0 - \Delta\tilde{\nu}_R)^4 \cdot I_0 \cdot N \cdot \frac{(J+1)(J+2)}{(2J+3)} \cdot g_i(J) \cdot \frac{\exp[-hcBJ(J+1)/kT_{\text{rot}}]}{Q_J} \tag{11.9}$$

Here $B = B_e - 1/2\alpha_e$, the other terms of (11.1) involving the quantum number J are neglected. The term $(J+1)(J+2)/(2J+3)$ results from the Placzek coefficient for the transition $\Delta J = +2$ [171]. All factors in (11.8) which refer to vibrational transitions are included in the proportionality constant C_4.

The rotational temperature is obtained by comparing the intensity of transitions from two different initial energy levels $E(v, J_1)$ and $E(v, J_2)$. Since the measurement of both transitions are performed under exactly the same conditions, the factors C_4, I_0, N, Q_J of (11.9) are identical for both transitions and therefore cancel each other:

$$\frac{I_{R,S}(J_1)}{I_{R,S}(J_2)} = \frac{(\tilde{\nu}_0 - \Delta\tilde{\nu}_R(J_1))^4 \cdot \frac{(J_1+1)(J_1+2)}{(2J_1+3)} \cdot g_i(J_1)}{(\tilde{\nu}_0 - \Delta\tilde{\nu}_R(J_2))^4 \cdot \frac{(J_2+1)(J_2+2)}{(2J_2+3)} \cdot g_i(J_2)} \cdot \left(\frac{\exp[-hcBJ_1(J_1+1)/kT_{\text{rot}}]}{\exp[-hcBJ_2(J_2+1)/kT_{\text{rot}}]}\right) \tag{11.10}$$

The resulting Raman shift $\Delta\tilde{\nu}_R$ from a rotational transition with $\Delta J = +2$ is $\Delta\tilde{\nu}_R = 4B(J+\frac{3}{2})$. If (11.10) is solved for the temperature T_{rot} this becomes:

$$T_{\text{rot}} = [(J_2)(J_2+1) - (J_1)(J_1+1)] \cdot \frac{h \cdot c \cdot B}{k} \cdot \left\{ \ln\left[\frac{I_{R,S}(J_1)}{I_{R,S}(J_2)} \cdot \left(\frac{g_i(J_2)}{g_i(J_1)} \cdot \frac{(2J_1+3)(J_2+1)(J_2+2)}{(2J_2+3)(J_1+1)(J_1+2)} \right) \left(\frac{\tilde{\nu}_0 - 4B\left(J_1+\frac{3}{2}\right)}{\tilde{\nu}_0 - 4B\left(J_2+\frac{3}{2}\right)} \right)^4 \right] \right\}^{-1} \tag{11.11}$$

The temperature measurement becomes more reliable, if it is performed for several transitions and then averaged.

In analogy of the equations for the Stokes transitions the equations for the Anti–Stokes transitions can be obtained, and measurements of the Anti–Stokes band intensities can be utilized for thermometry. Equation (11.10) can be rewritten in the form:

$$J(J+1) \cdot \frac{1}{T_{\text{rot}}} = \frac{k}{hcB} \cdot \left\{ -\ln\left[\frac{I_{R,S}(J) \cdot (2J+3)}{C_4 \cdot g_i(J) \cdot (J+1)(J+2)\left[\tilde{\nu}_0 - 4B\left(J+\frac{3}{2}\right)\right]^4} \right] \right\} \tag{11.12}$$

If for every measured transition the right side of (11.12) is plotted versus $[J(J+1)]$, the obtained points should lie on a straight line. The slope of this line is inversely proportional to the rotational temperature.

Figure 11.8 shows Stokes rotational Raman spectra of oxygen and nitrogen at ambient temperature (292 K) where especially the influence of the nuclear spin ($g_i(J)$) can clearly be seen. Figure 11.9 shows the corresponding plot according to (11.12).

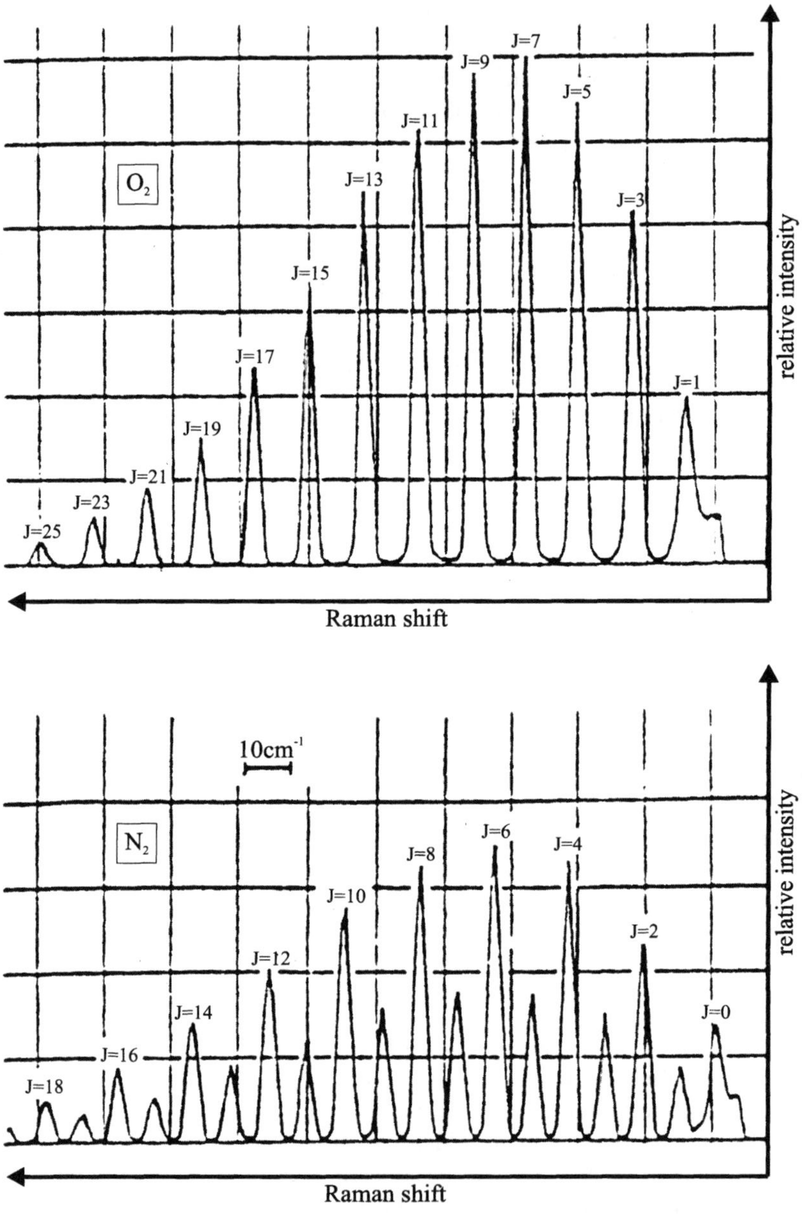

Fig. 11.8. Stokes rotational Raman spectra of O_2 and N_2 at ambient temperature. J is the quantum number of the initial state. Here, the influence of the nuclear spin, given by the term $g_i(J)$ in the equations, can be seen: $N_2 : g_i = 6$ for even values of $J, g_i = 3$ for odd values of $J; O_2 : g_i = 0$ for even values of $J, g_i = 1$ for odd values of J.

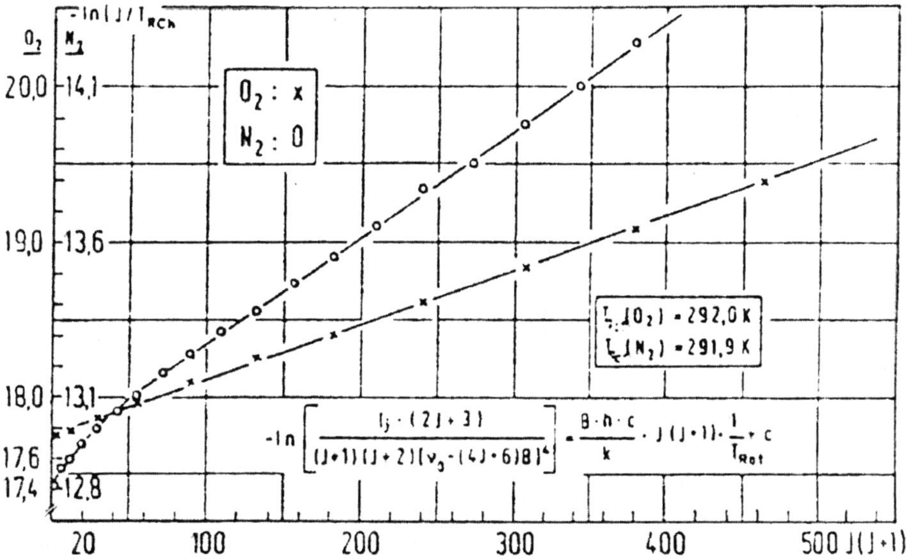

Fig. 11.9. Determination of rotational temperature from spectra of Fig. 11.8

Thermometry by Vibrational Raman Spectroscopy
The intensities of the Raman signals of the Stokes Q–branch ($\Delta v = +1, \Delta J = 0$) are given by (s. (11.1) and (11.8)):

$$I_R(v,J) = C_4(\tilde{\nu}_0 - \Delta\tilde{\nu}_R)^4 \cdot I_0 \cdot (v+1) \cdot \frac{J(J+1)}{(2J-1)(2J+3)} \cdot g_i(J) \cdot (2J+1) \cdot \frac{N \cdot \exp[-hc(\omega_e(v+\frac{1}{2}) + B\ J(J+1))/kT]}{Q_{vib} \cdot Q_{\mathrm{rot}}} \quad (11.13)$$

Again the energy term has been introduced into the exponent in a simplified form. It consists of a vibrational and a rotational term. Since the intensity of the Q–branch signals is about two orders of magnitude larger than that of the O– and S–branches, the Q–branch signals determine the shape of the recorded spectrum [168]. Due to the disharmonicities of the molecular vibrations the Raman shift of pure vibrational transitions decreases with an increasing initial vibrational quantum number. In addition, the (unresolved) rotational structure leads to a continuous spectrum as it is schematically shown in Fig. 11.2 and more detailed in Fig. 11.10.

As the rotational thermometry the vibrational thermometry is based on the temperature distribution of the molecules among the different vibrational states according to Boltzmann statistics. There are three basic vibrational thermometry methods: the contour fit method, the band–peak method of the 'hot–bands' and the Stokes/Anti–Stokes method.

Using the contour fit method means calculating the spectrum of the molecule under consideration for various temperatures and fitting it to the

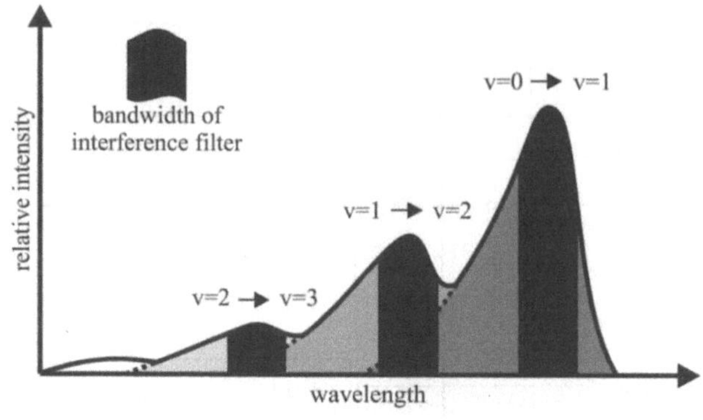

Fig. 11.10. Scheme of a typical vibrational Q–branch spectrum as it appears at higher temperatures. The shaded areas indicate the obtained intensity when the according bandwidth is observed

measured data as close as possible. In the calculations every parameter that influences the measurements have to be regarded. Beside the actual scattering process these parameters include specific parameters of the applied experimental set–up, such as the width of the registered band pass or the quantum efficiency of the reception unit. Figure 11.11 shows an example of a computer fit for a hydrogen molecule measured in flames. Due to the large constants for hydrogen, only the vibrational band from the ground state is observed. However, within this band the rotational lines can be resolved and compared to the theoretical curves.

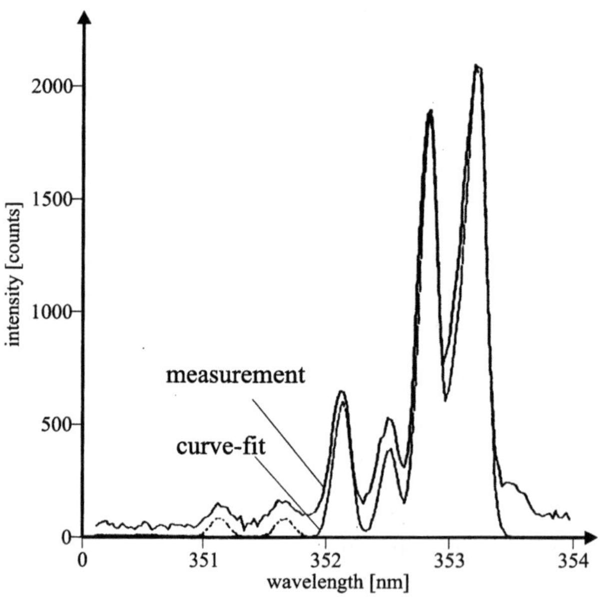

Fig. 11.11. Stokes rotational–vibrational Raman spectrum of H_2 in a lean H_2–air flame. Due to the large rotational and vibrational constants of hydrogen only the vibrational transition from the ground level is observed; however, the rotational transitions within the vibrational band can easily be resolved. The temperature obtained from the theoretical curve fit was 960 K

When applying the band–peak method the ratio of the Stokes Q–branch signals of different initial vibrational states is used to obtain the temperature of a given system. For the signal intensities of all transitions from two adjacent vibrational energy levels (11.13) leads to:

$$\frac{I_R(v)}{I_R(v+1)} = \frac{(\tilde{\nu}_0 - \Delta\tilde{\nu}_{R,v})^4}{(\tilde{\nu}_0 - \Delta\tilde{\nu}_{R,v+1})^4} \cdot \frac{(v+1)}{(v+2)} \cdot \exp[\frac{-hc}{kT} \cdot \omega_e[v-(v+1)]] \quad (11.14)$$

The expression $\omega_e[v-(v+1)]$ is the same for all vibrational transitions from adjacent initial vibrational energy levels and equals the value for the vibrational Stokes Q–branch $\Delta\tilde{\nu}_R$ of the considered species.

The expression $(\tilde{\nu}_0 - \Delta\tilde{\nu}_{R,v})^4/(\tilde{\nu}_0 - \Delta\tilde{\nu}_{R,v+1})^4$ is close enough to unity to be negligible. With the intensity measurement of two neighbored Q–branch transitions the temperature can be obtained from (11.15):

$$T = \frac{h \cdot c \cdot \tilde{\nu}_R}{k} \cdot \left(\ln \frac{I_R(v)}{I_R(v+1)} + \ln \frac{v+2}{v+1}\right)^{-1} \quad (11.15)$$

This equation holds true for transitions from any two adjacent vibrational energy levels. However, due to the low population of higher energy levels, measurements are generally limited to transitions from the ground and first level and for some molecules also the second upper level. As can be seen in Fig. 11.10 the intensity measured at the peak of the upper bands includes some scattered light from the transition from the ground level. This has to be considered before using (11.15). An example for the temperature as a function of the intensities of the Stokes Q–branch signals for nitrogen is given in Fig. 11.12.

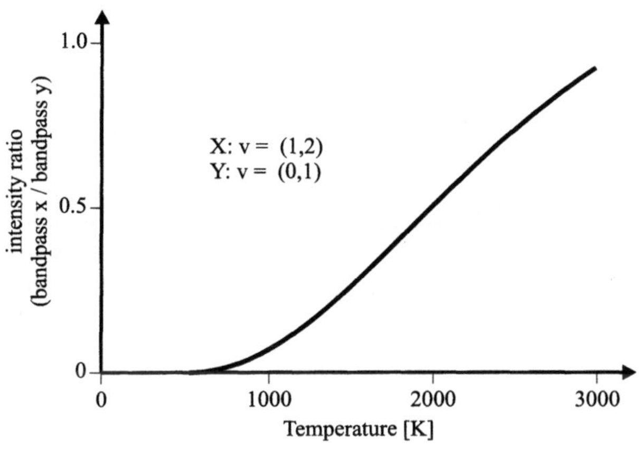

Fig. 11.12. Calculated temperature as a function of the intensity ratio of the transitions from the ground and first upper vibrational level for N_2. The bandpass (0.2 nm) of the collected signal was centered at the wavelength of the intensity peaks (s. Fig. 11.10)

A third method of vibrational thermometry is possible when the ratio of the intensities from the vibrational Stokes bands and the vibrational Anti–Stokes bands is taken to determine the temperature. For a conjugate pair of

Stokes or Anti–Stokes transitions (i.e. $v = 0 \rightarrow v = 1$ and $v = 1 \rightarrow v = 0$) the Raman shift $\Delta\tilde{\nu}_R$ is identical and the intensitiy ratio is similar to (11.14):

$$\frac{I_{R,S}(v)}{I_{R,aS}(v+1)} = C_5 \cdot \frac{(\tilde{\nu}_0 - \Delta\tilde{\nu}_R)^4}{(\tilde{\nu}_0 + \Delta\tilde{\nu}_R)^4} \cdot \exp[\frac{hc}{kT_{vib}} \cdot \Delta\tilde{\nu}_R] \tag{11.16}$$

The factor C_5 is introduced to consider any differences in detection efficiency at the two wavelengths, since they considerably differ. For the temperature (11.16) yields:

$$T = \frac{h \cdot c \cdot \Delta\tilde{\nu}_R}{k} \cdot \left(\ln \frac{I_{R,S}}{C_5 \cdot I_{R,aS}} + 4 \ln \frac{\tilde{\nu}_0 + \Delta\tilde{\nu}_R}{\tilde{\nu}_0 - \Delta\tilde{\nu}_R} \right) \tag{11.17}$$

Since the intensity ratio is constant for all pairs of transitions, this method is applicable with a still acceptable error when the vibrational bands are not resolved. Therefore, this method is suitable for all experimental conditions where a low signal intensity is expected, since the accumulated intensities of all transitions yield a higher level. Figure 11.13 shows the calculated intensity ratios as a function of temperature for some diatomic molecules, the resulting curves for other molecules are similar to the ones shown.

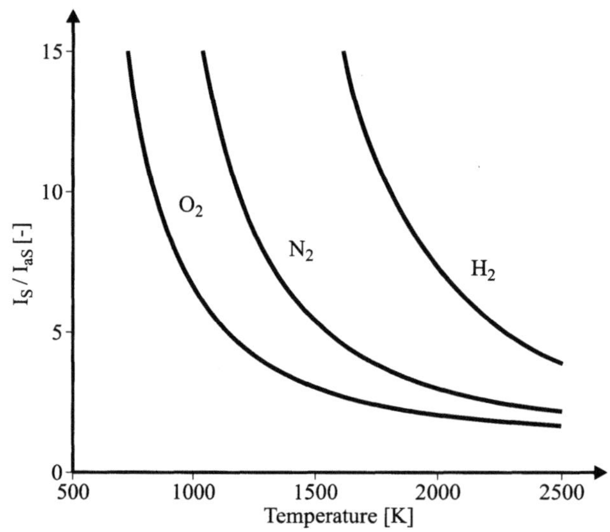

Fig. 11.13. Calculated ratio of the intensities of the Stokes and Anti–Stokes transitions as a function of temperature for H_2, O_2, N_2. The adjusted excitation wavelength is that of an excimer laser (XeCl) at $\lambda = 308\ nm$. For other light sources the curves are shifted to other values due to the $\tilde{\nu}_0$ influence in (11.16)

11.3 Experimental set–up

Even though not every applications of Raman spectroscopy and the relevant features of suitable equipment can be discussed in detail, this section attempts to give a general overview of possible experimental set–ups and the involved components. Beside the consideration of currently available components a short outlook will be given toward the most likely developments in each field.

Figure 11.14 shows a possible set–up for spot (quasi–zero–dimensional) measurements of temperature and species concentration with Raman spectroscopy.

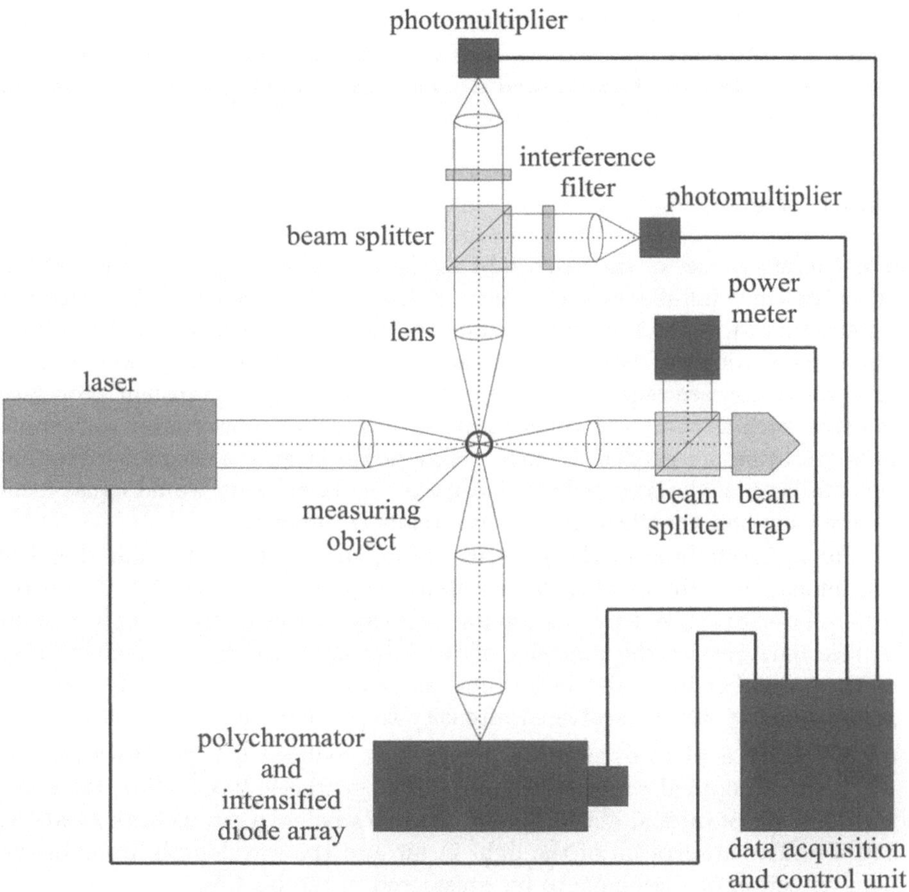

Fig. 11.14. Typical Raman set–up for point measurements. The two detection systems are only shown to demonstrate the different possibilities of arrangements, usually only one of the shown possibilities is chosen

The laser beam is focussed to a small spot, the diameter of which forms the cross section of the measuring volume, and is subsequently absorbed in a beam trap. The signal collection is usually arranged at an angle of 90° towards the direction of the laser beam. At this direction the signal intensity reaches a maximum and the size of the observed volume can best be determined. Even though for some specific applications the set–up of two separate signal collection units may be useful, generally only one unit is required to conduct the measurements. If the specific experimental apparatus containing the test–medium allows the installation of a convex mirror opposite to the signal collection lens, the obtainable signal intensity can be increased. In order to analyze the scattered intensities, the collected light has to be resolved spectrally. This is accomplished either by diffraction units (i.e. polychromators) or selective filters. The intensity of the light at the selected wavelength is then converted into electronic signals, digitized, and then numerically processed. Complete systems of the required data acquisition and control routines are commercially available.

11.3.1 Laser

The characteristics of the laser light as the excitation source for the Raman effect strongly influences the quality of the obtained signals. The selection criteria mainly depend on the specific conditions of the investigated medium. For investigations of liquid and steady flow fields only low power and no time resolution are required. For investigations of highly transient processes in gases, such as combustion phenomena, pulsed high–power laser with short pulse duration are needed. If such a high–power laser was used for investigations in liquids, the high energy density of the laser beam would cause local evaporation and therefore make measurements impossible.

For investigations in the gaseous phase, the scattered Raman signal is less intense than the incident laser radiation by about a factor 10^{-12}. Therefore, a laser optimized for maximum intensity of the scattered light should be selected. Besides the intensity of the laser light, the laser wavelength λ_0 or wave–number $\tilde{\nu}_0$ is also important, as can be seen in (11.8). The appropriate quantity for the scattered intensity is $\frac{I_0}{{\lambda_0}^4}$. For the number of photons registered by a photon counting device it is reduced to $\frac{I_0}{{\lambda_0}^3}$. However, for UV lasers all optical components must be of synthetic fused silica, since the transmission of optical glass rapidly decreases below a wavelength of about 350 nm. The absorption of the light in air and the wavelength limits of the light detection devices have to be considered in the far UV.

Beside high temporal resolutions the use of pulsed lasers can lead to improvements in the signal to noise ratio, especially with luminous backgrounds. However, a gas breakdown may occur, if the laser pulse has a high peak power due to the extremely high light intensities ($> 10^{10}\mathrm{W/cm^2}$). This breakdown is accompanied by strong light emission, that make Raman measurements almost impossible.

When choosing a suitable laser the possibility of interference of the Raman signals with laser induced fluorescence at certain laser wavelengths should also be considered. However, with increasing the availability of tunable lasers this problem becomes less important. Further criteria for appropriate lasers are the band width and the degree of light–polarization. The resolvability of the rotational lines of most molecules requires small band widths since the Raman effect involves a constant shift in wave–number rather than emission of signals at a specific wavelength. This means that the band width of the scattered light is at least as large as that of the laser beam. The band width of the laser should therefore be lower than the distance between the two neighboring lines that are to be resolved. The polarization of the laser light is important for the intensity of the Raman signal which is scattered in the direction of observation. The highest intensity is scattered in the direction perpendicular to both the direction of the laser beam and the plain of polarization.

Lastly it should be mentioned that different types of lasers require different amounts of maintenance and operational costs. Solid state lasers (i.e. Nd:YAG lasers) are comparatively easy to handle and require only electrical power, cooling water and sometimes flash–lamps. On the other hand need laser like excimer or dye laser frequent refills of the lasing medium but yield comparatively high energies at a low wavelength and tunable versions are available. Dye lasers are the most versatile lasers as far as tenability is concerned and emit light at wavelengths from 312 up to 985 nm when different dyes are used [163].

11.3.2 Focussing and Collection Lenses

In order to focus the laser beam several effects have to be considered. Due to spherical aberration, diffraction and divergence, the smallest spot of the laser beam if sent through a convex lens can never have a zero diameter. The beam converges to a minimum value and exhibits a more or less constant cross section for a certain length [171]. The smaller the waist the better is the spatial resolution of the system. Therefore, lenses with focal lengths as short as possible should be used, although in this case gas breakdown may occur (s. above). The beam can be manipulated by a set of lenses before it is sent into the measuring volume in order to optimize the shape of the focus spot. However, every surface that the beam passes through implies losses in energy. This significantly complicates the required alignment procedures. For practically every wavelength so called "best–form"–lenses are commercially available, which are designed to achieve minimal spot radii.

The collecting lens should be selected so that the solid angle of observation is maximized. This can be accomplished by short focal length or large lens diameters. However, the maximum lens diameter is limited by the smallest optical component in the light path following the collecting lens. Again, a set of lenses can be used to reach higher solid angles of observation, but the same restrictions occur as described above.

11.3.3 Spectral Frequency Selection

The selection of devices to spectrally resolve the scattered light depends on the quantity (concentration, temperature) to be obtained for which species, regarding any restrictions imposed by the investigated system. The goal should be to obtain sufficient resolving power with the highest possible signal intensities. Generally, the selective devices used for Raman scattering are interference filters or polychromators. It can be said that the polychromator is more versatile, but it yields lower signal intensities on the light detection device. If the intensity is to be measured at several different wavelengths, a set–up using filters becomes very complex as far as the alignment procedures is concerned. Additionally, the quantity of transmitted light is reduced with each surface and cut in half with each added beam splitter which results in low light levels.

If the goal of Raman measurements is exclusively to measure the temperature of a single species (preferably one with a large scattering cross section), a set–up with interference filters is most likely to be used. If, on the other hand, the system is supposed to be used flexibly for both concentration and temperature measurements with variable sets of species involved, a polychromator is the better choice. Filters are commercially available for virtually any center wavelength and bandwidth. They have to be selected for each specific application. The other possibility, mono– or polychromators, are very flexible but also very expensive. (Mono– and polychromators are basically the same devices. The term monochromator is used when a single frequency of observation is selected by an exit slit. In a polychromator no exit slit is installed and a broader section of the spectrum is observed.) The characteristic features of a polychromator are its focal length and the groove density of the installed diffraction grating. Spectral dispersion increases with the focal length and the grating groove density. Focal lengths range from about 0.25 m to 1 m. Grating groove densities can range from 20 $grooves/mm$ to a maximum of 3600 $grooves/mm$ which results in dispersions at the exit from about 200 nm/mm to 0.3 nm/mm. Modern polychromators can be equipped with two or three gratings of different groove densities, making the polychromators even more flexible. If further dispersion is required, it is also possible to arrange two or even three monochromators sequentially, commercially available as double or triple monochromators. The required resolution for a specific application of Raman spectroscopy has to be evaluated according to the equations given in the previous section.

11.3.4 Photon Converters

In order to convert the Raman signals into electrical signals either photomultiplier tubes, photodiode–arrays, or Charge–Coupled–Devices (CCD) are used. All three systems are photon counting devices with an output signal that is proportional to the number of photons received on the surface. While no spatial resolution is possible with photomultiplier tubes, photodiodes arranged in an array deliver one–dimensional resolution, and with CCD–cameras two dimensional resolution is achieved. The use of diode–array detectors is sensible if for the spectral dispersion of the Raman signals a polychromator is applied. In this case, each array of photodiodes is only exposed to a comparatively small width of wavelength and the intensity distribution over a wide range of wavelength is obtained in one simultaneous measurement. The spectral band width projected on a single diode is only determined by the used polychromator. A change of the diffraction grating immediately results in a different spectral resolution without any further change of the set–up.

Often diode arrays are equipped with light intensifiers, the so called multi–channel plates (MCP). With these MCPs several desired effects can be achieved. This means a dramatic increase in quantum efficiency, to about 10% (1 digital count for every 10 photons received), over a wide spectral range. Additionally, the MCP can be used to control the actual exposure time by applying the high voltage time resolved, which is needed to activate of the MCP. Minimum exposure times of about five nanoseconds are reached with fast getable units. Of course, the chosen exposure time must always be longer than the laser pulse duration in order to avoid signal intensity losses due to time jitter of both the laser pulse and the electronic control. Especially in luminous systems the background noise can significantly be reduced without any loss of signal. Due to the difference in the characteristic life time between Raman and fluorescence, the rejection of interfering fluorescence signals can be accomplished by an intensifier, if the laser and the detection system are properly tuned [164].

The more and more popular Charge–Coupled–Devices (CCD) are cameras which provide a high resolution in two spatial dimensions. When a CCD camera is used with a polychromator, one–dimensional measurements are possible instead of spot measurements (of course the laser beam has to be focused only in one dimension instead of two). If a filter is used instead of the polychromator two–dimensional measurements are possible. However, it should be considered that the signal intensity per evaluated unit decreases with resolution which results in a decreasing measurement accuracy.

11.3.5 Data Acquisition and Control

When Raman spectroscopy is conducted using cw–lasers and photomultiplier tubes, the necessary devices for data acquisition and control are not very complex. A voltage measurement device is the absolute minimum requirement. The application of pulsed lasers, diode arrays and MCPs requires more sophisticated equipment for process control and data acquisition. The triggering circuits for the exact activation with respect to time of each part of the system is arranged by electronic devices which in turn are controlled by (personal) computers.

Many experienced researchers compose their own systems, often integrating custom made units. However, for many applications complete systems including all necessary components for data acquisition and control are commercially available. With the increasing number of applications of light scattering spectroscopy more manufacturers offer such systems. They can include all the necessary software for control of the laser, the light detector (including MCP), the polychromator and even other components to control the investigated system. In combination with diode arrays the so called optical multichannel analyzers (OMA) are applied to process the large amount of data involved. They also include convenient routines for data acquisition, such as the automatic acquisition of two subsequent scans with and without a triggered laser. The difference between these two scans eliminates most of the background noise which is not related to the spectroscopic process. This results in a vast improvement in signal–to–noise ratio. They also supply common arithmetic routines to further process the raw data which facilitates the evaluation of the obtained data. All these routines help to make Raman spectroscopy more easily accessible for the inexperienced user.

11.4 Selected Applications

Raman spectroscopy is applied in a very wide variety of fields ranging from biology and chemistry for structural studies of molecules or reaction kinetic data, to engineering applications where the concentration and the temperature distribution in flow fields containing several species are simultaneously determined. Although the signal intensities decrease strongly with a decreasing species concentration, i.e. increasing temperatures, one of the most important applications of Raman scattering lies in the analysis of reacting processes

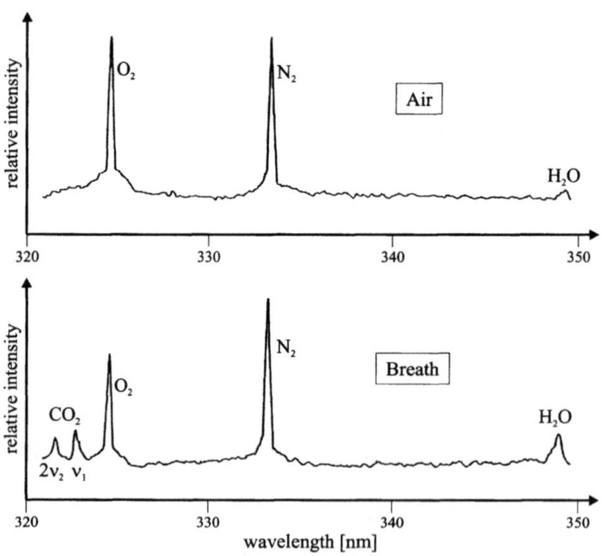

Fig. 11.15. Vibrational Raman spectra for room air and human breath

since conventional measurements often yield inadequate results.

In Fig. 11.15 a very simple yet vivid example of Raman investigations in gas flows with different species, namely the comparison of the composition of room air with human breath, is shown. The upper spectrum shows plain room air at a temperature of 21°C and a water vapor content of 1.5% corresponding to a relative humidity of 60%. The lower spectrum is that of human breath blown through a small tube into the measuring volume. Four effects can be observed:

- CO_2 appears at its two Raman–active vibrational modes. Its concentration is 8%, which corresponds to
- the decrease in O_2 concentration from 20% to 12%
- When assuming a constant relative nitrogen concentration the temperature increased from the initial 21°C to 29°C
- the steam content increased from 1.5% to 3.5%, corresponding to a relative humidity of 90% at the elevated temperature.

The next figures refer to combustion of H_2 in air. The experimental set–up was a stationary operated, closed tube type burner with a rectangular cross section of 20×30 mm.

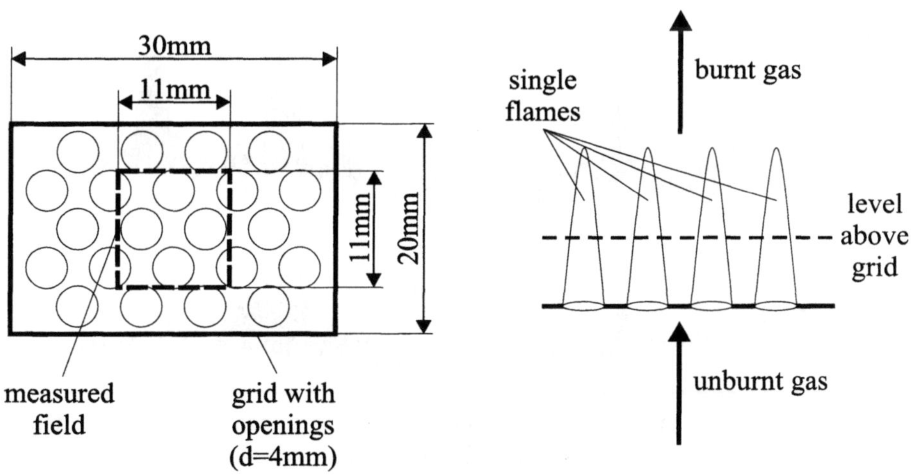

Fig. 11.16. Arrangement of stationary operated burner with metal grids as flame stabilizing devices. Behind each opening a separate turbulent flame is formed. The flames are self–similar. Raman measurements have been taken in a three–dimensional array, the results of concentrations for each species is plotted for horizontal levels above the grid

A metal grid was used to stabilize the flame. Figure 11.16 shows the dimensions of the grid and the conditions around the grid. The premixed, unburnt gases with a hydrogen concentration of 12 $vol.\%H_2$ approach the grid upwards with a velocity of 17 m/s. There is a separate turbulent flame stabilized behind each opening and the burnt gases leave the flame area upward. Raman point measurements have been taken in a three–dimensional array. The cross section of the array is represented by the dashed line in Fig. 11.16, in which measurements were taken at distances of 1 mm. The concentration distributions are depicted for each species in form of levels with various distances from the grid.

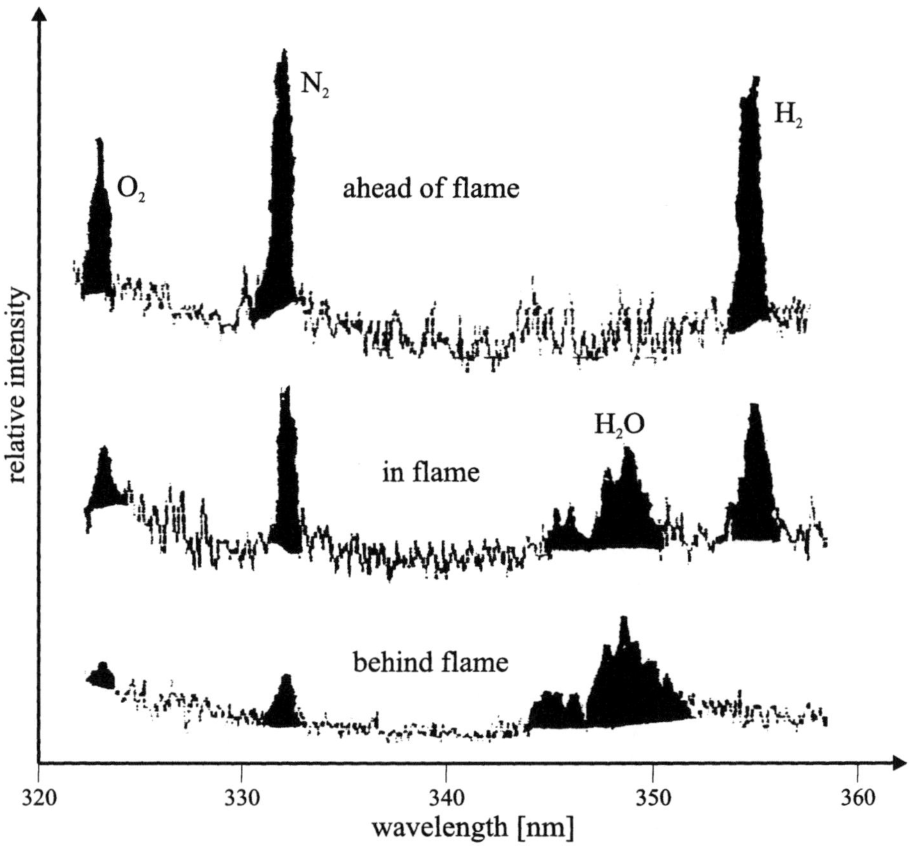

Fig. 11.17. Typical spectra from representative points in the burner. Top: unburnt mixture with 12 $vol.\%$ H_2 in air. Middle: spectrum from the turbulent reaction zone; the hydrogen has partially reacted with oxygen to form steam, every species is present. Bottom: completely burnt mixture. The under–stoichiometric combustion leaves surplus oxygen, but there is no hydrogen left

Figure 11.17 shows typical spectra of the characteristic zones. The first spectrum is taken before the grid; since no combustion had occurred, there is no steam present. The second spectrum is taken from the reaction zone.

Some of the hydrogen has reacted with parts of the oxygen to form steam as a product. All four species can be seen in the spectrum. The last spectrum is from a location above the reaction zone, the sub–stoichiometric combustion is completed. Only steam, nitrogen, and the surplus oxygen as species remain present. By assuming the nitrogen being inert, its concentration yields the temperature, Since the system is isobaric the law of the ideal gas can be applied. Figure 11.18 shows examplarily the nitrogen and hydrogen intensity distributions obtained from measurements in a level 17 *mm* above the grid.

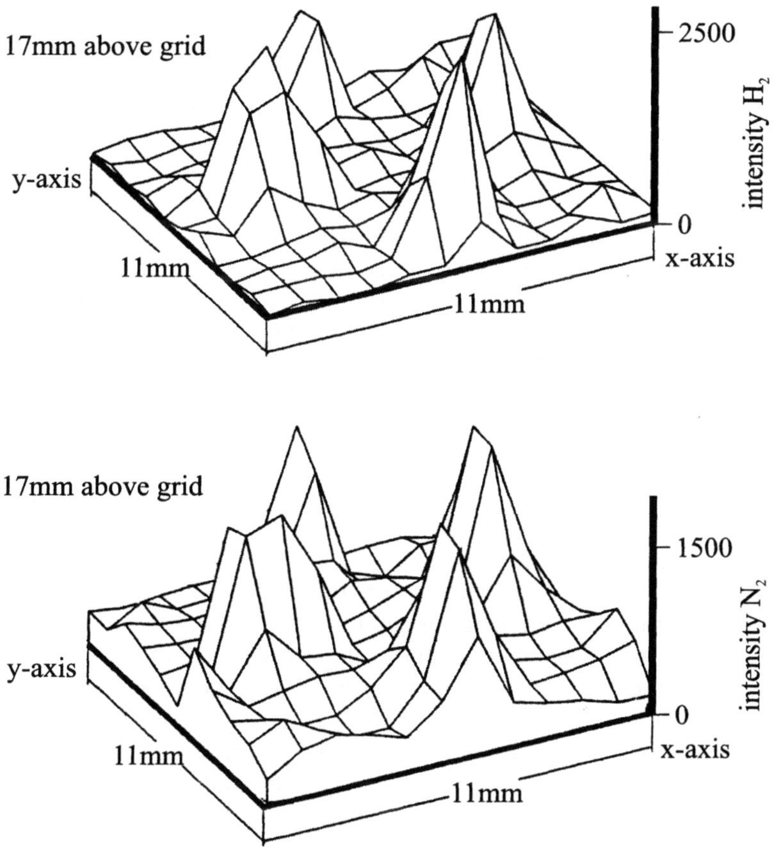

Fig. 11.18. Nitrogen and hydrogen intensity distributions (equivalent to concentrations) obtained from measurements in a level 17 *mm* above the grid. The structure of the single flames is clearly visible. The nitrogen concentration is inversely proportional to the system temperature. It can be seen that the nitrogen spikes are broader than the hydrogen spikes. This shows that the reaction mainly occurs at the surface of the conical flames

The integrated intensity from the spectra is proportional to the concentration of the regarded species. The structure of the single flames is clearly visible. In order to show the change in relative hydrogen concentration due to the combustion, the measured hydrogen concentration has to be divided by the values for nitrogen.

11.5 Concluding Remarks

Raman spectroscopy is a very versatile method to measure species concentrations and temperatures in various systems. It is applicable in all three phases, gaseous, liquid and solid and even applications in two phase systems have been accomplished [172]. Generally, Raman spectroscopy is applied when measurements with conventional measuring techniques lead to significant influences of the obtained results due to the interaction between the probes and the investigated system. Furthermore, very high resolutions with respect to both time and space can be achieved. One major advantage of Raman spectroscopy is the possibility to simultaneously measure both the concentrations of all the species involved and the temperature, even in highly reactive, turbulent processes.

Raman scattering is currently widely applied to combustion processes. In these systems the important mixing processes of fuel and oxidizer before the reaction can be investigated, as well as reaction kinetics in the flame zone and the species contained in the exhaust gases (i.e. pollutants). In such experiments the Raman scattering method is not only used for basic research but also applied to realistic technical set–ups (car engines, gas turbines, chemical reactors, etc. [165]).

The development of the components of a Raman probe rapidly continues. Lasers become more powerful with increasing versatility regarding tenability and spectral band width. Major developments are also expected in the field of light detectors and image intensifiers for which sensitivity, time resolution and scan rates are increased. Together with faster computers to store and to process the recorded data, more detailed information about the processes within the investigated system becomes available in less measurements. Therefore, the field of applications of Raman scattering is rapidly enlarged. Attention has also been given to the development of mobile Raman systems. In addition to this, with decreasing costs of simpler Raman set–ups, the thought of utilizing this technique outside laboratories, i.e. for quality control of production processes, is not too remote.

12 Laser induced Fluorescence

Peter Andresen

12.1 Introduction

Among the newest measurement techniques are those based on laser induced fluorescence. The first successful application of fluorescence imaging was reported in 1982. Despite this short history LIF techniques have become very powerful and are the most widely used non–intrusive techniques for gas measurement. This is mainly due to their high signal strength as compared to the main competitors (i.e. Rayleigh scattering and Raman scattering) as shown in Chap. 7 (Techniques Based on Light Scattering) of this book. This high signal strength make two–dimensional measurements with excellent time and space resolution possible. The principal applications of LIF are the measurement of minority species concentrations and temperatures. In special arrangements of the LIF probe, however, pressure and velocity distributions have also been obtained by fluorescence measurements.

LIF measurement is based on the natural fluorescence of molecules and atoms. Since this fluorescence occurs in transitions from weakly populated excited electronic energy levels the signals are as weak as the population density. The effect of the laser induction process is to promote a considerable number of particles, molecules or atoms, from the densely populated lower energy levels to the excited levels. Therefore subsequent emission of radiation is strongly enhanced by fluorescence. The intensity of the fluorescence signal is used to determine the absolute concentration of the regarded molecular species, sometimes with considerable problems in the conversion. The temperature can be obtained by different techniques based on the temperature–dependent equilibrium population distribution among the energy levels, either by exciting different transitions and observing the total light emission or by spectral analysis of the emitted light after broadband excitation.

Although all LIF methods are based on laser excitation followed by natural fluorescence of the particles, different methods of LIF are distinguished according to the characteristics of the actual transition processes involved and the configuration of the LIF probe. In addition to the standard linear laser induced fluorescence, laser induced saturated fluorescence [LI(S)F] and laser induced predissociation fluorescence [LI(P)F] are to be distinguished. The latter techniques provide better measurement accuracy, yielding higher evaluation and interpretation of the scattered intensities. For LI(P)F this is obtained at the cost of significant loss of signal intensity.

LIF methods are generally applicable to a large number of molecular species. Due to its large signal intensities and the high achievable spatial resolution, the main field of application of LIF is the two–dimensional imaging of reacting flows. Here the intermediate products formed in the separate steps of the overall reaction process and residuals appearing with rather low concentrations are of main interest. With the rising importance of environmental aspects in all fields of industrial production, energy consumption and emission of pollutants the detectability of low concentration molecular species will be in great demand. LIF methods provide powerful tools for monitoring these types of compounds.

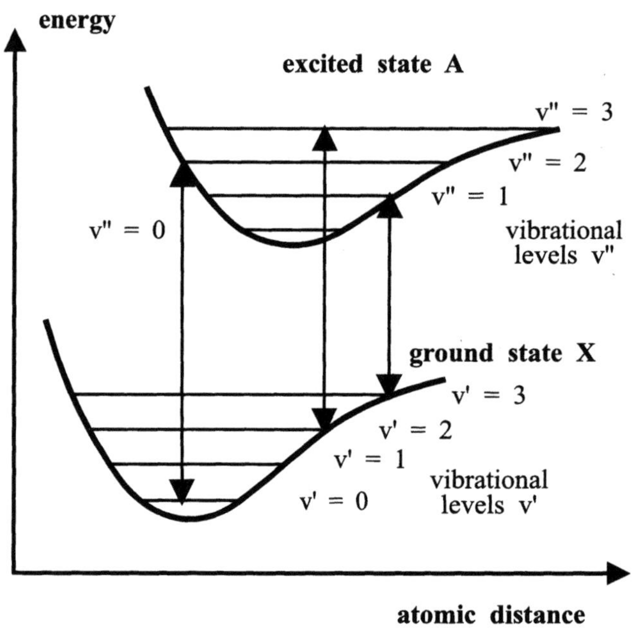

Fig. 12.1. Energy diagram of electronic levels and their vibrational levels. The rotational levels within each vibrational level are not shown in this schema, although they do play an important role in LIF thermometry measurements. Consult Chap.s 7 and 8 for more details on the ro–vibronic structure of molecules. Each potential curve exhibits a specific minimum indicating its stability. The spacing of the atoms in the molecule during the vibrational movement varies between the intersections of the line of the vibrational level and the potential curve. According to the Franck–Condon principle, transitions from the end points are more intense than from the middle of vibrational lines

12.2 Basic Principles of Laser Induced Fluorescence

12.2.1 General Considerations

Before the physical processes involved in LIF measurements are discussed, the simplified energy level model of molecules presented in Chap. 7 of this book is outlined. It is depicted in simplified form in Fig. 12.1 and shows two electronic energy levels with their inherent vibrational levels. Generally the lower energy level regarded is the ground electronic state designated by the letter X; only this state is considerably populated. The upper level here is designated by the letter A, i.e. the first excited state. However, in other cases this may well be a different state $(B, C, ...)$. The total energy content E_{tot} in units of $[\text{cm}^{-1}]$ of a molecule in the electronic level e (X, A, ...), the vibrational level v (0, 1, ...) and the rotational level J (0, 1, ...) is approximately given by

$$E_{\text{tot}} = E_{\text{e}} + E_{\text{v}} + E_{\text{J}} = T_{\text{e}} + \omega(v + \frac{1}{2}) + BJ(J+1) \tag{12.1}$$

The constant values for the term energies T_{e} as well as the vibrational and rotational constants, ω and B respectively, of the regarded electronic level can be found in e.g. Huber/Herzberg [179].

Generally, fluorescence is a transition from an upper to a lower energy level in conjunction with a spontaneous emission of a photon. The energy of the photon is equal to the energy difference of the two levels involved. According to Planck's law this energy is directly proportional to the frequency of the emitted light. Transitions within the electronic ground state involve low energy differences and therefore appear in the infrared region of the electromagnetic spectrum. Transitions between electronic states, on the other hand, are generally associated with emission frequencies in the visible (green or blue) or the UV region.

There is usually an extremely low population in electronically excited states (the population of the A state of OH at T =2500 K is roughly nine orders lower than the population of the ground state X). The reverse process of emission–absorption – is used to efficiently populate the upper electronic level and obtain higher signal strength. Typical lifetimes, meaning the time the molecule stays in the upper electronic level, are of the order of 10^{-8} s. For molecules in the excited electronic levels there are different subsequent processes possible, shown schematically in the energy level diagram of Fig. 12.2:

- the molecule can undergo a laser–stimulated transition back to the ground state (emitting a photon of light at the laser frequency in the direction of the laser beam)
- the molecule can absorb an additional photon of the incident light and go to an even higher energy level (including ionization)

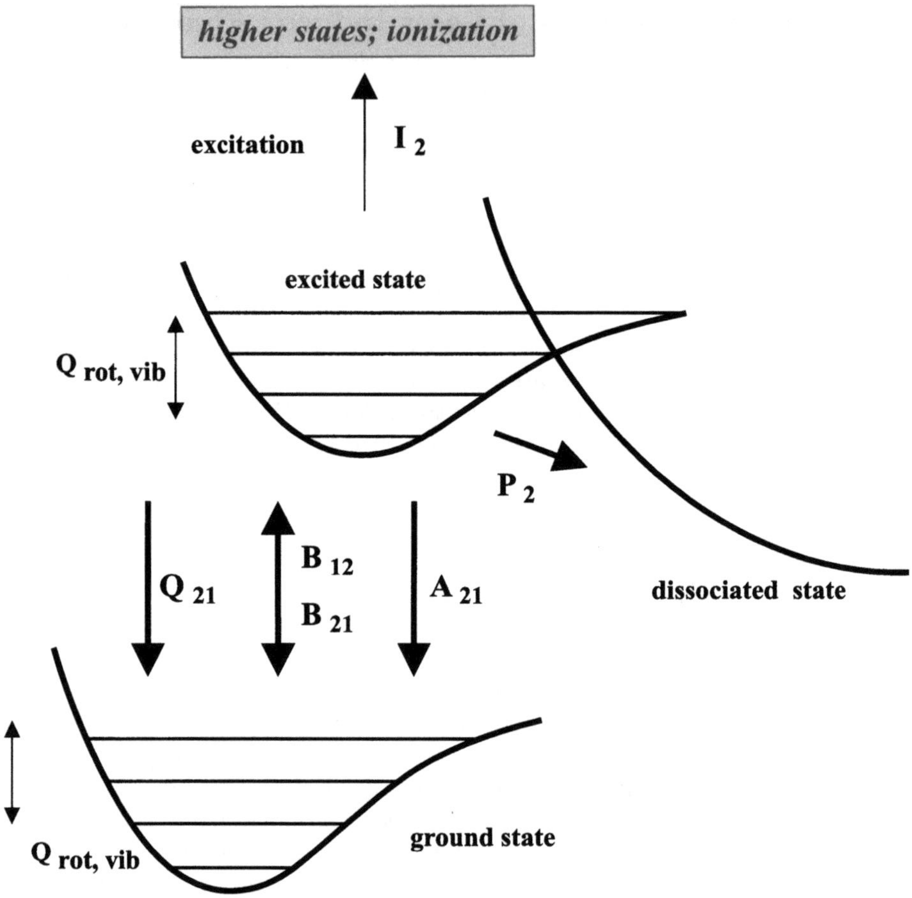

Fig. 12.2. Possible events for a molecule in the upper electronic state specified by the rate coefficients used in the equations. The intermolecular collisional quenching rate Q_{21} includes processes involving two electronic states. Those involving only ro–vibronic levels within the same electronic state are denoted by $Q_{rot,vib}$. The fluorescence transitions A_{21} do not necessarily occur between the identical ro–vibronic levels involved in the laser excitation process, but rather follow the "natural" transition probabilities determined by the Franck–Condon selection rules. The laser–stimulated processes between the regarded states E_1 and E_2 are determined by the Einstein coefficient B in both directions B_{12} and B_{21}, which are coupled by the degeneracies g_J of the corresponding energy levels. Processes leading to ionization are denoted by I_2. The actual transition rate is obtained by multiplying the Einstein coefficient by the spectral intensity of the laser. For some molecules the potential curves of stable upper electronic states have an intersection point with an unstable electronic state (repulsive molecular configuration without minimum in potential curve) leading to dissociation of the molecule

- collisions of the molecules with other molecules may cause a transition to a lower electronic state without emission of light; this effect is referred to as "collisional quenching"
- interactions within the molecule may lead to dissociation; if the dissociation caused by a shift from a stable configuration to an unstable electronic configuration (showing a repulsive energy potential curve without specific energy minimum), the process is called "predissociation"
- the molecule undergoes a fluorescence transition to the ground electronic state involving light emission
- collisions with other molecules may cause vibrational and rotational transitions within the excited state with subsequent fluorescence

Since time scales of these processes are of vital importance for interpretation of the actually observed intensity of fluorescence, the time dependent equation of the population of the different energy levels involved has to be established. The population $N_j(t)$ of a specific energy level j as a subset of the set of levels i is defined by the differential equation

$$\frac{\mathrm{d}N_j(t)}{\mathrm{d}t} = \sum_{i \neq j} N_i(t) R_{ij} - N_j(t) \sum_{i \neq j} R_{ji}, \tag{12.2}$$

where R_{ij} is the integral rate coefficient in units of $[\mathrm{s}^{-1}]$ for all processes yielding an increase in the population of the level j, and R_{ji} which in analogy represents all processes depopulating the level j.

If the regarded set of energy levels E_i is reduced to two levels E_1, E_2, coupled by the energy content of a photon of monochromatic radiation of the corresponding frequency ν_{21}

$$\Delta E_j = E_2 - E_1 = h\nu_{21} \tag{12.3}$$

with j=2 as the upper level, 12.2 yields

$$\frac{\mathrm{d}N_2(t)}{\mathrm{d}t} = N_1(t) R_{12} - N_2(t) R_{21}. \tag{12.4}$$

If the integral coefficients R are explicitly stated, (12.4) extends to

$$\frac{\mathrm{d}N_2(t)}{\mathrm{d}t} = N_1(t)[B_{12} I_\nu] - N_2(t)[Q_{21} + B_{21} I_\nu + A_{21} + I_2 + P_2]. \tag{12.5}$$

The coefficients Q, B, A, P and I appear in place of the coefficient R. Collisional excitation Q_{12} is usually negligible. Since the photo–ionization rate is usually comparatively small, this term shall be neglected in the following considerations. The rate of predissociation depends mainly on the species investigated, but is nearly independent of the environment. For the following considerations, a non–predissociated species or electronic state shall be regarded (i.e. $P_2 = 0$); the case of predissociation will be discussed separately later. The first Einstein coefficient A_{2f} for a transition from the upper level 2 (in the electronically excited state) to the lower levels f (in the electronic ground state) is equal to the reciprocal value of the characteristic lifetime τ of the excited energy level E_2 of the species s

$$\sum_f A_{2f} = \frac{1}{\tau_{2,s}} \tag{12.6}$$

where f denotes a state in the lower electronic state. The distribution of the transitions among the lower states f occurs according to the corresponding Franck–Condon factors for the transition. The values of the lifetimes and the Franck–Condon factors again are to be taken from Huber and Herzberg [179] and Eckbreth [176].

The second Einstein coefficient B represents laser stimulated processes, both for population and for depopulation. It is coupled to the coefficient A by the frequency of the laser light ν

$$B_{21} = \frac{c^3}{8\pi h}\frac{1}{\nu^3}A_{21} \tag{12.7}$$

and has to be multiplied by the laser spectral intensity I_ν. This is a function of the laser irradiance I_L at the considered frequency ν of the transition and the spectral bandwidth $\Delta\nu$

$$I_\nu = \frac{2I_L}{\pi c \Delta\nu} \tag{12.8}$$

The coefficients B for stimulated population and depopulation are coupled by the degeneracy of the two levels involved

$$g_1 B_{12} = g_2 B_{21} \tag{12.9}$$

Assuming that the total population $N_1(t) + N_2(t)$ of the two energy states remains constant, and in addition that the laser pulse is long compared to the time constant τ of the solution of the differential (12.5), the system reaches a steady state with

$$N_2(t \gg \tau) = \text{constant} = N_1(0)\frac{B_{12}I_\nu}{[B_{12}I_\nu(1 + g_1/g_2) + Q_{21} + A_{21}]} \tag{12.10}$$

yielding the rate of fluorescence FR

$$FR = A_{21}N_2(t \gg \tau) = A_{21}N_1(0)\frac{B_{12}I_\nu}{[B_{12}I_\nu(1 + g_1/g_2) + Q_{21} + A_{21}]}. \tag{12.11}$$

In order to capture the influence of the spectral intensity of the laser I_ν and the population/depopulation ratio given by the term $(1 + g_1/g_2)$ (12.11) is arranged in the form

$$FR = N_1(0)B_{12}I_\nu\left(\frac{A_{21}}{A_{21} + Q_{21}}\right)\left(\frac{1}{1 + I_\nu/I_\nu^{\text{sat}}}\right) \tag{12.12}$$

The newly introduced term I_ν^{sat} is called the saturation intensity of the laser light and is defined as

$$I_\nu^{\text{sat}} = \frac{A_{21} + Q_{21}}{B_{12}(1 + g_1/g_2)} \tag{12.13}$$

Is the laser intensity well below this saturation limit ($I_\nu \ll I_\nu^{\text{sat}}$ or $B_{12}I_\nu \ll Q_{21} + A_{21}$), (12.12) yields

$$FR_{\text{lin}} = N_1(0)B_{12}I_\nu \frac{A_{21}}{A_{21} + Q_{21}}. \tag{12.14}$$

This is called the linear fluorescence signal because it is proportional to the laser intensity. The factor $A_{21}/(A_{21} + Q_{21})$ is often referred to as the Stern–Vollmer factor. The problems arising from linear LIF are the often unknown quenching rates Q_{21}, which depend on temperature, pressure, and composition of the gas under investigation and are generally significantly higher than the rate of the A_{21}.

If the laser intensity is well above the saturation level ($I_\nu \gg I_\nu^{\text{sat}}$ or $B_{12}I_\nu \gg Q_{21} + A_{21}$), fluorescence becomes independent of Q_{21}. The fluorescence rate for this case is

$$FR_{\text{sat}} = N_1(0)\frac{A_{21}}{1 + g_1/g_2}. \tag{12.15}$$

and the technique is often called laser induced saturated fluorescence [LI(S)F] [183]. The assumption that no molecules are lost, i.e. $N_1(t) + N_2(t) =$ constant, holds strictly only for the first excited states of atoms. For molecules, collisions as well as radiation usually leads to other states $i \neq 1, 2$. In this case it has to be assumed that no chemical reactions occur and, secondly, the collisional redistribution is fast enough to maintain Boltzmann equilibrium. In most cases the stationary state is not obtained at the beginning and the end of the laser pulse (time) as well as in the wings of the beam profile (space).

Another possibility to avoid the problems of linear LIF in near atmospheric pressure investigations is the use of the so–called laser induced predissociated fluorescence [LI(P)F]. Here the excitation transition is chosen such that predissociation of the upper state occurs at a high rate and the P term in (12.5) becomes important. Introduction of P_2 in (12.14) yields

$$FR_{\text{pre}} = N_1(0)B_{12}I_\nu \frac{A_{21}}{A_{21} + Q_{21} + P_2}. \tag{12.16}$$

The schematics of the potential curves for predissociation and the processes involved in LI(P)F are shown in Fig. 12.2. Usually the predissociation rate P_2 is significantly larger than both the fluorescence rate A_{21} and the quenching rate Q_{21}

$$P_2 \gg Q_{21} \gg A_{21} \tag{12.17}$$

therefore the proportionality factor for the fluorescence yield becomes approximately

$$\frac{A_{21}}{A_{21} + Q_{21} + P_2} \simeq \frac{A_{21}}{P_2} \tag{12.18}$$

Although the high predissociation rate significantly decreases the overall fluorescence intensity, the quantitative interpretation of the signals is more accurate. In contrast to LI(S)F, that is based on sufficiently fast redistribution, the effect of collisions in the electronically excited state is completely eliminated in LI(P)F. In LI(P)F (as well as in LIF and LI(S)F) care has to be taken to avoid depletion of the ground state, otherwise collisions within the electronic ground state repopulate the level and make the signal sensitive to the gas composition.

Whereas the errors introduced by collisional quenching in the excited state are well known and extensively discussed in recent reviews [174,182,185] this RET in the electronic ground state has been completely neglected even in the very recent years. Corrections for the quenching in the electronically excited states become rather reliable if gas composition and temperature are only roughly known, as, for example, in the exhaust gases of premixed flames with known stochiometry. For some of the most widely used molecules, like OH and NO, even rovibrationally resolved quenching data are available today for several electronically excited states.

The collisional effects in the electronic groundstate can also cause severe errors in both density and temperature measurements at laser intensities often used in real experiments. This is because higher pump rates are used to obtain better S/N [175,186]. Errors increase with an increasing laser intensity because the population in the probed state decreases strongly with the consequence that the state is refilled by rotational energy transfer within the ground state. In this way molecules that originally could not be excited by the laser field (because they were in other states) are now excited and contribute to fluorescence, thus fluorescence increases. The RET rates in the ground state are by no means constant. They depend on J and usually decrease with increasing J which implies that the low J states are refilled more efficiently. LISF, which operates at the highest laser intensities is most affected by RET in the ground state. For example, the ratio of the population in two quantum states which depend on nothing but temperature, was measured to depend on laser intensity. The measured ratio was found to vary as much as a factor of two with laser intensities easily achieved on sheets, which implies very large temperature errors [175,186]. Unfortunately, in contrast to the excited state quenching, there is presently too little data on the RET in the electronic ground state to apply appropriate corrections. Large additional errors may also result at low densities from polarization effects because these are not included in the line strength factors. Therefore, LIF detection should be done under the magic angle.

12.2.2 Concentration Measurement

The concentration of a species is coupled to the fluorescence rate by (12.11). The term related to the concentration is the population of the lower laser–coupled state $N_1(0)$ before the start of the excitation process. In a volume V the number of molecules $N_{S,1}$ of the regarded species S in the energetic state defined by the quantum numbers for vibration v_1 and rotation J_1 is

$$N_{S,1} = \gamma_S N_{\mathrm{tot}} V \cdot \left(\frac{g_1(2J_1+1)}{Q_{\mathrm{rot}} Q_{\mathrm{vib}}}\right) \cdot \exp\left(-\frac{hcE_1(v,J)}{kT}\right) \tag{12.19}$$

with γ_S as the mole fraction of the species, N_{tot} the total number density of molecules, $E_1(v,J)$ the molecular energy of the state in $[\mathrm{cm}^{-1}]$, k as Boltzmann's constant and T the system temperature in $[K]$. $Q_{\mathrm{rot,vib}}$ are the sum of all vibrational and rotational states and are stated explicitly in Chap. 11 "Raman scattering". The total number density of molecules is also dependent on the temperature. Assuming the applicability of the equation for the perfect gas at constant pressure, it is inversely proportional to the temperature,

$$N_{\mathrm{tot}} \simeq N_L \frac{273}{T} \tag{12.20}$$

where $N_L = 6 \cdot 10^{19} \mathrm{cm}^{-3}$ is Loschmidt's number.

The fluorescence signal obtained and processed in an experiment is the number of photons n_{LIF} reaching the detector from a considered volume of the measurement field. For the example of the linear fluorescence signal (12.14), (12.19), and (12.20) yield

$$n_{\mathrm{LIF}} = \left(\eta_c \frac{\Omega}{4\pi}\right)\left[\gamma_S N_{\mathrm{tot}} V \left(\frac{g_1(2J_1+1)}{Q_{\mathrm{rot}} Q_{\mathrm{vib}}}\right) \exp\left(\frac{hcE_1(v,J)}{kT}\right)\right]\left(B_{12} I_\nu \frac{A_{21}}{A_{21}+Q_{21}}\right). \tag{12.21}$$

The first factor in (12.21) is necessary to correct for the losses in the collection of the photons. Since the fluorescence photons are emitted into the full solid angle of 4π, only the part covered by the collection solid angle Ω can be captured. The detection efficiency η_c is determined by the arrangement and includes losses at the various elements in the optical path of collection. The number of captured photons can again be converted to energy or intensity values at the detector surface by Planck's law.

The difficulties in determining the quenching rates Q_{21} have already been considered above and also apply to (12.21). For LI(S)F and LI(P)F the last factor has to be replaced according to (12.15) and (12.16) respectively.

For a given set–up, apart from the strong influence of quenching effects, the signal obtained in a measurement is dependent upon the concentration of the species and the temperature. The temperature influence on the population density of the lower laser–coupled state 1 depends primarily on the constants of the molecule under investigation and can be neglected for atoms. In many cases it is possible to find a transition which does not show a strong temperature dependence in the temperature range expected in the experiment. For example, in combustion environments the change in molecular species concentration of the intermediate reaction products exhibits much stronger changes than the change in temperature. Therefore the obtained fluorescence signals give a very good qualitative picture of the concentration of these intermediate species, yielding ample information on the location and fluctuations of the reaction zone. In any case, the selection of the excitation levels is a very important step in the design of the LIF probe and depends on the conditions of the system under investigation.

12.2.3 Temperature Measurement

Temperature measurements are possible by either exciting different transitions and observing the total emissions at a these excitation wavelength or by observing the spectral distribution of the emissions induced by a fixed transition.

If the light source can be tuned to stimulate electronic transitions from different single rotational levels within a vibrational level, the observed fluorescence intensity I_{LIF} can be obtained from (12.21). The variables characterizing the experimental set–up, i.e. η_c, Ω, I_ν and V, are identical for excitation of spectrally close lower states. Assuming furthermore that the coefficients for emission A_{21} and quenching Q_{21} do not vary with excitation frequency, which is not a really valid assumption, (12.21) can be written in the logarithmic form

$$\ln\left(\frac{I_{\mathrm{LIF}}}{B_{12}g_1(2J_1+1)}\right) = -\frac{hcE_1(v,J)}{kT} + \mathrm{C}, \tag{12.22}$$

C: constant, and the solution for the temperature becomes:

$$T = \frac{hcE_1(v,J)}{k} \cdot \left[-\ln\left(\frac{I_{\mathrm{LIF}}}{B_{12}g_1(2J_1+1)}\right) + \mathrm{C}\right]^{-1}. \tag{12.23}$$

Experimental scanning through the spectral range of transitions of the regarded species usually require more time than the characteristic timescales of the system under investigation allow without change of said system. Therefore it is very suitable to use two distinct lines for excitation and to compare the broadband fluorescence signals. Assuming a transition from two rotational states a and b (i.e. $J_{1,a}$ and $J_{1,b}$) with $J_{1,a} < J_{1,b}$ from the same vibrational level of the ground state to the same ro–vibronic level v_2, J_2 in the upper electronic state, the following equation evolves

$$\frac{I_{\mathrm{LIF},1}}{I_{\mathrm{LIF},2}} = \frac{B_{12,1}g_{1,a}(2J_{1,a}+1)}{B_{12,2}g_{1,b}(2J_{1,b}+1)} \cdot \exp\left[\frac{hc\left(E_1(v,J_{1,b}) - E_1(v,J_{1,a})\right)}{kT}\right] \tag{12.24}$$

However, this equation is only applicable if the laser spectral intensity and the detection efficiency are equal for both transitions. Otherwise slight modifications of (12.24) must be done according to (12.9) and (12.21).

The excitation method requires light sources that are both tunable and able to exhibit very narrow bandwidths in order to avoid simultaneous excitation of several rotational levels in the lower laser–coupled state. In addition, because fluorescence is often observed in the same spectral range, the lasers and detection devices have to be triggered sequentially, with time delays below the characteristic timescales of the investigated system. Therefore, the fluorescence spectroscopy technique may be advantageous in some applications. In this technique excitation occurs with a single laser pulse, which may have a wider bandwidth exciting several transitions within a vibrational band. The lifetimes in the upper electronic state of the molecules undergoing spontaneous emission are assumed to be longer than the time it takes the molecules to reach rotational thermal equilibrium, i.e. a distribution among the rotational levels J_2 according to the Boltzmann equation. Therefore, the fluorescence spectrum from the electronically excited state can be analysed for its rotational structure to obtain the temperature. Due to the necessity of thermal equilibrium this method is sometimes referred to as thermally–assisted fluorescence. Equation (12.21) yields for the fluorescence spectroscopy method with the spontaneous emission rate $A_{21}(J_2)$ from the upper rotational level J_2

$$\ln(\frac{I_{\mathrm{LIF}}}{A_{21}(J_2)\ g_2\ (2J_2+1)}) = -\frac{hcE_2(v,J)}{kT} + \text{constant} \tag{12.25}$$

The spectral selection of the observed signal must be very fine to distinguish the different rotational levels. This can be achieved with high resolution. However, this is not possible for planar imaging applications.

In the above equations for temperature measurement the fluorescence from the same location and the same vibrational level of the upper laser–coupled state was used. Thus the assumption of identical collisional quenching conditions is applicable. However, if fluorescence induced by monochromatic light is to be monitored two–dimensionally with one light collection channel, the temperature dependence of collisional quenching Q_{21} has to be taken into account. This technique is called absolute fluorescence because the temperature is obtained from the absolute fluorescence intensity distribution. A simple approach, introduced by Hanson [178], to model the collisional quenching coefficient Q_{21}, is

$$Q_{21} = N_{\mathrm{tot}}\sigma_c c\bar{v}(T), \tag{12.26}$$

where σ_c is the collisional cross–section and $c\bar{v}(T)$ is the mean velocity of the molecule as a function of temperature. If the collisional cross–section is assumed not to be temperature dependent and the mean molecular velocity is proportional to the square root of the temperature, the factor for collisional quenching becomes

$$Q_{21} \sim N_{\mathrm{tot}}T^{1/2}. \tag{12.27}$$

Since in every regarded location the same transition is used, all factors in (12.21) except for the collisional quenching, the exponential expression and the mole fraction of the observed species are constant. Assuming $Q_{21} \gg A_{21}$, (12.21) and (12.26) give

$$I_{\mathrm{LIF}} \sim \gamma_S T^{1/2} \exp(\frac{hcE_1(v,J)}{kT}). \tag{12.28}$$

If the mole fraction γ_S of the regarded species does not change within the observed volume, the obtained fluorescence can be scaled to the temperature very easily with the above equation, needing only one reference point of known temperature. This is also possible in reacting flows, if inert particles are added to the flow and used for LIF.

12.2.4 Tracer LIF

Although LIF, as discussed up to now, works for a considerable number of molecules there are many "dark" compounds that do not fluoresce at all and thus cannot be detected by LIF, e.g. most fuels, N_2, CO_2. In other cases LIF can be applied but is too difficult to quantify, e.g. H_2O, O_2. The spatial and temporal distribution of dark compounds can nevertheless still be analyzed using a modification of LIF, i.e. a method called tracer–LIF.

In this method LIF is done with trace amounts of fluorescent molecules that are homogeneously mixed to the dark compound of interest, for example to fuel or to air. As in normal LIF, the resulting signal intensity is proportional to the density of the dark compound. With the set–up shown in the Fig. 13.3. spatial distributions of dark compounds can be visualized in this way.

Various tracer molecules have been successfully used to visualize and quantify fuel distributions, both for liquid droplets and fuel vapor. Usually UV–lasers are used in tracer LIF because most popular tracer molecules can only be excited at short wavelengths. A detailed discussion about fuel tracers and problems resulting in their use can be found in the review by Zhoa [188].

Among the most popular tracer molecules for liquid fuels are ketones, aldehydes, and various dyes. If the fuel evaporates the tracer molecules may evaporate as well and may be detected by LIF in the gas phase and might be used to visualize the fuel vapor. Special compounds are available that allow to distinguish fuel droplets from fuel vapor via their emission spectra. Also, the regular fuels from gas stations usually contain additives, e.g. aromatic hydrocarbons, that fluoresce and can be used to locate both liquid droplets and vapor. The correct choice of tracer for a given fuel depends on many details. The LIF signal is for example dependent on quenching, pressure, and temperature. In evaporation studies, the vapor pressure of the tracer should match the vapor pressure of the fuel otherwise the tracer concentration is enhanced or depleted in the liquid or the gas phase.

A clear illustration of the importance of tracer–LIF today is its use in spray flames, gasoline and Diesel direct injection engines. In these commercial combustion devices liquids are atomized, small fuel droplets generated and evaporated so that the fuel vapor will subsequently mix with the surrounding turbulent air flow to supply a mixture that gives efficient combustion and low pollutant formation. Because the intelligent design of the fuel air mixing device is crucial to the quality of combustion, laser techniques are used extensively today to optimize the performance of these combustion devices.

There is a strong need for sensitive and quantitative measurements of the density of both fuel and air. Unfortunately LIF does not work for most fuels (e.g. all saturated hydrocarbons), neither for the liquid droplets nor the fuel vapor. Quantitative measurements of air densities by LIF are also difficult. Although O_2 has been successfully detected via LIPF and used as a measure for air densities in non–reacting flows, this is not very practical because of the short wavelength required and difficulties in the quantitative conversion to densities.

Acetone and NO become popular tracer molecules to measure the density and the motion of the air. Tracer–LIF is used to visualize and quantify the 2D–distributions of liquid droplets and fuel vapor and also to analyze the air motion in the presence of liquid droplets. Combining tracer–LIF and MIE–scattering with the appropriate choice of tracers allow to determine size and velocity distribution as well as the temperature of droplets.

12.3 Experimental Setup and Procedures

12.3.1 Experimental Setup

Fig. 12.3 shows a typical set–up for planar LIF measurements. The laser beam is formed into a thin light sheet by a set of lenses. The lenses would typically consist of a spherical convex lens with a long focal length followed by two cylindrical lenses making up a one–dimensional telescope. The achievable size of the light sheet depends strongly on the effective absorption cross section of the LIF process. The effective cross section again depends on the following: the molecular species undergoing the induced fluorescence process, the selected transition and the chosen technique (either linear, saturated or predissociated fluorescence). As a guideline a value of 50 mm may serve for both the height of the light sheet and the width of the observed area, which depends on the angle of divergence of the light sheet and the camera used for observation having a fixed side–length ratio. The typical thickness of the sheet which determines the spatial resolution in this direction, is of the order of a few hundred μm. In order to suppress background light the beam is trapped in a light absorber (beam trap) after it has passed the observation volume.

The fluorescence is captured by a collecting lens and imaged onto the detector surface. In the collection path combinations of interference and cut–off

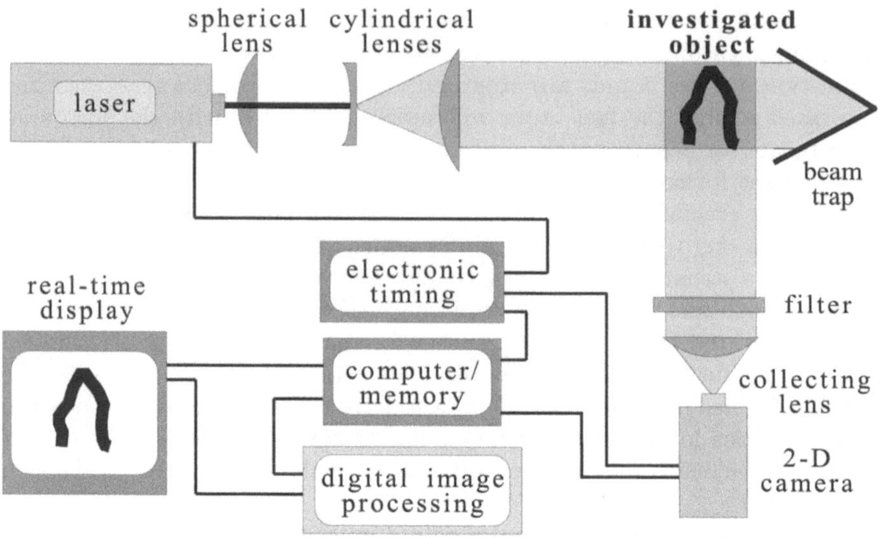

Fig. 12.3. Typical set–up for planar laser induced fluorescence measurements. The laser beam is formed into a thin light sheet by means of a long focal length spherical lens and a pair of cylindrical lenses acting as a telescope. The emitted light is captured by collection optics and imaged onto the surface of a solid state detector. In order to limit the light captured by the camera to the desired fluorescence signal, optical filters are used. The data obtained by the detector are transferred to the acquisition and control unit, where they are processed and stored. The control unit is also responsible for the timing requirements of the laser, the camera and, if required, also the experimental system under investigation

filters are positioned, so the fluorescence signal alone is actually observed by the detector. The detector is generally a solid–state camera equipped with a gateable image intensifier. The digitized intensity data of the two–dimensional image is transferred from the camera to the acquisition and control unit installed in or coupled to a computer. This control unit is also responsible for the timing of the laser, the camera and the experiment under investigation. The data of the images are stored in grey scale values according to the dynamic range of the camera and the intensity of the emitted light. These digital values can be processed by standard or custom designed procedures for evaluation of concentration and temperature fields. In order to show fine intensity differences the processed image is finally displayed in false color pictures, where a specific color is associated with a distinct intensity range.

LIF measurements are not always performed as two–dimensional imaging experiments, but may be integrated in other spectroscopic techniques as well. In this case the set–up differs from the one shown above, resembling instead the set–up for Rayleigh scattering and Raman scattering (s. Chap. 11). The application of LIF in a "spectroscopical set–up" will be shown in one of the examples below.

The modern developments in laser and camera technology as well as digital image processing in the recent years have caused LIF applications to become increaslingly more powerful and simultaneously easier to use. Although these more technological aspects are interesting, they are too complicated to be discussed in detail here; consequently only selected topics are discussed below. Due to the experience of LIF applications acquired in chemically reacting systems, i.e. combustion processes, these processes also present the basis for a discussion of the equipment.

LIF applications in combustion processes require tunable, intense, pulsed lasers in the deep UV range. Tunability is required because of the resonant nature of the excitation process. Intensity is required for high excitation efficiency, which is the origin for the high sensitivity of LIF. The lasers have to be pulsed to distinguish the laser induced emission from the emission arising naturally in combustion. With short pulsed lasers, in combination with gated fluorescence detection, the natural emission is suppressed by orders of magnitude. Operation in the deep UV is required because the spectroscopy of combustion species is almost exclusively in this frequency range.

In most LIF applications Nd:YAG or excimer lasers are used to pump tunable dye lasers in the visible part of the spectrum. Frequency doubling or mixing is typically used to generate the desired frequencies in the deep UV range. Pulse durations of such laser systems are in the range of 10 ns. Bandwidths in the UV around 0.2 cm^{-1} are relatively easy to achieve. In the recent past most LIF applications were performed with rather moderate pulse energies of μJ to a few mJ at maximum. Since the invention of BBO crystals, the pulse energies available in the deep UV ($210 - 250$ nm) became much higher (several mJ per pulse), which in many cases is sufficient for planar LIF imaging on considerably large laser sheets.

An alternative to such fully tunable deep UV lasers are tunable excimer lasers. These lasers have the highest spectral brightness and pulse energies in the deep UV (a few 100 mJ per pulse within less than a wave number). Although tunable excimer lasers operate in a narrow frequency range (KrF: $193 - 194$ nm, ArF: $248 - 249$ nm) a variety of different LIF techniques can be used to detect important combustion species (such as OH, O_2, H_2O, NO, CO). Additionally these lasers offer more advantages for laser diagnostics. They are easily frequency shifted by stimulated Raman scattering in high pressure hydrogen cells, opening up other tuning ranges. Because of their deep UV wavelength, "tracer molecules", added to the flow for visualization purposes or added to hydrocarbon fuels to measure fuel air ratios can be excited efficiently. The lasers can also be used – with slight modifications – to pump dye lasers and generate fully tunable UV radiation. In addition, due to their extreme power in the deep UV range, the lasers are ideally suited for Rayleigh– and Raman–diagnostics. In this way many different laser diagnostic techniques can be employed with tunable excimer lasers alone, yet the lack of full tunability can be a considerable disadvantage in a number of applications.

In the recent years there have been several important developments in laser and camera technology which facilitate LIF considerably. Firstly, tunable narrow band laser radiation in the deep UV is now commercially available from full solid state lasers on the basis of OPO technology without any use of dye lasers. Secondly, two dimensional picosecond LIF has become possible recently due to the commercial availability of both narrow band picosecond lasers in the deep UV and ultrafast cameras with gate length as short as 80 *psec*. Although the technology is still expensive, it has been demonstrated that single shot psec–LIF images of OH can be recorded in atmospheric flames. This technology removes all problems in the quantification of LIF because both quenching in the excited state as well as RET in the ground state become negligible [173,184].

On the detection side – only a few years ago – photmultipliers had been used almost exclusively for the detection of fluorescence. Today, modern camera technology has greatly increased the potential of LIF methods. The development of short gated photometric camera systems with ultimate sensitivity (matching that of phototubes), in combination with powerful digital image processing opens up new areas for LIF applications. It has been mentioned that cameras in combination with laser sheets are used for two–dimensional laser induced fluorescence imaging from spatially well defined planes and that spatial scanning in the third dimension will yield species and temperature distributions in all three dimensions. In the example discussed below, it is demonstrated that even qualitative information from such images allows details about mixing and combustion to be understood.

Two–dimensional imaging is not the only application for cameras for LIF processes. First, a camera mounted behind a spectrograph or polychromator yields an optical multi channel analyzer which is ideal for a survey of spectroscopy. Second, the same set–up can be used to simultaneously record spectroscopic and one–dimensional spatial information, as shown in an example below. In addition, some cameras can be modified to work in a "slow streak mode" to obtain time resolved spectroscopic information, generating hundreds of spectra delayed only by microseconds.

Although the most severe errors in LIF measurements are introduced by the laser sources, great care has also to be applied in the selection of the detection units, in particular with cameras. The most popular CCD–cameras based on home video standard are sufficient for TV and qualitative visualization purposes, but are not precise measuring ("photometric") instruments. Apart from the often missing linearity, the digitization depths (dynamic range) with only 8 bit (256 intensity values) and the noise of typically 3 bits (8 intensity values) yield signal strengths that exceed the noise by a factor of $30 - 50$ at best. Less errors are introduced by carefully selected high quality cameras, although also here the photometric features have to be carefully tested. One problem that arises for pulsed laser experiments with pulse durations in the nanosecond to microsecond range even in some high quality camera systems is that the output signal does not increase linearly with input illumination above some fluorescence intensity. This linearity, however, is clearly required for photometric operation. If too many photons reach the camera in a few nanoseconds, some intensified cameras become non–linear even though they are strictly linear when the same number of photons reach the camera in a longer time interval (e.g. some milliseconds). The problem in such pulsed applications arises from the image intensifier. The wall of a channel in a micro channel plate has only a small capacity, i.e. it stores only a finite charge. The charge is lost if too many electrons are amplified in the same channel and will only be replaced in milliseconds at the large resistance of micro channel plates. As a consequence, the voltage across the channel breaks down, the channels saturate and the linearity of the output signal with input illumination is no longer warranted, i.e. the cameras are not photometric devices. Because both LIF and Rayleigh experiments can yield rather high emission intensities in a single laser shot, this “single shot linearity” should extend over a wider range. Today cameras with a single shot linearity up to 14 bits are availible.

The lens chosen to project the image onto the photo–sensitive surface is another important design consideration in imaging experiments. LIF diagnostics in combustion are primarily done in the deep UV. The only commercially available lens for this region of the spectrum is the UV–Nikkor from Nikon. Although this lens features excellent characteristics with high spatial resolution, it is not designed for weak light collection, i.e. it has poor collection efficiency. Despite the fact that LIF is famous for its high sensitivity, the accuracy of LIF measurements depends crucially on the light collection efficiency. At the cost of some spatial resolution, the collection efficiency can be increased easily by a factor of 10 with specially designed UV lens combinations with a spatial resolution still below 100 μm. Since this resolution is of the same order as the resolution given by the thickness of the laser sheet, it represents no real deficiency. The use of a carefully selected imaging lens is therefore by far the most inexpensive way to improve the measurement precision.

12.3.2 Experimental and Evaluation Procedures

The laser induced fluorescence phenomena which may occur in combustion processes are numerous, particularly if intense UV lasers are used for excitation. The actual appearance of possible emission phenomena in a given combustion device depends not only on laser features (i.e. wavelength bandwidth, power, or focussing conditions) but also on the type of flame and on the location within the flame (i.e. temperature and gas composition, radicals, soot, and unsaturated hydrocarbons). For tightly focussed high power lasers in particular, where multi photon processes become probable, many of the emission phenomena are poorly understood and can not be used for quantitative diagnostics.

In the selective detection of some given species, great care has to be taken to avoid interferences with emissions from other species occurring at the same excitation wavelength. Very often spectral filtering is a necessity. The precise knowledge of the spectral composition of the emission in a given combustion device is a prerequisite for the use of a given diagnostic method. As an example, consider the selective detection of NO via the $D(v'=0)-X(v''=1)$ state around 193 nm by a tunable ArF excimer laser, which interferes heavily with O_2 emission. If the probe volume contains not only NO but also hot oxygen molecules, there will be simultaneous excitation of O_2 and NO at almost any suitable excitation wavelength in the tuning range of the ArF laser. There are only a few exceptions (e.g. the line $R_1(26.5)$) where only NO is excited. In order to selectively measure NO without interference from O_2 it is necessary to carefully filter the emission (e.g. around 208.5 nm), in particular in planar LIF applications. It should be emphasized that many other interfering emissions, e.g. from soot, aromatic hydrocarbons, Mie, Rayleigh, and Raman scattering, have to be avoided as well. The interfering emissions must be analyzed for each combustion condition before any quantitative measurements can be reformed.

An elegant approach to characterizing the various laser induced emission phenomena in an unknown (combustion) object within a very short time is to record excitation/emission spectra with an optical multi channel analyzer. This is demonstrated in a laminar methane–oxygen flame operating close to stochiometric conditions. The procedure to obtain a full set of emission spectra is as follows. The laser is set to a certain frequency at one end of the tuning range, here 249 nm(40180 cm^{-1}). The laser beam in this set–up is formed into a narrow line rather than a sheet and passes through the combustion system. The laser line, or rather a section of it, is imaged onto the entrance slit of a low resolution spectrograph to cover the emission from a large spectral range, here from 220 to 310 nm. An intensified digital camera, mounted in the image plane of the spectrograph, records the emission. The emission from 100 laser shots is integrated "on chip". The emission intensities at a given excitation wavelength is integrated and stored as the first line of a new image, which represents an emission spectrum for the initially selected laser frequency. Now the laser is tuned by a small increment (0.1 cm^{-1}) to the next excitation frequency generating a second emission spectrum which is stored in the second line of the new image. The procedure is repeated until the end of the tuning range (40375 cm^{-1}) is reached. In the actual procedure, the computer activates the camera, tunes the laser to the next frequency, fires the laser 100 times at this frequency, integrates the emission for the same emission wavelength, and stores the emission spectra in the new image. The scan takes less than 10 minutes. The resulting image, containing

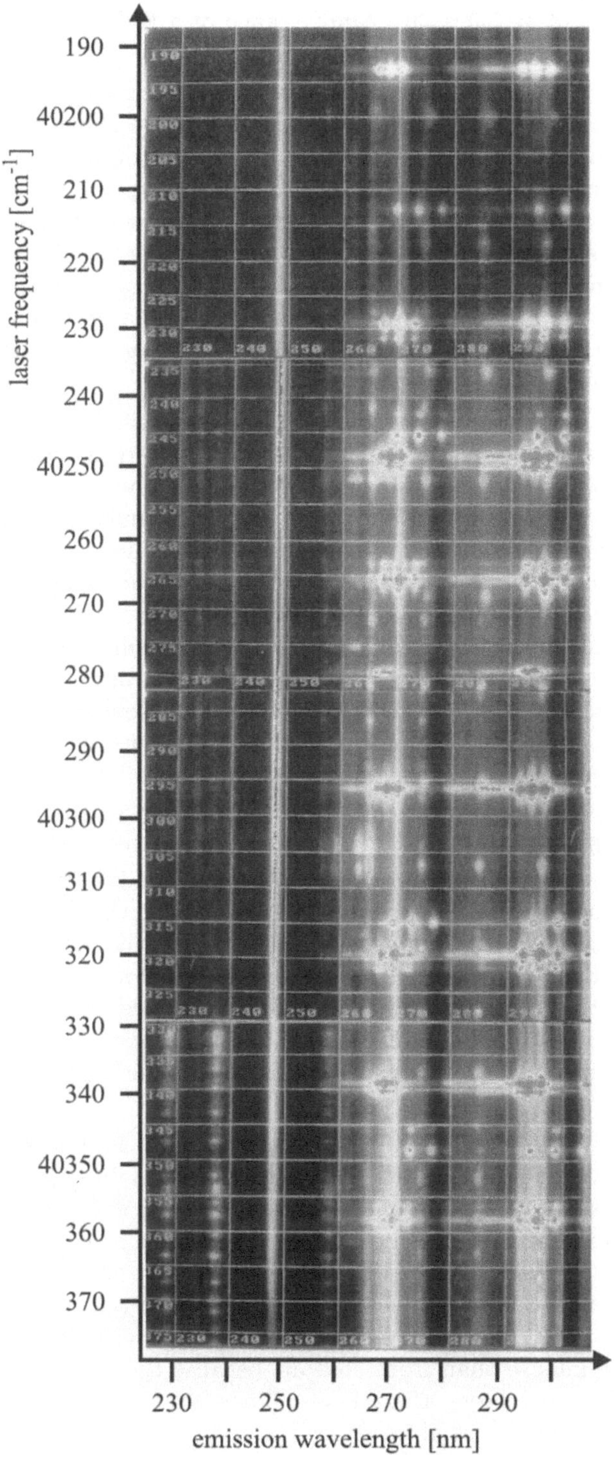

Fig. 12.4. Complete set of emission spectra in the range between 220 and 310 *nm* for all excitation frequencies of a tunable KrF laser

the complete emission spectra of the entire tuning range of a KrF laser, is shown in Fig. 12.4.

Subsequent lines in Fig. 12.4 contain emission spectra for increasing laser frequency. The excitation/emission spectrum shown in Fig. 12.4 shows all the emission phenomena occurring at all excitation frequencies of the KrF laser. It provides a spectral map that can be used as a fingerprint for a given combustion condition. It can be used to identify interference of different emission phenomena. Horizontal profiles yield emission spectra for a specific excitation wavelength, vertical profiles yield excitation spectra for well defined emission wavelength. Excitation/emission spectra for other combustion situations may be generated by varying stochiometries or scanning spatially through the flame.

The combination of a camera with a spectrograph also yields an attractive alternative to planar LIF applications, in which spatially resolved information along a line (one–dimensional measurement) is recorded simultaneously with the emission spectrum. The method allows simultaneous multispecies and temperature measurements along a line, which is of great interest for applications in turbulent combustion. In this method the laser beam is again formed to a line and the emission imaged to the spectrograph as before. In this way a single spectral line in the image plane of the spectrograph contains spatially resolved information from the selected section of the laser line that is projected to the entrance slit. The achievable spatial resolution depends not only on the type of spectrograph but also on the quality of the lens imaging the laser line to the entrance slit (e.g. achromatic features).

Figure 12.5 shows an example of a measurement in a laminar methane–oxygen flame that was performed in the above described manner, integrating 200 laser shots. The emission extends from 220 to 300 nm. The schematic on the left of the figure shows the path of the laser line traversing the flame. Along the section of the laser line that is imaged to the entrance slit (marked as black bar) different combustion conditions are met. From 0 to 2 mm (compare y axis in the image with the schematic) the emission is sampled from the central cold core of the flame with only premixed reactants). The combustion region extends from 2 to 6 mm. The cold air surrounding the flame is probed from 6 to 8 mm.

The emission from the cold reactants is in the upper part of the image. Three emissions are observed: Rayleigh scattering at 248.5 nm, vibrational Raman scattering from methane at 269 nm and vibrational Raman scattering from O_2 at 258 nm. N_2 Raman scattering is not observed because a methane–oxygen flame is used. The emission induced by the laser in cold air in the lower part of the image contains Rayleigh scattering and Raman scattering from N_2 and O_2 as well as a little bit of H_2O. The Rayleigh scattering from the cold reactants is more intense than from cold air because the Rayleigh cross section for methane is much larger than for air. There are no problems with interfering emissions in these spatial locations.

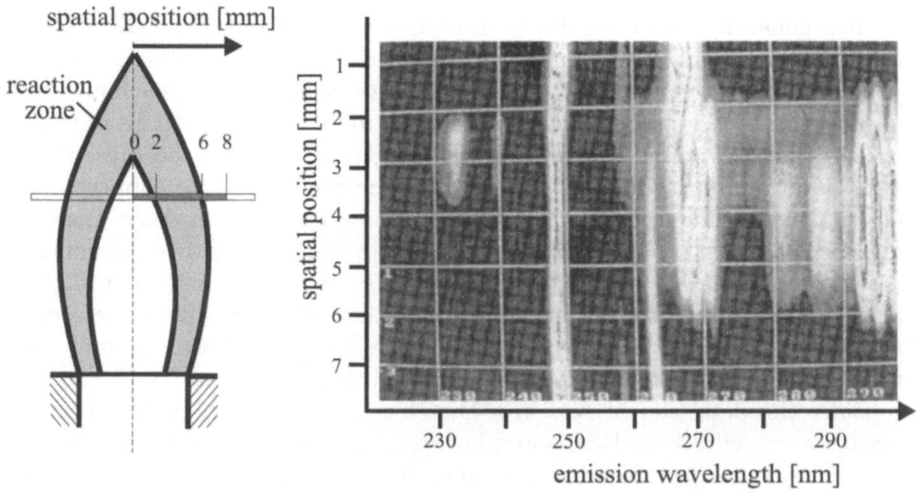

Fig. 12.5. One dimensional emission spectrum recorded from a laminar methane–oxygen flame. The spectrum contains emissions from Rayleigh, Raman and LIF processes. The laser frequency is tuned to the $P_1(8)$ line of the $A(v' = 3)-X(v'' = 0)$ absorption band of OH

The central part of the image shows the emissions induced by the laser in the combustion region. The two very strong emissions at 2704 and 4300 nm at the right, each consisting of three single lines, originate from selective excitation of OH. The maximum of the OH emission is at 4 mm, roughly at the midpoint between the reactants and the cold air. Compared to the cold reactants and the air, Rayleigh scattering is weak in this region because the temperature is high and the corresponding density is low. The intensity of the O_2 Raman line is low because of its consumption in the combustion process. The H_2O Raman line at 270 nm is not seen at all. It is dominated by the intense interfering OH emission at the selected resonance frequency of OH. The still rather intense N_2 line in this region of the flame may be a little surprising. An interesting observation is the considerably intense broad-band emission between the cold reactants and the region with maximum OH. Although the precise origin of this emission is not understood, it is most probably due to LIF of thermally decomposing hydrocarbons. There is obviously another, rather intense, narrow band emission at shorter wavelength around 230 nm. Because of these interfering emissions great care has to be applied in this region in the quantitative interpretation of Rayleigh, Raman and LIF, demonstrating the necessity of knowing the spectral composition of light. From such images temperatures and densities of major species have been derived for different flame heights and stochiometries using appropriate calibration procedures. For more details the reader is referred to Reckers [187].

It is generally recommended to use these spatially and spectrally resolved measurements along a line for spectroscopically difficult conditions. In most realistic combustion devices pressure and temperature can become high which implies that spectroscopy can become even more complex. Many unknown species such as aromatic hydrocarbons, PAH's or soot may be present and yield additional laser induced emissions which are difficult to suppress with filters. Due to line broadening the selective excitation of a desired species may become very difficult. In such spectroscopically unknown environments it is preferable to do spatially and spectrally resolved measurements along lines instead of two dimensional imaging. Then the full emission spectrum is available at every spatial position and can be used later to analyze if there are additional strong interfering laser induced emissions. As an example, consider the detection of NO in a Diesel motor at very high pressure and temperature with soot, PAH's and hot oxygen present. At these pressures a sufficiently selective excitation is no more possible because of line broadening. Although the emission spectrum becomes very complex, it still contains a full vibrational A–X progression which can be used to extract the density of NO.

An additional advantage of line measurements is that Rayleigh scattering and Raman scattering can be recorded simultaneously to LIF. Polarization techniques became a powerful technique to discriminate the usually unpolarized LIF from spontaneous Raman– and Rayleigh scattering [180].

After the LIF data are recorded and stored, they have to be processed in order to obtain detailed information on the investigated process. The conversion of raw LIF data to densities and temperatures is complicated by the effect of quenching collisions (s. above). Methods to eliminate this systematic error have been discussed above. However, this is not the only source of errors in LIF applications.

Lasers are probably the most critical experimental devices in cases where quantitative application of LIF is desired and yield the largest errors in measurements. For almost every pulsed laser source, the spatial profile of the laser beam will contain areas with higher and lower intensities. Similarly the laser intensity is not constant over the time of the laser pulse; instead intensity increases to a maximum and decreases again. The varying intensity, both with respect to space and time, may lead to partial saturation, which may cause problems. For partial saturation the intensity is not determined by the Hönl–London and Franck–Condon factors anymore. On the contrary, when saturated LIF is applied, the spatial wings of the laser beam and the low intensity at the beginning and end of the pulse contain less intensity and will lead to incomplete saturation. A problem with intensity arising in all LIF methods described above (LIF, LISF, LIPF) is that the ground state population may be depleted at too high laser intensities, again leading to errors in the conversion of data. For LIF measurements the frequency and bandwidth stability is also very important. In particular for temperature measurements via two–line thermometry, the tuning to the excitation line center is of utmost importance, because small deviations in the ratio of the observed intensities lead to errors of several hundred Kelvin.

One of the most straightforward ways to quantify LIF data is based on calibration procedures. The calibration procedure is explained for the case of OH LIPF detection in the oil burning furnace described in the example below. Since the combustion in laminar flames is both theoretically and experimentally well studied and understood, such a flame can be used to predict both the temperature and the absolute OH density in the flame and thereby serve as a calibration reference.

In the overall calibration procedure the spatial OH distribution is first measured in the oil burning furnace with a given experimental set–up of camera and laser sheet. This measurement yields fluorescence intensity fields $S_1(x,y)$ and $S_2(x,y)$ for the two rotational states 1 and 2 probed by $Q_1(11)$ and $P_2(8)$, respectively. The oil burner is then removed from its actual measurement position and replaced by the one–dimensional laminar methane–air flame. The resulting signal is measured and yields fluorescence intensity fields $S_{10}(x,y)$ and $S_{20}(x,y)$ at known density fields $n_{10}(x,y)$ and $n_{20}(x,y)$ for the two probed quantum states 1 and 2. The unknown density fields $n_1(x,y)$ and $n_2(x,y)$ for the two quantum states at each location (x,y) in the furnace are then simply obtained by:

$$n_1(x,y) = \frac{S_1(x,y)}{S_{10}(x,y)\; n_{10}(x,y)} \tag{12.29}$$

$$n_2(x,y) = \frac{S_2(x,y)}{S_{20}(x,y)\; n_{20}(x,y)} \tag{12.30}$$

From these absolute densities in the two quantum states the temperature is determined according to the Boltzmann law as described in Sect. 12.2. The described calibration procedure eliminates several potential errors because the calibration and the actual measurement are done under identical conditions. Because the signal for LIPF does not depend on the collisional environment, calibration can be done once. For normal LIF and LISF, calibration is more involved. However, RET in the ground electronic state remains a problem.

12.4 Selected Applications

Because of its great selectivity and sensitivity, the laser induced fluorescence method has been used extensively at low densities in the field of laser spectroscopy. Although there are also numerous applications of LIF to combustion problems (s. review by Hanson [177]), most of them remained qualitative visualization experiments, because the conversion of measured fluorescence intensities to densities and temperatures is more difficult at the elevated densities in real combustion problems (collisional quenching problems). With more recent LIF methods, such as saturated LIF or LIPF, the problems in the conversion are more relaxed.

The extreme selectivity in combination with the high sensitivity of LIF makes it ideally suited to the important detection of minority species in complex combustion processes. In the following two recent examples for LIF applications are discussed.

The first demonstrates the application of different LIF processes to a commercial oil burning furnace (manufactured by Viessmann), which is sketched in Fig. 12.6. Two different burner configurations, differing in the resulting amount of NO_x content in the exhaust gases, are investigated to understand the origin and reasons of the achieved NO_x reduction. The example serves to demonstrate that even the qualitative visual information from raw fluorescence data allows one to understand many details both about fuel air mixing and the subsequent combustion.

The NO_x reduction from 70 *ppm* in the conventional burner configuration to 45 *ppmNO* in the improved burner configuration is achieved by a slight modification in the fuel air mixing device (s. inset in Fig. 12.6). The mode and amount of fuel injection through the nozzle is identical in both mixing devices; only the air supply is modified. Two modifications have been implemented. First, the diameter of the cylindrical air slit between the baffle plate and the outer rim of the wall is decreased from 4 to 3 *mm* by moving the baffle plate forward. Second, the diameter of the hole in the middle of the baffle plate was decreased from 5 to 3 *mm*. Because the fuel injection remained unchanged, the air pressure behind the slits had to be increased to admit the same amount of air in order to keep the average stochiometry the same.

The results of the LIF experiments are shown in the images in Fig. 12.7. In the following, only the major results shall be discussed; for details the article by Koch [181] should be consulted. Image 1 at the upper left side shows a normal photograph of the natural flame emission together with a coordinate system for the identification of the dimensions. All images are drawn to the same scale to help the eye compare the spatial location of the different monitored species relative to each other in the flame.

Images 4–7 in Fig. 12.7 can be interpreted qualitatively as "averaged spatial distributions" of different species. The images on the left show the results for the conventional configuration, the images on the right for the NO_x reduced configuration. Image 4 shows the unburnt fuel, image 5 the distribution of "hot" O_2, image 6 the OH distribution and image 7 the distribution of NO. These species distributions are obtained by using different LIF processes at different excitation wavelengths in the tuning ranges of the ArF and KrF lasers. NO is probed with the ArF laser via LIF using the line $R_1(26.5)$ in the $D(v'=1)-X(v''=0)$ band. O_2 is probed with the ArF laser via LIPF using the line $R(25)$ of the $B^3\Sigma_g(v'=10)-X^3\Sigma_g(v''=2)$ band. OH is probed with the KrF laser via LIPF using the $P_1(8)$ line of the $A^2\Sigma(v'=3)-X^2\Pi(v''=0)$ band. The unburnt fuel distribution is obtained

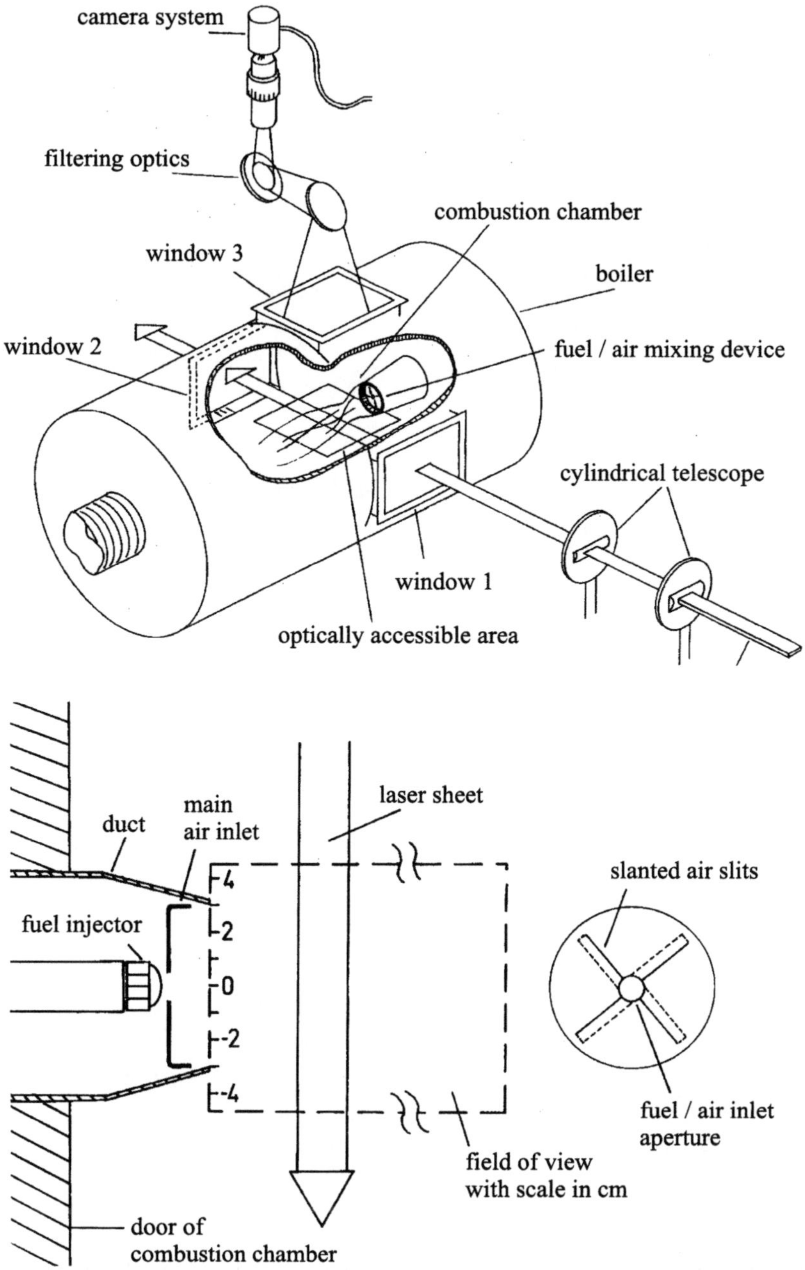

Fig. 12.6. Commercial oil burning furnace investigated by LIF methods for reduction in NO_x emissions. The upper picture shows the overall arrangement of the furnace and the LIF probe, the lower picture shows the details of the fuel/air mixing device as well as the field of view.

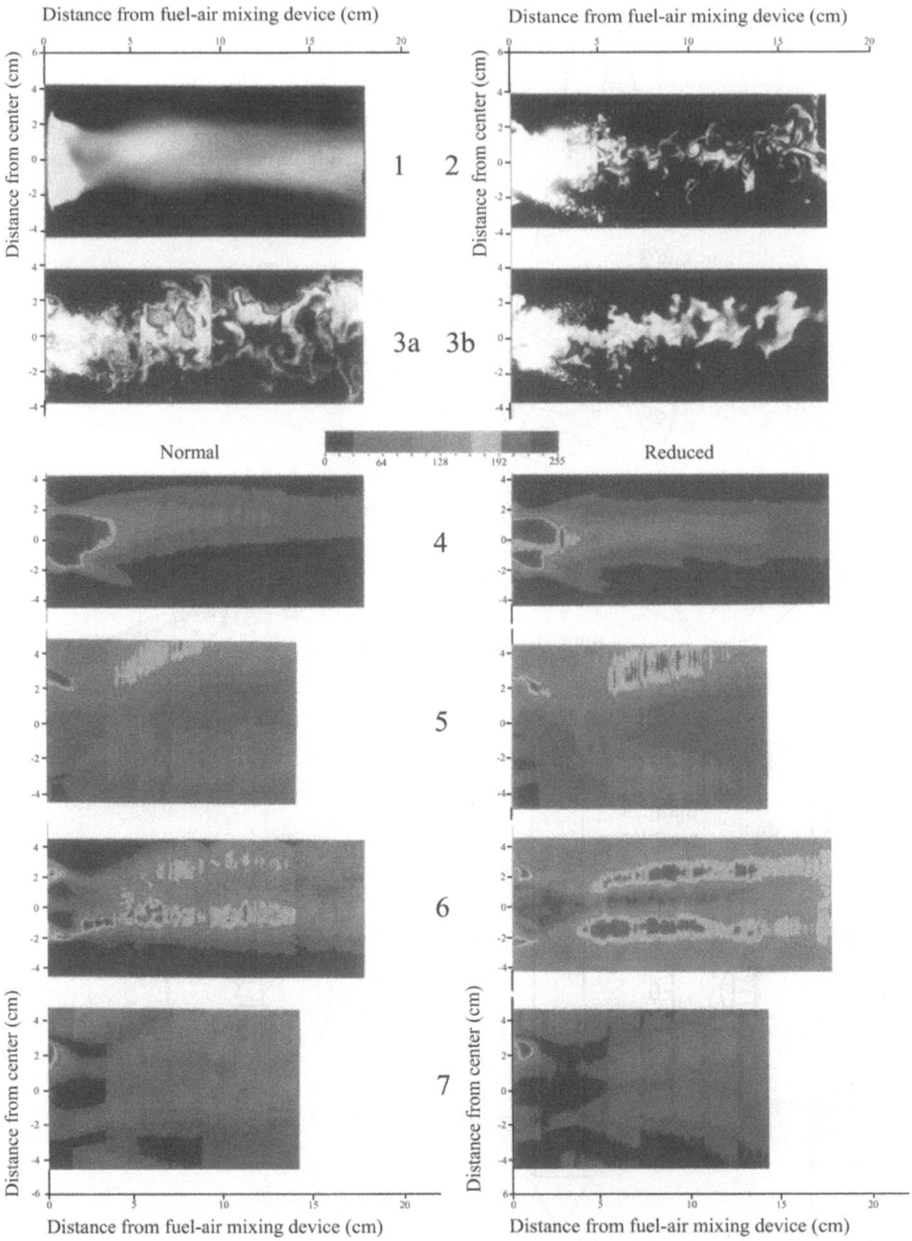

Fig. 12.7. Results of measurements in the commercial oil burner. 1: photograph of natural flame emission; 2: elastic scattering (Rayleigh and Mie); 3: fluorescence intensities with KrF laser off (3a,left) and on (3b,right) resonance; 4: unburnt fuel; 5: hot O_2; 6: OH; 7: NO. For images 4–7, the data on the left are for the conventional and on the right for the reduced burner configuration

by a non–resonant LIF process with the KrF laser, which is not understood in detail.

Image 2 in Fig. 12.7 is recorded in a single laser shot of the KrF laser without any spectral filtering or blocking. It shows elastically scattered light (Rayleigh– and Mie–scattering). Since the Rayleigh cross sections are much larger for the hydrocarbon molecules of the fuel, the image really represents the distribution of unburnt fuel (liquid droplets and regions with high gaseous fuel content). The image shows clearly the highly turbulent structure that is typical for the mixing of fuel with the surrounding air.

Images 3a and 3b are shown to explain how the averaged spatial distributions in images 4–7 are obtained. Image 3a is obtained with the KrF laser tuned off any resonance and blockage of the elastically scattered Rayleigh and Mie light with suitable filters. Although the broadband fluorescence obtained here is not understood in detail it can be shown to originate from unburnt fuel. The emission extends from 250 to 600 nm and is independent of the excitation wavelength. It shows up as a background in all other images. The only difference between images 3a and 3b is that for the photograph on the right (3b) the laser is tuned to the OH resonance $P_1(8)$. Obviously there is an additional weaker emission in image 3b, located in most cases further outside of the strongest emission area. Images 4–7 are now obtained by the following procedure. First 256 single images are summed up with the laser "on resonance" for the probed molecule. Then 256 images are summed up "off resonance". In the third step the averaged "off resonance" image is subtracted from the averaged "on resonance" image. This procedure yields the images 4–7. Because the non–resonant contributions are subtracted, the images show selectively the averaged spatial species distributions probed by the corresponding resonance. It should be mentioned that apart from the subtraction of non resonant contributions the images represent raw data.

In contrast to the highly turbulent structure in the single shot images (2, 3a, 3b), there remains a much smoother but clearly inhomogeneous spatial structure appearing in each of the distributions. This structure is highly reproducible and depends sensitively on details of the fuel–air mixing device. A closer comparison between the images at the left and right side reveals considerable differences for the conventional and the NO_x reduced burner configuration. A detailed discussion of the results is beyond the scope of this review article and can be found in the corresponding literature. Nevertheless a brief discussion is given here.

Certain features in the images are common to both burner configurations and easily understood. Unburnt fuel is found mainly in the middle of the combustion chamber as expected from the injection in this direction. O_2 is found further outside, since most air is injected through the cylindrical slits at $y = \pm\ 3\ cm$. The mixing takes place in the area between the peak concentrations of O_2 and fuel. Qualitative information on local stochiometries might be derived from comparisons of the two separately obtained distributions. It is not surprising that OH is formed in this mixing area (s. 6a, 6b). The OH indicates the active locations of the burner in which the combustion reaction actually takes place. The sharp structures in the OH images at the left (x =0 cm, $y = \pm\ 3$ cm) originate from burning vortices which stabilize the flame. That the peak NO concentrations appear roughly at the same spatial locations as OH, with a slight shift downstream is intuitively clear since these are the regions where the highest temperatures, leading to NO_x, appear.

The comparison between the images on the left (conventional configuration) conventional configuration and the right side (improved low NO_x configuration) yields several interesting observations, which are understood straightforwardly by the corresponding differences in the two fuel air mixing devices:

1. All distributions on the left side show some asymmetry towards positive y–values. On the right side all distributions are less asymmetric, peaked more in the forward direction. This results from the higher injection velocities due to the reduced area of the injection ports at constant mass flow, leading to a more directed flow.
2. The penetration of fuel inside the burner is deeper in the images on the right side than in those on the left side. The corresponding OH distribution on the right side indicates that for the NO_x reduced burner configuration almost no burning takes place in the middle, consistent with the presence of unburnt fuel in this area. This is explained by the faster overall air flow: the mixing only yields stochiometries suitable for combustion further downstream. This observation is supported by the O_2, NO, and OH distributions which are also shifted 2 cm downstream in the images on the right side.
3. The NO distribution shows roughly 30% more fluorescence intensity for the conventional burner configuration shown on the left side. This corresponds roughly with the NO reduction measured conventionally in the exhaust gases. The distribution on the left shows more NO in the middle near the central axis than in the images on the right side. This is expected from the corresponding OH distribution at the right side image showing almost no combustion in the middle area.
4. The OH distribution on the right is more confined. This is interpreted as a result of the higher degree of turbulence introduced by the increased air flow velocity in the burner. The higher turbulence enhances mixing and consequently leads to a faster combustion. This results in lower residence times of the gases at high temperatures and therefore the amount of NO formation is reduced.

The final conclusion on the investigations regarding the reduction is based on the larger volume of high temperature (burning along the middle axis) in the conventional configuration as opposed to the faster mixing and combus-

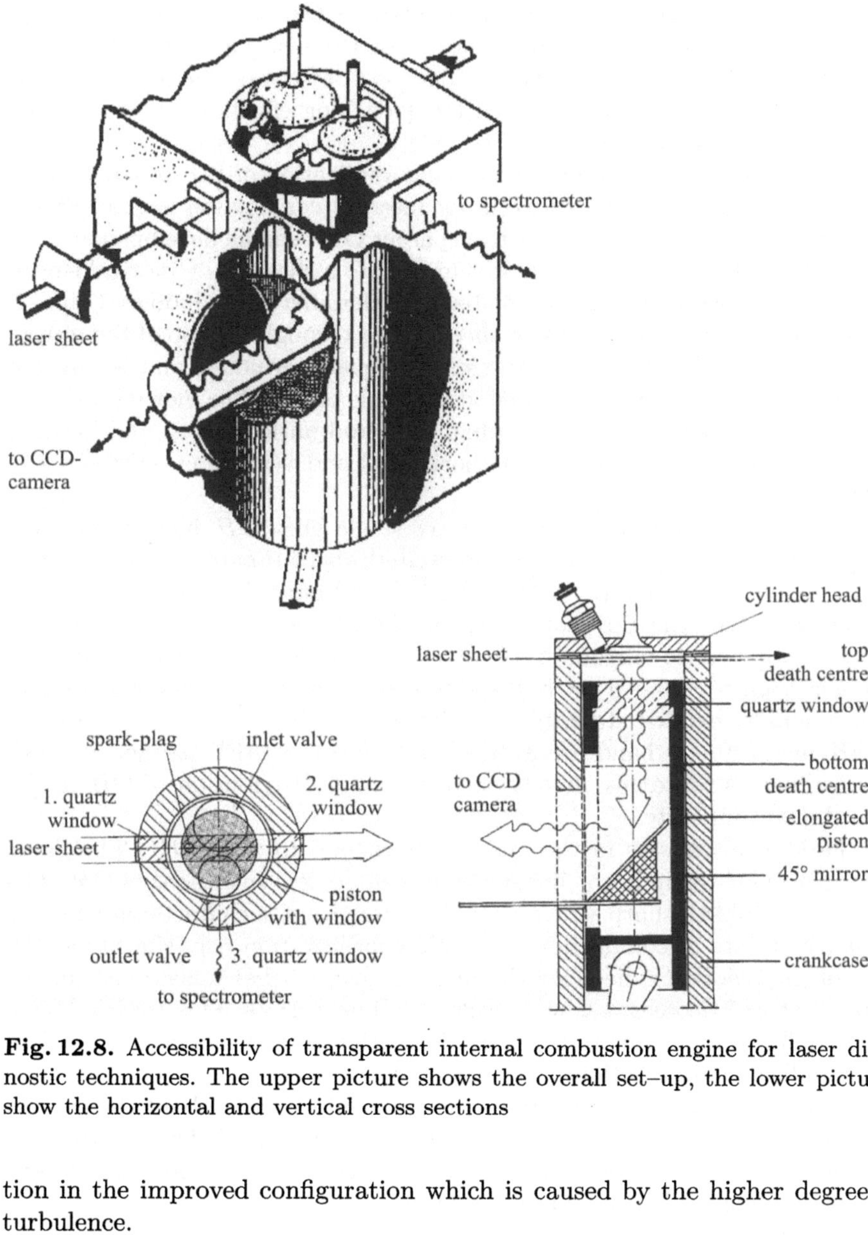

Fig. 12.8. Accessibility of transparent internal combustion engine for laser diagnostic techniques. The upper picture shows the overall set–up, the lower pictures show the horizontal and vertical cross sections

tion in the improved configuration which is caused by the higher degree of turbulence.

Regarding the quantitative conversion of data, the different LIF processes involved in the measurements have to be distinguished carefully. The distribution of fuel is most difficult to quantify since in this case not even the precise origin of the emission is understood [181]. The O_2 and OH distributions represent correct spatial distributions of these species because they are obtained by the LIPF method which is not affected by quenching effects.

However, because the images represent O_2 and OH distributions in single quantum states, absolute number density for these species can only be calculated if the temperature is known. The NO detection scheme is based on normal LIF and has to be corrected for quenching effects in the conversion to absolute concentrations.

A second example of a very interesting application of LIF to combustion processes is the detection of OH in an internal combustion engine. Figure 12.8 shows the optical access to a transparent Otto engine developed at VW. The light of a tunable excimer laser is formed to a sheet by cylindrical lenses. This sheet passes through the combustion chamber via windows 1 and 2. The emitted fluorescence from the sheet passes through the top of the piston, which is equipped with an UV transparent quartz window, and is directed to the camera by a mirror mounted at an angle of 45°. The viewing range of the camera, the location of spark plug, inlet and outlet valves, and location of the laser sheet are shown in the horizontal and vertical cross sections in the lower part of Fig. 12.8.

Figure 12.9 shows an example of six single shot OH images obtained shortly after ignition. The OH is again excited with a tunable excimer laser at the $P_1(8)$ line in the $A^2\Sigma(v' = 3) - X^2\Pi(v'' = 0)$ band. i.e. the LIPF method is used to detect the OH concentration distributions. The fluorescence is filtered via dielectric mirrors to selectively detect the collisionally unaffected emission from $v' = 3 - v'' = 2$. Because this transition is weak and ends in a vibrationally excited state, the well known problems with self absorption of OH in the $0 - 0$ band are avoided. Because the LIPF method is used, the raw images represent already correct spatial distribution of OH in the probed quantum state.

The high intensities in the images correspond to more OH. The flame front clearly shows up in the images as maximum concentrations of OH. The irregular, wrinkled, sharp structure of the flame front is characteristic for the turbulent nature of combustion. The OH concentration rises sharply at the front of the reaction zone acing the unburnt gas, reaches a maximum in the flame front and decreases slower toward the burnt gases. This effect of "OH super equilibrium" is due to the faster formation than decomposition of OH in the radical chain reactions.

Using the described set–up and procedures information about the process of combustion in internal combustion engines can be gained. This information is vital for further improvement of current engines with respect to efficient combustion, engine knock and pollutant emission.

12.5 Concluding Remarks

Laser induced fluorescence methods are extremely powerful tools, mainly applied to concentration and species measurements. For planar imaging, the strength of the fluorescence signals is often many orders above that of Ra-

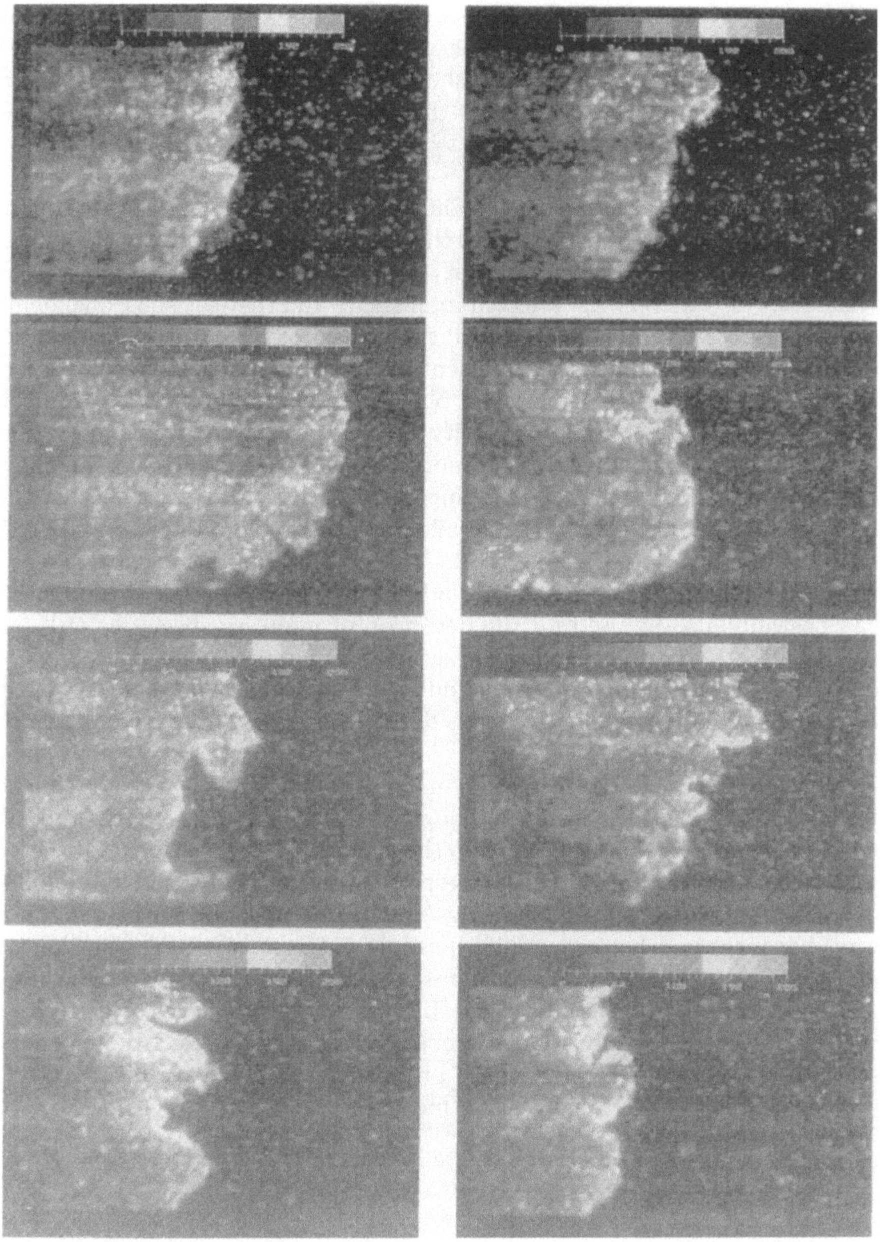

Fig. 12.9. *OH* fluorescence images obtained shortly after ignition in the transparent internal combustion engine

man scattering and Rayleigh scattering techniques. Even for large areas of observation of several centimeters side length the achievable spatial resolution is very good and the resolution in the third dimension, defined by the thickness of the exciting laser light–sheet, is of the order of 100 μm. Temporal resolution is mainly determined by the excitation source. For pulsed lasers typically used for LIF measurements it is usually in the lower nanosecond range.

One of the most important advantages of LIF– compared to most other laser diagnostic methods is its extreme selectivity and sensitivity. The extreme selectivity of LIF results from the fact that molecules can be excited resonantly only at very few selected frequencies and that the subsequent emission represents a highly characteristic fingerprint of the emitter. The extreme sensitivity originates from the efficient excitation by pulsed deep UV lasers and the highly sensitive detection of photons. These factors make LIF ideally suitable for the detection of minority species in complex reaction systems, especially combustion processes. Minority species, such as OH or CH, play a central role in our basic understanding of combustion processes because they represent sensitive probes of the detailed reaction dynamics and indicate, for example, the positions of flame fronts. Most pollutants, such as NO, NO_2 or CO, are minority species that can only be detected by LIF. Although LIF can basically also be used for the detection of majority species as well (e.g. O_2, H_2O), its main domain is the detection of minority species. It allows one to selectively probe even single quantum states of molecules, e.g. in real internal combustion engines, which is difficult with any other laser diagnostic method.

One disadvantage of LIF techniques is the necessity of exact frequency tuning of the exciting lasers. Although there are lasers available to cover nearly all spectral ranges necessary, the handling of some of these lasers is sometimes uncomfortable. However, especially in this field, rapid development of solid–state, easy to use lasers that are tunable over wide ranges, is on its way. Advances in other aspects, such as increasing sensitivity, resolution and read–out speed of solid–state cameras, together with more powerful lasers will enhance the applicability of the LIF techniques for large scale research such as in aircraft combustion chambers or large furnaces. LIF techniques seem to be the most dominant and promising non–intrusive techniques for flow field imaging and quantitative measurements in reacting flows for the future.

13 Absorption

Volker Ebert and Jürgen Wolfrum

13.1 Introduction

Absorption spectroscopy is one of the oldest techniques of non–intrusive investigation of gaseous media. In this technique the radiation emitted by a source towards a radiation detector is absorbed along its way to the detector. This loss is monitored and analyzed for its dependence on wavelength. Absorption is widely used for industrial gas analysis because of its simplicity, low cost and effectiveness. Nevertheless it is still an area of active research. Up to the late sixties most spectroscopic data was gained only with the help of broadband sources and monochromators or filters, however, with the development of lasers, research gained an important tool to improve the quality of the data and, more importantly, to access completely new areas for the application of non–intrusive optical measurements. These new possibilities are based on the superior properties of laser radiation which include high spectral resolution, very high spectral power density and directivity of the radiation. Chemical analysis and determination of temperature by optical methods in some cases has already been made possible or could be carried out much more specifically with these features, while signal to noise and sensitivity could be enhanced dramatically.

In most cases when absorption measurements are employed, the subject of interest is either the absolute concentration of one or more chemical species present in a mixture with several other components, or the purpose is to measure the temperature of this mixture. In the first case one has to find a spectral window where the subject of study is the only or at least the dominant absorber. In the second case, in addition to finding a spectral window, the absorption feature therein has to be chosen by having a maximized temperature dependence. One of the major advantages of the laser is that due to the high spectral resolution, the spectral windows can be very narrow.

The quantity measured in an absorption experiment is the spectral transmission T. Absorption A can be only measured directly by a few specialized methods. These quantities are defined as:

$$T(\nu) = \frac{I(\nu)}{I_0(\nu)}; \quad A(\nu) = \frac{I_0(\nu) - I(\nu)}{I_0(\nu)}; \quad A(\nu) = 1 - T(\nu) \tag{13.1}$$

with: I_0 the initial light intensity and $I(\nu)$ the light intensity behind the absorbing medium.

The macroscopic description of the absorption process is realized by Lambert–Beers law:

$$I(\nu, z) = I_0 \cdot exp(-k(\nu) \cdot z) \tag{13.2}$$

with: $k(\nu)$ the linear absorption coefficient and z the thickness of the absorbing medium. A homogeneous medium (in partial pressure, total pressure and temperature) and a monochromatic light source is assumed. Another commonly used quantity is the spectral absorption cross section $\sigma(\nu)$. In the picture of geometrical optics σ is a measure of the area blocked by each molecular absorber. σ is related to the linear absorption coefficient k(ν) by:

$$\sigma(\nu) = \frac{k(\nu)}{c}; \quad c = \frac{\#absorbers}{unitvolume} \tag{13.3}$$

For high incident power levels the absorption will not be linear any more. The linearity will only be maintained as long as the initial number of absorbers in the ground state is not changed by the incident light. As life times of the excited states increase dramatically with increasing wavelength, care has to be taken of non–linearities especially towards the mid and far infrared spectral region or at high laser power densities.

13.2 Line spectra

The quantum mechanical description of atoms and molecules explains the existence of discrete energy levels. This quantization holds not only for electronic transitions (e.g. in atoms) but also for the excitation of vibrational and rotational transitions in molecules. If the temperature and the pressure are not too high this will result in absorption spectra consisting of separated lines. These lines can be characterized by four properties:

1. Spectral position
2. Shape
3. Width
4. Line strength

While position and strength are very much intrinsic properties of the molecule itself, width and shape are very strongly influenced by the physical boundary conditions such as pressure, temperature and chemical composition of the gas mixture.

13.2.1 Position

The theoretical calculation of absolute line positions is still prone to large errors, particularly for molecules, so that it is best to rely on experimental data. What can be calculated accurately are the relative positions of molecular lines with quantum mechanical descriptions of models of the rigid rotor and harmonic oscillator. The level separations for the quantum mechanical rotor and oscillator are:

$$E_{rot} = \frac{1}{2I} J(J+1)h^2 \tag{13.4}$$

with J = rotational quantum number and I = moment of inertia of the molecule; and including corrections for deviations from rigidity:

$$\begin{aligned} E_{rot} &= B_\nu J(J+1) - D_\nu J^2(J+1)^2 \quad \text{with} \\ B_\nu &= \frac{h^2}{2I \cdot 100ch}; \quad D_\nu = \frac{4B^3_{\nu=0}}{\omega_e^2} \quad \text{and} \end{aligned} \tag{13.5}$$

$$E_{vib} = h\nu_0(v + \frac{1}{2}) \tag{13.6}$$

with v = vibrational quantum number, ν_0 = classical vibrational frequency, c = velocity of light; and including anharmonic corrections:

$$\begin{aligned} E_{vib} &= \omega_e\left(v + \frac{1}{2}\right) - X_e\omega_e\left(v + \frac{1}{2}\right)^2 \\ \omega_e &= \frac{\nu_0}{100c}; \quad X_e \approx +0.01 \end{aligned} \tag{13.7}$$

(For more details see books mentioned under general reading in the reference section to this chapter.)

13.2.2 Shape and width

As mentioned above, the shape and width of an absorption line changes substantially with changing boundary conditions. Increasing pressure tends to increase the line width and therein limits the spectroscopic resolution which can be achieved. The theoretical treatment of isolated absorption lines with center frequency ν_0 can be simplified by splitting up the linear absorption coefficient $k(\nu)$ in a product of a frequency independent line strength factor S and a line shape function $\phi(\nu - \nu_0)$

$$\begin{aligned} k(\nu) = S \cdot \Phi(\nu - \nu_0); \quad \int_{-\infty}^{\infty} \Phi(\nu - \nu_0) d\nu = 1, \quad \text{with} \\ S = \int_{-\infty}^{\infty} k(\nu) d\nu \end{aligned} \tag{13.8}$$

centered on ν_0 the line center frequency. The area under ϕ is normalized to unity (In most cases! In some papers ϕ is normalized to $\sqrt{\pi}$ leading to confusing appearances of this factor). This normalization explains the alternative definition of S as an integrated absorption coefficient. With ν in units of wave numbers (cm^{-1}) and k in units of cm^{-1}, ϕ has to have units of cm and therefore S is in cm^{-2}. The introduction of the line strength S allows one to treat the broadening effects separately via the dedicated line shape function ϕ and its width γ. In this text γ is always used as half width at half maximum (HWHM). Below the three basic broadening mechanisms, their shape functions and width γ will be introduced and explained.

A. Natural line broadening

The finite lifetime of all excited states in atoms or molecules results in radiative line broadening. This can be understood by using Heisenbergs uncertainty principle $\Delta E \cdot \Delta t \cong \hbar$ to relate the uncertainty in energy of an excited state to its finite lifetime τ_j. An absorption process accessing such a "widened" energy state will therefore show a distribution of transition frequencies with a width γ_{nat} related to the life times of the connected states.

$$\frac{1}{\tau_{21}} = \frac{1}{\tau_1} + \frac{1}{\tau_2} \rightarrow \frac{1}{\pi \cdot \tau_{21}} = \gamma_{nat}[Hz] \rightarrow$$

$$\gamma_{nat} = \frac{A_{21}}{\pi} = \frac{16}{3} \cdot \frac{\pi^3}{\varepsilon_0 h c^3} \cdot \frac{g_{initial}}{g_{final}} |R_{21}|^2 \, {\nu_{21}}^3 \tag{13.9}$$

Using the Einstein coefficient for spontaneous emission A_{21}, which has a cubic dependency on the transition frequency ν_{21}, to express the lifetime τ_{21} it becomes obvious why we find such a dramatic increase in the natural line width moving from the far infrared to the ultraviolet. The shape function (s. Table 13.2) used to describe radiative broadening is called Lorentzian. It can be derived from classical electrodynamics using a damped oscillator model as well as from quantum mechanics. A special feature of a Lorentzian is its slowly falling wings, which reach the 1% level of the peak height after nearly 10 half widths (compare to Doppler broadening).

B. Doppler broadening

In a gas in thermal equilibrium at temperature T the thermal motion of the absorbers with mass $m(kg)$ respectively $M(AMU)$ is described by a Maxwell–Boltzmann distribution. At room temperature the most probable speed $v_p = (2kT/m)^{1/2}$ is in the order of 500 m/sec. The Doppler effect predicts a velocity dependent shift in transition frequency caused by the moving absorber, so that any molecule seems to have a different absorption frequency in the inertial system of the light source. This frequency distribution has the same shape as the Maxwell–Boltzmann distribution and can be described by a Gaussian shape function with a width $\gamma_{Doppler}$ (the shape function is shown in Table 13.2):

$$\gamma_{Doppler} = \left(\frac{\nu_0}{c}\right)\sqrt{\frac{2kTln2}{m}} = 3.581 \cdot 10^{-7}\nu_0\sqrt{\frac{T}{M}} \tag{13.10}$$

Compared to the Lorentzian the Gaussian wings are much steeper and reach the 1% level of the peak height already at 2.6 $\gamma_{Doppler}$. For clarification table 13.1 shows some examples of Doppler line widths for several molecules at different wavelengths.

Table 13.1. The typical lifetimes of electric dipole transitions and the corresponding natural, Doppler–and collisional line widths using the $\nu^3 - law$ are shown. The long lifetimes and the narrow natural and Doppler widths for microwave (MW)– and far infrared (FIR)–transitions are usually not accessible because of the onset of saturation broadening at already low incident power levels and the dominance of collisional effects.

transition type	spectral range	wave-length μm	lifetime typical sec	γ_{nat} cm^{-1}	γ_{Dop} $T = 300\ K$ $M = 30\ amu$	$\gamma_{coll}/\gamma_{Dop}$ @1 bar ~ 0.1 cm^{-1}	energy/kT $T = 300\ K$
rotational	MW	1000	10^{+3}	1.06E-14	1.13E-5	8830.711	0.048
	FIR	100	10^{-1}	1.06E-11	1.13E-4	883.071	0.480
vibrational	MIR	10	10^{-1}	1.06E-08	1.13E-3	88.307	4.796
	NIR	1	10^{-7}	1.06E-05	1.13E-2	8.831	47.960
electronic	Vis	0.55	10^{-7}	6.38E-05	2.06E-2	4.857	87.199
	UV	0.1	10^{-9}	1.06E-02	1.13E-1	0.883	479595

C. Collisional broadening

Perturbations by the surrounding medium during the interaction between photon and absorber can have the same consequence as the shortening of the lifetime of an excited state and therefore also cause a broadening of the transition. However, this mechanism depends very much on the type and strength of the molecular interaction (short/long range forces) and the quantum state of both perturber and absorber, so that all quantities describing these effects become functions of these parameters. A complete and exact analytical description of collisional broadening with one single line shape function has not been possible up to now. Theoretical treatments have been developed first by Lorentz [223], then by Van Vleck and Weiskopf [247] and Herbert [213]. An extensive description of the subject can be found in Sobel'man et. al [243]. In the simplest but nevertheless commonly used form by Lorentz this broadening is described by the Lorentz shape function, which is identical to the one describing the radiative broadening (s. Table 13.2). The pressure and temperature dependence of the half width can be empirically expressed as:

$$\gamma_{Lorenz} = \gamma_{0L}\left(\frac{p}{p_0}\right)\left(\frac{T_0}{T}\right)^s = \gamma_L^0 \cdot p \cdot \left(\frac{T_0}{T}\right)^s \tag{13.11}$$

with p_0, T_0 standard p, T actual pressure and temperature respectively. ${\gamma_0}_L$ denotes the half width at 1 atm and is often combined with p_0 to the pressure broadening coefficient γ_L^0. The temperature dependence does not necessarily have to be a power law, but it most frequently works out to be sufficient with $S =$ 0.5. To take into account the different collision partners separate coefficients are introduced for *foreign gas broadening* (perturber and absorber are different species) and *self–broadening* (perturber and absorber are the same species) and combined to give.

$$\gamma_{Lorenz} = \left(\gamma^0_{L_{self}} \cdot p_{self} + \gamma^0_{L_{foreign}} \cdot p_{foreign}\right)\left(\frac{T_0}{T}\right)^s \tag{13.12}$$

The self–broadening coefficient is quite high for molecules with a high permanent dipole moment ($NH_3 : 1\ cm^{-1}atm^{-1}$) whereas for light, homonuclear molecules it can be very low ($H_2 < 0.01\ cm^{-1}atm^{-1}$). Compared to self broadening, foreign gas broadening is usually considerably smaller. As an example H_2S with N_2 as perturber has a high foreign broadening coefficient with 0.15 $cm^{-1}atm^{-1}$.

D. Voigt function–Mixed line shapes

At intermediate pressures, when none of the above mentioned mechanisms dominates or in high resolution experiments, neither a Lorentzian nor a Gaussian profile is adequate for a precise description. The experimental line shape is then described by a convolution of Gauss and Lorentz profiles, the so called Voigt function g_{Voigt}. Unfortunately the Voigt shape function cannot be expressed in an analytical form, however, it can be gained via approximate methods as developed by Pierlussi [231] or Humlicek [215] or via tabulated values of the Voigt function, e.g. by Abramovitz and Stegun [397] whose tabulated function $V(x, y)$ is related to the shape function through

$$g_{Voigt}(\nu) = \frac{V(x,y)}{\gamma_{ED}\sqrt{\pi}} \quad \text{with}$$

$$x = \frac{(\nu - \nu_0)}{\gamma_{ED}}; \quad y = \frac{\gamma_{Lorentz}}{\gamma_{ED}}; \quad z = \frac{\gamma_{Doppler}}{\sqrt{ln2}} \tag{13.13}$$

where y, often denoted "a", is a measure of the Doppler content and x is the normalized frequency separation from the line center. $V(x, y)$ describes for $y = 0$ a Gaussian and for large y a Lorentzian profile. In general Voigt line profiles should be used if the points of interest are more than a few line widths (γ_{ED}) away from the line center or if the collisional width is not more than three times larger than the Doppler width (γ_{ED}) that is if $y < 3$ or $x > 5$. The Voigt function is Lorentz–like in the wings, while in the central part the profile it is predominantly Gaussian. For comparison Fig. 13.1 depicts a Lorentzian, a Gaussian and a Voigt shape function all normalized to unit area and plotted on a relative frequency scale x.

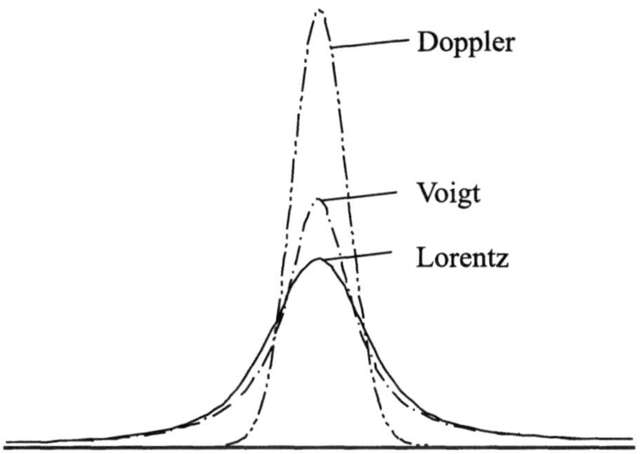

Fig. 13.1. Line shape functions for a Gaussian, a Lorentzian and a Voigt profile (with $y = 1$). All profiles normalized to unit area and plotted on a relative frequency scale.

Fast approximations for the line center absorption coefficient, $k(\nu_0)$ or the line width γ_{Voigt} (HWHM) can be made with the equations shown by Olivero and Longbothum [227].

$$k(\nu_0) \cong S\left[\left(\frac{\beta}{\gamma_{ED}\sqrt{\pi}}\right) + \left(\frac{1-\beta}{\pi \cdot \gamma_L}\right)\right] \quad \text{with}$$

$$\beta = \frac{\gamma_{ED}}{(\gamma_L + \gamma_{ED})}; \quad \gamma_{ED} = \frac{\gamma_{Doppler}}{\sqrt{ln 2}} \quad \text{and} \tag{13.14}$$

$$\gamma_{Voigt} = 0.5346 \cdot \gamma_{Lorentz} + \sqrt{\left(0.2166\gamma_{Lorentz}^2 + \gamma_{Doppler}^2\right)} \tag{13.15}$$

While the Voigt function might suit most applications, deviations can be found in high precision or high resolution experiments achievable with tunable diode laser spectroscopy. Here the Voigt profile reveals inaccuracies of the order of -10% in peak height and up to +1% in width. In addition line shifts of the order of 10^{-3} $cm^{-1}atm^{-1}$ can be measured, so that more advanced models have to be used. Whereas only state perturbing collisions – leading to collisional broadening – have been included so far these new models take into account influences of velocity changing elastic collisions on the profile. Depending on the assumed absence or presence of a correlation between the different collision types these models use so called Galatry profiles G or generalized Galatry profiles denoted H to describe the line shape [207]. G can describe the narrowing of a Doppler broadened line under increasing pressure also known as Dicke narrowing [197]. This phenomenon can be roughly understood on the basis that the Doppler width reflects the molecular velocity distribution. If the collision rate (pressure) is high enough to change the velocity of the absorber several times during the absorption process, then the Doppler effect monitors only the average velocity, which tends towards zero for increasing collision frequency. If not hidden by excessive collisional broadening, Dicke narrowing can be measured in several gases and can create sub Doppler line widths [229,232]. With the generalized profile H line narrowing and line shifts can be modeled. Since neither G nor H can be expressed analytically the reader is referred to the literature [228,248] for approximations or tables.

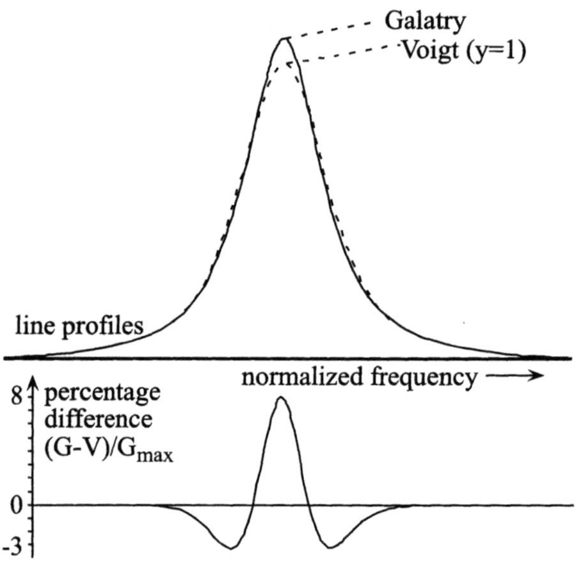

Fig. 13.2. Top: Comparison of normalized Voigt and Galatry profiles; Bottom: Percentage difference between Galatry and Voigt shape function normalized to the peak value of the Galatry profile.

In Fig. 13.2, normalized Voigt and Galatry profiles are shown as an example. The characteristic frequency dependence of the relative difference (Fig. 13.2, bottom) can serve as a good indication when to apply Galatry instead of Voigt line shape functions.

13.2.3 Line strength

The efficiency of the absorption of light in a gaseous medium is characterized by the wave number dependent linear absorption coefficient $k(\nu)$. As

Table 13.2. Analytical expressions for the different line shape functions ϕ_i, the line width γ_i (HWHM $[cm^{-1}]$) and the peak height of the linear absorption coefficient k(ν_0)

Cause		Shape function	Line width γ_i HMHW $[cm^{-1}]$	Peak height k (ν_0)
Radiative Broadening	Lorentz	$\frac{\gamma_{nat}/\pi}{(\nu-\nu_0)^2+(\gamma_{nat})^2}$	$\frac{16\pi^3 g_n}{3\varepsilon_0 h g_m}\|R_{nm}\|^2\nu_{nm}^3$	$\frac{S}{\gamma_{nat}\cdot\pi}$
Doppler Broadening	Gauss	$\frac{\sqrt{ln2}}{\gamma_D\sqrt{2}}\cdot exp\left[\frac{-ln2(\nu-\nu_0)^2}{\gamma_D^2}\right]$	$\frac{\nu_0}{c}\sqrt{\frac{2kTln2}{m}}$	$\frac{S}{\gamma_D}\sqrt{\frac{ln2}{\pi}}$
Pressure or Collisional Broadening	Lorentz	$\frac{\gamma_L/\pi}{(\nu-\nu_0)^2+(\gamma_L)^2}$	$\gamma_L^2\cdot p\cdot\left(\frac{T_0}{T}\right)^S$	$\frac{S}{\gamma_L\cdot\pi}$
Pressure and Doppler Broadening at Intermediate Pressure	Voigt	$\frac{\gamma_{ED}^{-1}}{\sqrt{\pi^3}}\int_{-\infty}^{\infty}\left[\frac{y\cdot exp\left(-t^2\right)}{y^2+(x-t)^2}\right]\partial t$	$\cong\frac{0.5346\gamma_L}{\sqrt{0.2166\gamma_L^2+\gamma_D^2}}+$	$\cong S\left[\frac{1-\beta}{\pi\cdot\gamma_L}+\frac{\beta}{\sqrt{\pi}\cdot\gamma_{ED}}\right]$

mentioned before when dealing with line spectra, it proves very useful to introduce a line strength S defined as:

$$S=\int_{-\infty}^{\infty}k(\nu)d\nu \tag{13.16}$$

$$k(\nu)=S\cdot\phi(\nu-\nu_0);\quad \int_{-\infty}^{\infty}\phi(\nu-\nu_0)d\nu=1 \tag{13.17}$$

The area underneath ϕ is normalized to unity, so that the line strength S can also be expressed through $k(\nu)$, which explains why S is also called the integrated absorption coefficient. With these expressions the Lambert–Beer law can be written as:

$$I(\nu-\nu_0,z)=I_0\cdot exp\left[-S\cdot\phi\left(\nu-\nu_0\right)\cdot z\right];\quad [S]=\left[cm^{-2}\right] \tag{13.18}$$

Unfortunately the designation *line strength* is used for various quantities derived from the integrated absorption coefficient. This can cause a substantial amount of confusion very easily. For various applications S (in units of cm^{-2}) is normalized to the pressure p, the concentration of absorbers c or other quantities. The result are sometimes misleading expressions, which give S in different units, e.g. by:

$$\frac{S}{p}=S' \text{ to set to } \left[\frac{cm^{-2}}{atm}\right] \text{ or} \tag{13.19}$$

$$\frac{S}{c}=S' \text{ to set to } \left[\frac{cm^{-2}}{\frac{molecules}{cm^{-3}}}=\frac{cm}{molecules}\right] \tag{13.20}$$

Conversely S is also related to the dipole transition element $|R_{21}|^2$ introduced in quantum mechanics, the Einstein coefficient for spontaneous Emission A_{21}, or the oscillator strength f_{12} known from classical electro–dynamics, which defines the number of classical oscillators needed to produce the equivalent absorption found for the transition investigated. The numerical and analytical conversion expressions between some of these quantities are shown in Table 13.3. The table is calculated in SI system with N_1 denoting the concentration of absorbers in $[m^{-3}]$, g_n the degeneracy of the initial state and ν the frequency of the transition in $[Hz]$ – not in $[cm^{-1}]$! The table is always read from the left:

$$S_{12} = \frac{e^2}{4\epsilon_0 m_e c} N_1 \cdot f_{12} \quad \text{or} \quad f_{12} = \frac{3.773 \cdot 10^{-5}}{N_1} \cdot S_{12}$$

Table 13.3. Numerical and analytical conversion factors between commonly used quantities to express the "line strength" of a transition.

	S_{12} $[m^{-1}]$	f_{12} $[-]$	A_{21} $[s^{-1}]$	$\lvert R_{12}\rvert^2$ $[cm^2]$
S_{12}	1	$\frac{e^2}{4\varepsilon_0 m_e c} N_1$	$\frac{N_1 g_2}{g_1 \nu^2} \cdot \frac{c^2}{8\pi}$	$\frac{N_1}{g_1} \cdot \frac{2\pi^2}{3\varepsilon_0 ch} \nu$
f_{12}	$\frac{3.773 \cdot 10^5}{N_1}$	1	$\frac{g_2}{g_1 \nu^2} \cdot \frac{4 m \varepsilon_0 c^3}{8\pi e^2}$	$\frac{1}{g_1} \cdot \frac{8\pi^2 m_e}{3e^2 h} \nu$
A_{21}	$\frac{g_1 \nu^2}{N_1 g_1} 2.8 \cdot 10^{-16}$	$\frac{g_1 \nu^2}{g_2} 7.42 \cdot 10^{-22}$	1	$\frac{1}{g_2} \cdot \frac{16\pi^3}{3\varepsilon_0 c^3 h} \nu^3$
$\lvert R_{12}\rvert$	$\frac{g_1}{N_1 \nu} 2.673 \cdot 10^{-37}$	$\frac{g_1}{\nu} 7.095 \cdot 10^{-43}$	$\frac{\nu^3}{g_2} 1.05 \cdot 10^{-22}$	1

After these frequently confused relations between different expressions for the line strength, which mostly stem from experimental procedures, we finally discuss the dependencies of S in a microscopic picture. To calculate a rovibrationally resolved electronic transition we start from the quantum mechanical model of the molecule based on nuclear and electronic wave functions Ψ which lead us to discrete energy levels E_m.

With the Boltzmann distribution law we determine by the equation below the fractional population F_m; that means the number of absorbers N_m out of an ensemble of N in the initial energy level m and also introduce the partition function Q:

$$F_m = \frac{N_m}{\sum\limits_j N_j} = \frac{g_m \; e^{(-E_m/kT)}}{\sum\limits_j g_j \; e^{(-E_j/kT)}} = \frac{g_m}{Q} \; e^{(-E_m/kT)} \tag{13.21}$$

$$Q = \sum_j g_j \; e^{(-E_m/kT)} \tag{13.22}$$

To compute the complete expression we therefore have to know all energy levels E_i (at least the thermally accessible ones) and their degeneracy g_i. For a molecule this is further complicated by the fact that electronic, vibrational and rotational excitations are possible and that the nuclear spin has to be included. Fortunately, because of the large relative differences in energy and in the characteristic time scales for these excitations, the wave function $\Psi_{total} = \Psi_{electr} \cdot \Psi_{vib} \cdot \Psi_{rot}$ and the total fractional population F can be separated and expressed as a product $F = F_{electr} \cdot F_{vib} \cdot F_{rot}$. This is also known as the Born–Oppenheimer approximation. The Q_i for vibrational and rotational transitions are:

$$Q_{vib_{tot}} = \prod_{J=1}^{3N-6} \sum_n e^{(-nhc\nu_j/kT)} = \prod_{J=1}^{3N-6} \frac{1}{1 - e^{(-hc\nu_j/kT)}} \tag{13.23}$$

$$Q_{rot_{tot}} = \sum_j (2J+1) e^{(E_j/kT)} = \frac{2IkT}{h^2\sigma} = \frac{kT}{B\sigma} \tag{13.24}$$

$$Q_{rot_{tot}} = \frac{1}{\sigma}\sqrt{\frac{\pi(kT)^3}{ABC}} \tag{13.25}$$

They are gained under the assumption of describing the vibration of a molecule consisting of N atoms by $3N-6$ (if $N \geq 3$) independent harmonic oscillators with Eigenfrequencies $\nu_i(cm^{-1})$ ($3N-5$ if $N=2$). The total vibrational partition function $Q_{vib_{tot}}$ is therefore a product of $Q(\nu_i)$ for each normal mode j. The rotational partition function is derived from a rigid rotor model with three different moments of inertia I_a, I_b, I_c and rotational constants $A, B, C = h^2/2I_{a,b,c}$. σ is a symmetry factor [255] and the term $(2J+1)$ stems from the degeneracy of each rotational level with quantum number J. The last term to evaluate the line strength is the electric dipole transition moment $|R_{21}|^2$. For simplification $|R_{21}|^2$ is calculated with fixed nuclei positions ignoring vibration and rotation and is associated with the electronic transition only. However, because of the slow movement of the nuclei compared to the electron movement, the effect of vibration and rotation is more a distribution of the transition moment over a large number of rovibrational sublevels than a change in its strength. Applying the Born–Oppenheimer approximation to factorize the total wave function we can split $|R_{21}|^2$ in three terms, each of which depends only on the quantum number and the spatial coordinates associated with the different transitions.

$$\left|R^{12}_{rot,vib}\right|^2 = \left|R^{12}_{electr}\right|^2 \cdot \left|R^{\nu_{initial}}_{\nu_{final}}\right|^2 \cdot \left|R^{J_{initial}}_{J_{final}}\right|^2 \quad \text{with} \tag{13.26}$$

$$\left|R^{12}_{electr}\right|^2 = \left|\int \Psi^{*final}_{electr} \mu \Psi^{initial}_{electr} dr_{electr}\right|^2 \tag{13.27}$$

$$\left|R^{\nu_{initial}}_{\nu_{final}}\right|^2 = \left|\int \Psi^{*final}_{\nu ib} \Psi^{initial}_{\nu ib} dr_{\nu ib}\right|^2 \quad \text{Frank} - \text{Condonfactor} \tag{13.28}$$

$$\left|R_{J_{final}}^{J_{initial}}\right|^2 = \left|\int \Psi_{rot}^{*final} \Psi_{rot}^{initial} dr_{rot}\right|^2 \quad \text{Hönl - Londonfactor} \tag{13.29}$$

for a vibrational transition within an electronic level

$$\left|R_{rot,\nu ib}^{12}\right|^2 = \left|R_{\nu ib}^{12}\right|^2 \left|R_{J_{final}}^{J_{initial}}\right|^2 \tag{13.30}$$

$$\left|R_{\nu ib}^{12}\right|^2 = \left|\int \Psi_{\nu ib}^{*final} \mu \, \Psi_{\nu ib}^{initial} dr_{electr}\right|^2 \tag{13.31}$$

$$\left|R_{J_{final}}^{J_{initial}}\right|^2 = \left|\int \Psi_{rot}^{*final} \mu \, \Psi_{rot}^{initial} dr_{rot}\right|^2 \quad \text{Hönl - Londonfactor} \tag{13.32}$$

The Hönl–London and the Franck–Condon factor can be evaluated if the vibrational and rotational wave functions are known. Since both factors are the squares of integrals evaluating the spatial overlap between the initial and the final wave function, the total transition probability will be higher the less the distribution of the nuclei is changed during the transition. For light molecules or high J values deviations caused by the non–rigidity of the molecule have to be included. This is often done by a product term to the Franck–Condon factor called Hermann–Wallis factor $F_{\nu r}$, which for linear molecules can be expressed as: $F_{\nu r}(m) = 1 + am + bm^2....$ where $m = -J^{final}$ for the P–Branch ($\Delta J = -1$), $m = J^{final} + 1$ for the R–Branch ($\Delta J = +1$) and a, b are molecular constants. So finally we can express the rovibrational resolved electronic transition strength S as:

$$S = \frac{8\pi^3}{4\pi\varepsilon_0 3hcQ_{\nu ib}(T)Q_{rot}(T)} \nu \cdot e^{(-hc\nu/kT)} \cdot \left(1 - e^{(-hc\nu/kT)}\right) \cdot \tag{13.33}$$
$$\left|R_{electr}^{12}\right|^2 \cdot \left|R_{\nu_{final}}^{\nu_{initial}}\right|^2 \cdot F(m)(2J^{final} + 1) \cdot \left|R_{J_{final}}^{J_{initial}}\right|^2$$

from which we assume Q_{electr} to be unity, which is true for most molecules except for radicals like NO, which can have very low lying electronic energy levels. Two terms have to be explained: the $(2J + 1)$ term corrects the degeneracy in the final rotational level, while the $(1 - e^x)$ term corrects reduced absorption due to stimulated emission. The influences of nuclear spin statistics have been omitted. The reader is referred to standard literature. To compare line strengths measured at temperature T with values referenced to a temperature T_0 or vice versa we can simplify the above equation under omission of the stimulated emission term to:

$$S = S(T_0) \left(\frac{Q_{\nu ib}(T_0) \cdot Q_{rot}(T_0)}{Q_{\nu ib}(T) \cdot Q_{rot}(T)}\right) \cdot exp\left(\frac{h\nu}{k}\frac{T - T_0}{TT_0}\right) \tag{13.34}$$

13.3 Experimental techniques

13.3.1 Overview

In preparing and performing an absorption experiment the first question is "Where are the spectral and spatial absorption regions of the species under investigation?" Since the experimental spectral data on absolute line positions and strengths are still much more accurate than the theoretical calculations it is best to check specific journals such as Journal of Molecular Spectroscopy, Journal on Quantum Spectroscopy and Radiative Transfer, Journal of Chemical Physics, Applied Optics, Applied Physics and others. A short introduction to the nomenclature and a substantial set of spectroscopic and molecular data can be found in [234]. Another possibility is to try to get access to line data banks like JPL [230,233], GEISA [217,218], or HITRAN [238,239] which provide extremely extensive data sets (with several million transitions) for most atmospheric majority species and trace gases in the spectral range from the microwave region to the UV. For CO_2, H_2O and others there is even a high temperature version of HITRAN called HITEMP in preparation [239]. Some of the data sets–sometimes including software to calculate synthetic spectra–are available on CD–ROM or via the internet. The next important decision to be made is the choice of the light source. Since the advent of the laser the use of broadband sources is in many cases only acceptable for industrial tasks where the financial aspects are dominating over ultimate sensitivity or resolution. The high spectral power density of the laser dramatically reduces the problem of detector noise, one of the major limits of working with broadband sources and monochromers, rendering it to a more minor difficulty.

The optimum instrument for high sensitivity and high resolution studies during the last one and a half decades remains the lead salt diode laser. Its emission wavelength matches the region of the strongest molecular absorption lines or the so called "Finger print region" from $3 \sim 20\ \mu m$ corresponding to fundamental rovibrational transitions. The combination of the high spectral resolution ($10^{-4}\ cm^{-1}$ laser line width) and the low Doppler width of the IR–absorption lines results in very high sensitivity and specificity. A basic experimental set–up is shown in Fig. 13.3. It consist of a light source, absorbing region and detector arrangement. To apply the Lambert–Beer law the initial intensity I_0 has to be probed. In addition a relative and absolute frequency calibration is necessary. The first is performed with an etalon with a low free spectral range to produce a high resolution, periodic intensity change on a detector when the laser output wavelength is tuned. The absolute frequency is measured by a wavelength meter or a reference cell filled with a gas with known absolute transition frequencies. For high sensitivities, i.e. to measure absorption below 1%, a noise reduction technique is essential. Noise reduction can be achieved with four widespread techniques:

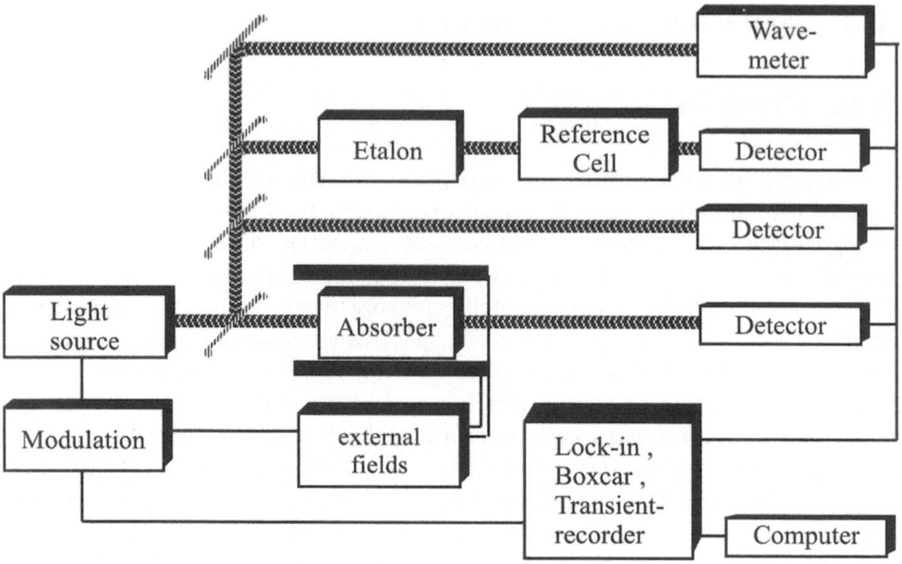

Fig. 13.3. Principal experimental set–up of an absorption experiment

- Amplitude or wavelength modulation with harmonic detection by lock–in amplifiers
- High frequency modulation and detection with a homodyne signal technique
- Sweep integration techniques by using transient recorders
- Time gated signal averaging with Boxcar or gated amplifiers

The techniques to be used depend on the tuning scheme of the experiment. The most simple technique for the necessary modulation for harmonic detection is amplitude modulation of the detector signal with a chopper (mechanical or optical). The modulation used in the harmonic detection scheme shifts the absorption signal from DC to higher harmonics of the modulation frequency where the signal can be demodulated by a lock–in amplifier. This acts as an extremely narrow electronic filter, which detects only the periodic signal content with the correct frequency and phase. This concept is also applicable to fixed frequency lasers. Compared to a measurement at a single wavelength, an improved noise reduction and much higher system stability can be expected if more spectral information is extracted by recovering the complete absorption line shape. One way to achieve this is to use a tunable laser, so that a periodic wavelength modulation combined with slow scanning of the wavelength is providing the amplitude modulation of the signal to use harmonic detection. As an alternative it is possible to modulate the position of the absorption line itself (!) by applying external electrical or magnetic fields, so that wavelength modulation and harmonic detection can be combined with fixed frequency lasers. Absorption on the order of 10^{-3} can be detected on a routine basis, while under optimum conditions less than 10^{-5} have been detected this way [200,202,235]. High frequency modulation techniques using modulation frequencies in the hundred MHz range are capable of shot noise limited detection and can reach sensitivities of 10^{-8} relative absorption [242] (and references therein) [251,252]. At these high frequencies the demodulation cannot be done by standard lock–in amplifiers any more, so that traditional mixers and other HF–components have to be used.

If wavelength tuning is possible an alternative to the harmonic technique is provided by sweep integration [193,194,199,203,204]. Direct absorption spectra obtained from successive scans are stored and added onto each with a transient recorder. This reduces the noise because the statistical noise amplitude in any wavelength channel only increases with the square root of the scan number while the signal increases linearly. About the same sensitivity as with the Lock–in technique can be achieved this way, but the feature to subtract reference spectra more easily opens up new possibilities in discriminating between interfering species. The same instrument also offers the possibility of resolving fast time dependent concentration changes by working on a "single shot" basis. Time resolutions of the order of 20 μsec can be achieved.

Pulsed lasers, like excimer or dye lasers, which are very frequently used for UV absorption measurements, are most often used with other noise reduction schemes due to their short pulse duration in the order of nanoseconds. This is the domain of time gated noise reduction techniques, where the spectroscopic signal is averaged over several pulses producing a similar noise reduction efficiency like the sweep integration technique [195].

Another application of time gated signal averaging is called integrative spectroscopy. There the laser is repeatedly scanned over one single isolated absorption line and the recorded absorption signal is integrated [198,201]. If a linear frequency scan can be assured the time integration of the signal corresponds to a frequency integration of the absorption. This value is then proportional to the line strength, or if this quantity is known the result is a direct measurement of the absolute concentration, if most of the absorption line can be covered.

$$\int_{\substack{\text{gate}\\\text{time}}} -ln\left(\frac{I(t)}{I_0}\right)\frac{d\nu}{dt}dt = \int_{\substack{\text{absorption}\\\text{line}}} -ln\left(\frac{I(\nu)}{I_0}\right)d\nu = S \cdot z \cdot p_{O_2} \quad (13.35)$$

This data processing scheme has several advantages:

- If the gate time is chosen correctly the concentration signal will be very insensitive to fluctuations of the total pressure.
- By repetitive averaging of the integrated signal an improvement in the S/N proportional to the square root of number of sampled gates can be achieved.
- If the gates are analyzed on a "single shot" basis, then the time resolution can be below 100 μs.

- If additional leading and trailing gates are also sampled, it is possible to correct for a linear sloping [209,236] and quadratic background and etalons [201].
- By placing two more gates symmetrically in the steepest parts of the absorption line signal an error signal for a line locking scheme [209] and a correction for pressure broadening [201] can be derived.

Techniques used in absorption spectroscopy

An overview of the different approaches to laser absorption spectroscopy is shown in Fig. 13.4. Explanations of these techniques will be given here but ample information can be found in the associated literature. For fixed frequency lasers there are two possibilities to adapt modulation techniques for noise reduction. Instead of tuning the laser the absorption line itself can be continuously tuned via the Zeemann effect, with an external magnetic field, so called **Laser magnetic resonance (LMR)** [196], or with an external electric field, known as **Stark spectroscopy** [246,254], if the absorbers have a significant magnetic or electric dipole moment [250]. LMR for instance allows very sensitive detection of radicals, since it discriminates between majority species, if they do not react on the external magnetic field. For lasers with discrete but very densely laying lines like CO_2, HF and CO a differential absorption technique has been developed. Here a pair of interference free laser lines have to be found in a way that one line is on resonance and the other is off resonance, so that the concentration of the absorbers can be measured by comparing the **differential absorption** on and off the absorption line. An example with a powerful CO_2 laser for the analysis of industrial stack gases is shown below.

A wide field of applications has been opened by the introduction of **tunable infrared diode lasers** which spurred the development of **modulation techniques** [193–195,235,242] because of their versatile tuning possibilities. Some examples for that will be given later on.

Finally there is a variety of **indirect absorption detection techniques** where the signal is not derived from photon detection but transferred to a quantity which can be measured with higher accuracy or/and better resolution. The absorption of photons and subsequential relaxation of the excited molecules induces a temperature change in a gaseous medium and hence a pressure change in a closed absorption cell that can be detected with a sensitive microphone. This **photo–acoustic effect** [214] combined with harmonic detection allows sensitivities in the sub–ppb range at pressures ranging from about $10^3 \sim 10^5$ Pa.

Opto–thermal spectroscopy [208] uses the long lifetimes of rovibrationally excited molecules and atoms in very low pressure media to detect the excitation. The energy stored in the absorber is thereby released on impact onto a liquid nitrogen or helium cooled target (bolometer) and the resulting temperature change of the target is measured depending upon the wavelength of the excitation. Combined with the strong temperature dependency of super conductivity, thermal powers of 10^{-12} W can be detected.

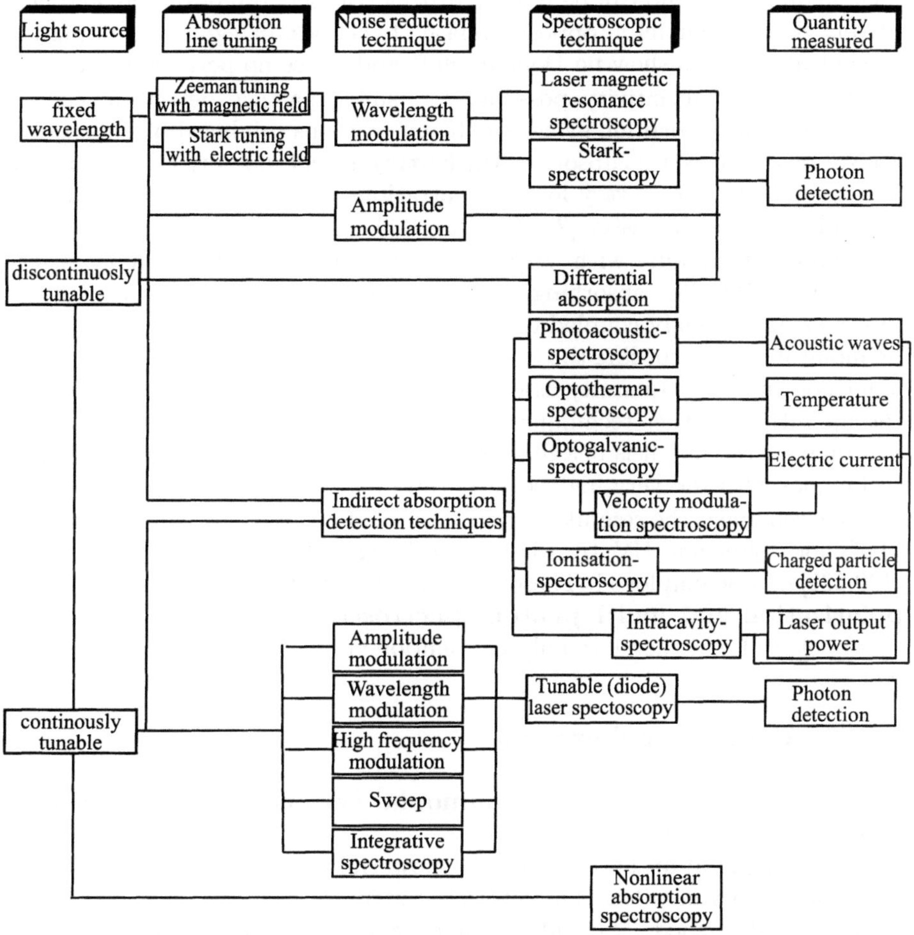

Fig. 13.4. Overview showing some of the different experimental techniques in laser absorption spectroscopy

A powerful method to analyze the chemistry of gas discharges and flames is the **opto–galvanic spectroscopy** [245], where the change of the discharge current is monitored while tuning the wavelength of light sent through the plasma. The absorption of the incoming light changes the population density of a certain state and thereby changes the ionization probability and the discharge current respectively. This method is especially valuable to investigate unstable, highly reactive or intermediate species, which can be permanently produced in the discharge. The selectivity of this method can be greatly enhanced by the introduction of the **velocity modulation spectroscopy** [210], where a periodic modulation of the discharge current is applied. The periodic Doppler shift caused by this velocity modulation is extracted with lock–in amplifiers and can be used to distinguish between positively and negatively charged particles (opposite phase of the lock–in signal), while signals of neutral particles show no Doppler shift and hence no periodicity at all.

The highest sensitivity possible, e.g. single particle detection, can be achieved if the absorption event is detected via ions produced by subsequent ionization of the excited absorber. **Ionization spectroscopy** is hence a valuable method to detect very low concentrations of radicals and is often used to analyze flame chemistry [211,216,221,253].

Another technique with very high sensitivities is **intra–cavity spectroscopy** [212]. The absorbing medium is placed inside the laser cavity or into a cavity strongly coupled to it. The result of an absorption process can be monitored by recording the laser output power with variable output wavelength or any other quantity associated to it. By combining the intra–cavity set–up with the other techniques like opto–galvanic, photo–acoustic or ionization spectroscopy it is possible to combine the specific advantages of these techniques with the very high sensitivities of the intra–cavity set–up.

To circumvent the resolution limit imposed by Doppler broadening particularly for visible and UV–transitions **nonlinear absorption spectroscopy** [220,222] can be employed. These very powerful techniques like **saturation, polarization and multi–photon spectroscopy** are beyond the scope of this book and will therefore only be mentioned here.

13.3.2 Experimental examples

A. In situ measurements of ammonia concentration in industrial combustion systems

As an example of an application of a powerful fixed frequency laser in the mid–infrared, an *in–situ* gas monitoring system to measure concentrations of ammonia is described. The addition of ammonia or other nitrogen containing chemicals to the flue gas is the most promising way to reduce the emission of nitric oxide (NO). NO can be converted to nitrogen and water either in a catalytic process or by direct reaction with the NH_2–radical. In both cases the consumption of ammonia has to be as complete as possible to avoid severe damage to the power plants by ammonia sulfates otherwise produced and emission of NO or NH_3. Thus a fast responding *in–situ* ammonia sensor is needed to optimize the NO removal and NH_3 consumption by an active process control system. The species concentration is determined by a differential absorption measurement with a stepwise tunable $^{13}CO_2$ wave guide laser. The laser can be rapidly switched between two adjacent emission lines by modulating the laser resonator length at 2 kHz using a piezo element. The high spectral resolution of the laser suppresses interferences by other species.

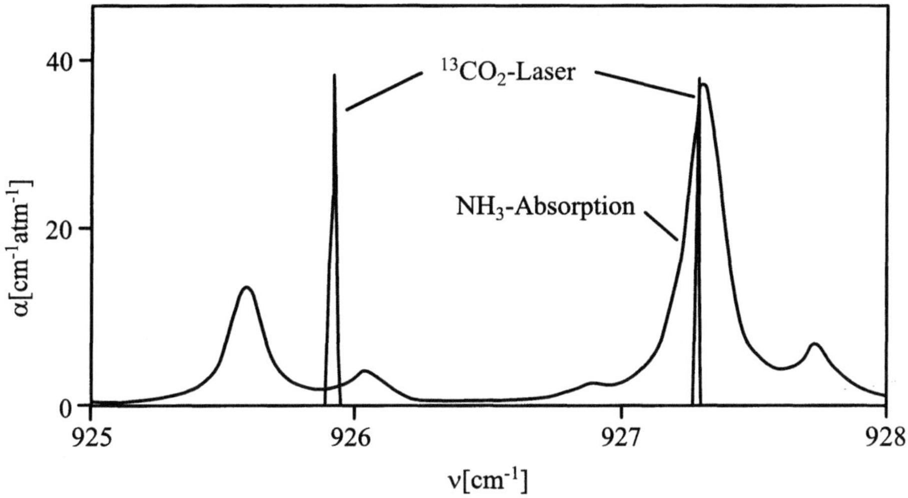

Fig. 13.5. Computer generated absorption spectrum between 925 cm^{-1} and 928 cm^{-1} of a mixture of 15% H_2O, 15% CO_2, 200 ppm N_2O and 1 ppm NH_3 diluted in pure nitrogen at STP

As depicted in Fig. 13.5 none of the two lines suffers from interference by water, while interference by $^{12}CO_2$, unavoidable in a combustion process, is suppressed completely by using an isotopic ^{13}C–based laser gas. This way one laser line measures the gas absorption, while the other is used as a reference to correct for dust induced transmission fluctuations. The $R(18)$ $^{13}CO_2$ laser line coincides with the strong a $Q(6,6)$ NH_3 absorption line, whereas the $R(16)$ laser line shows only small NH_3 absorption. The determination of species concentration by differential absorption techniques is based on the Lambert–Beer law which can be converted to:

$$c = \frac{-1}{\alpha_{NH_3} \cdot z} \; ln \left[\frac{I\left(\nu_1\right)}{I\left(\nu_2\right)} \cdot \frac{I_0\left(\nu_2\right)}{I_0\left(\nu_1\right)} \right] \tag{13.36}$$

where the concentration c only depends on the ratio of the initial intensities $I_0(\nu_1), I_0(\nu_2)$, the transmitted intensities $I(\nu_1), I(\nu_2)$, the absorption path length z and the absorption coefficient α_{NH_3}, which is temperature and pressure dependent. The high switching frequency ensures that the referencing is fast enough to completely compensate even very quick dust induced transmission fluctuations, which are common in *in–situ* applications, so that this system is robust enough to work even in rough industrial environments. The high average CO_2–laser power of about 1 W emitted in a single mode (compared e.g. to 100 μW of a IR–diode laser) enables long absorption paths even in dusty gases. Since only the peak absorption is measured with this device, it is necessary to experimentally determine the temperature dependence of the line strengths and the line broadening parameters to correct for line shape effects and to allow an application of this device at various gas temperatures. This was done at temperatures of up to 660 K with a tunable IR–diode laser system [226]. This Laser–In–Situ–Ammonia–sensor, LISA, which is commercially produced by ZEISS, Germany, could be successfully applied in the chemical industry, power plants and waste incinerators. A detection limit of 1 ppm NH_3 at atmospheric pressure could be reached [224] and *in–situ* path lengths of up to 10 m could be monitored even under severe dust loads. The experimental set–up is shown in Fig. 13.6.

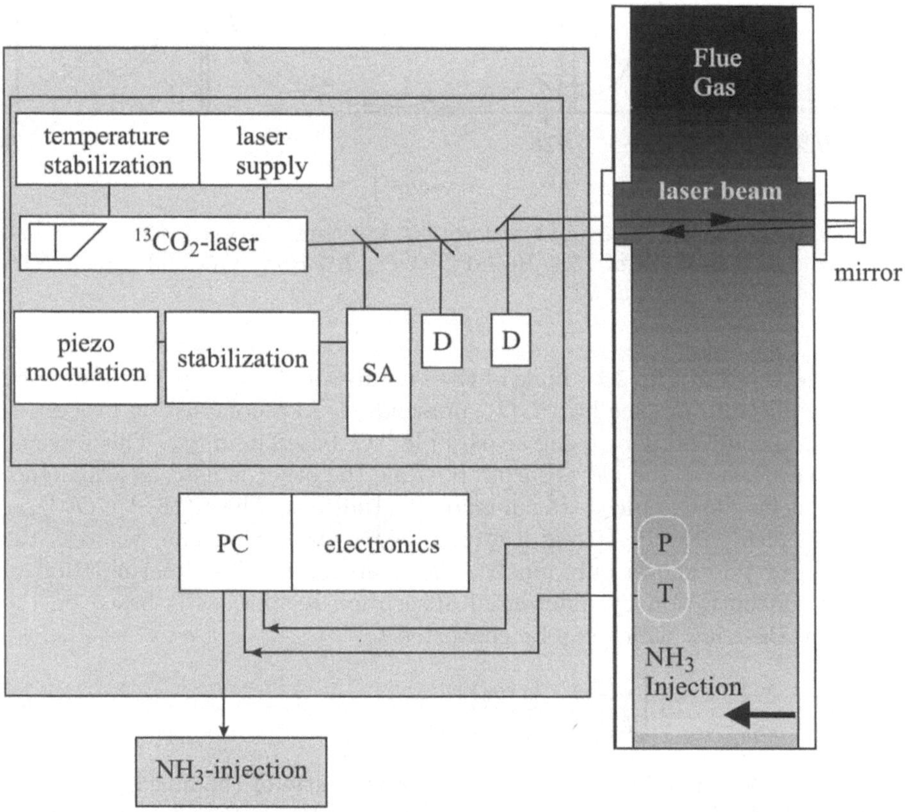

Fig. 13.6. Experimental set up of the Laser–In–Situ–Ammoniamonitor (LISA)

The $^{13}CO_2$ laser, a spectrum analyzer (SA), the detectors (D) and optical elements, such as mirrors (M) and beam splitter (BS) are mounted on a compact solid platform. The laser beam is split into three parts. One is directed into the spectrum analyzer, which is used to spectrally stabilize the laser. The second beam serves as an intensity reference yielding $I_0(\nu_1)/I_0(\nu_2)$ whereas the third beam is used as the *in–situ* "sensor" which probes the measurement volume in a double path configuration, being reflected by a mirror mounted on the opposite chimney wall. Regarding the optical set–up, special care had to be taken to avoid etalon effects, which strongly influence the intensity ratios. The pre–amplified detector signals are corrected for transmission fluctuations synchronously to the piezo modulation frequency by analog electronics. The final calculation of the absolute concentration is carried out by a computer.

In several applications in industry it was found that the distribution of ammonia in the flue gas channel is very inhomogeneous. This was the motivation for a two dimensional measurement with the LISA–system (Fig. 13.7).

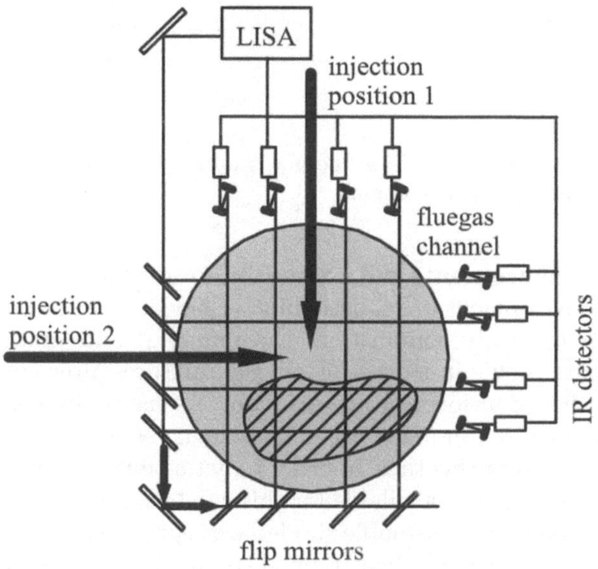

Fig. 13.7. Experimental set–up for two–dimensional NH_3–detection

The laser beam successively traverses the channel in eight different paths with the help of folding mirrors. The mechanism of the folding mirrors had to be very stable and robust to work in rough industrial environments. Thus the misalignment over long periods had to be better than 1 *mrad*. After laboratory tests a measurement was possible during reduction experiments in a coal burner. After a full cycle of absorption measurements in the different paths the ammonia distribution was calculated by the PC with a simple projection algorithm. The application of very complex and time intensive tomographic calculations is not necessary since only eight projections under two different angles have been used.

As shown in Fig. 13.8 the distribution of ammonia was very inhomogeneous and strongly dependent upon the geometrical configuration of the injection nozzles. It was a further proof that a single point measurement in the flue gas channel can be very misleading and is far from being representative.

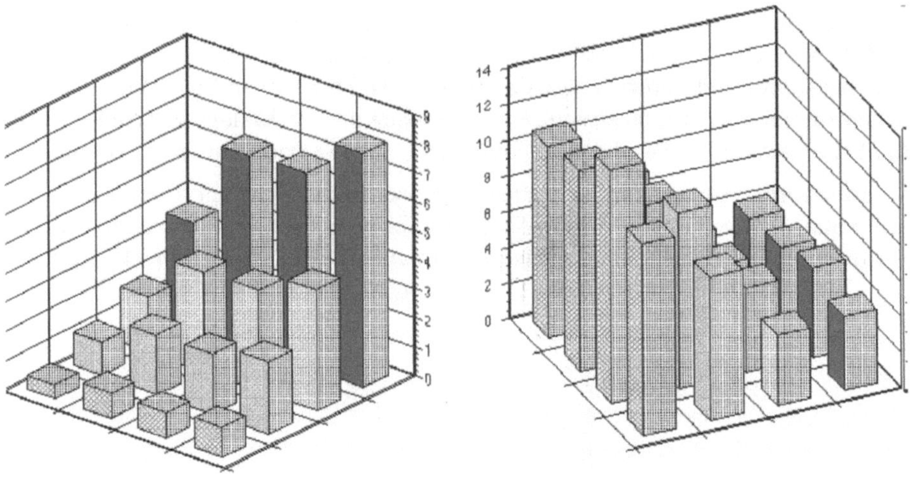

Fig. 13.8. Measured ammonia distribution in a flue gas channel during reduction experiments with different urea injection positions

B. Fast temperature measurements with tunable diode lasers

In order to support the development of realistic kinetic models for ignition and flame propagation in unsteady combustion, the temporal and spatial evolution of the temperature profile in methanol–oxygen mixtures after ignition can be measured by infrared absorption spectroscopy. As the combustion process of the mixture takes place in some ten milliseconds, some hundred spectra have to be recorded within this time in order to get a time resolution suitable to test the model calculations available for this combustion scenario. This difficult task can be solved with tunable diode laser spectroscopy. For this purpose the temperature dependent absorption of two rovibrational transitions of a single species have to be captured. The ratio of the logarithms of the transmissions T_i of the two suitable lines is independent of the species concentration:

$$\frac{ln(T_1)}{ln(T_2)} = \frac{c \cdot z \cdot \Phi_1 \cdot S_1(T)}{c \cdot z \cdot \Phi_2 \cdot S_2(T)} \approx \frac{S_1(T)}{S_2(T)} \tag{13.37}$$

and in good approximation equal to the ratio of the individual line strengths S_i, as the shape functions ϕ_i are very similar for transitions of the same species. Carbon monoxide as a natural component of most combustion systems is a good candidate for a temperature indicator since its absorption spectrum is easy to interpret, its spectral position at about 4.7 μm is isolated from that of other interfering molecules and furthermore mid IR diode lasers are available, which can be continuously tuned with very high speeds over a spectral range where several dozens of useful combinations of absorption lines are accessible. The temperature dependence of the line strengths of the rovibrational transitions of CO can be evaluated according to the relationships given in the theoretical section. For these measurements the lines $0P2$ ($v = 0,\ J = 2 \Rightarrow v = 1,\ J = 1,\ 2135.3131\ cm^{-1}$) and $1R4$ ($v = 1,\ J = 4 \Rightarrow v = 2,\ J = 5,\ 2135.5466\ cm^{-1}$) were chosen. The temperature dependence of the strengths of these lines and their ratio is shown in Fig.13.9.

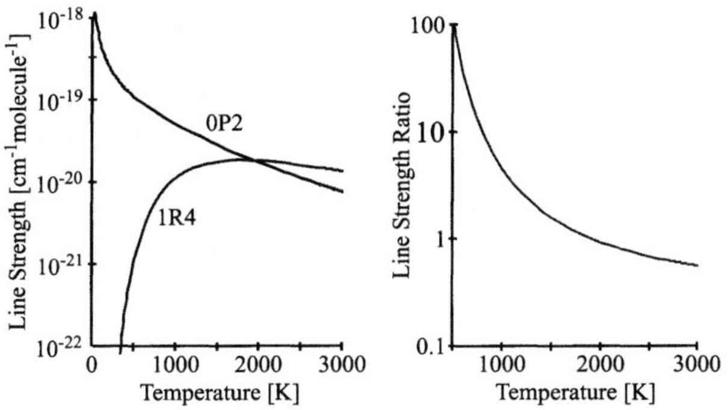

Fig. 13.9. Left: Line strengths of the CO absorption lines $0P2$ and $1R4$; Right: Ratio of the line strengths of the CO absorption lines $0P2$ and $1R4$

Due to the signal–to–noise ratio of the absorption measurement and the variation of the ratio with temperature, temperatures in the range from 800 K to 2800 K can be measured with this line–pair. The experimental set–up is shown in Fig.13.10.

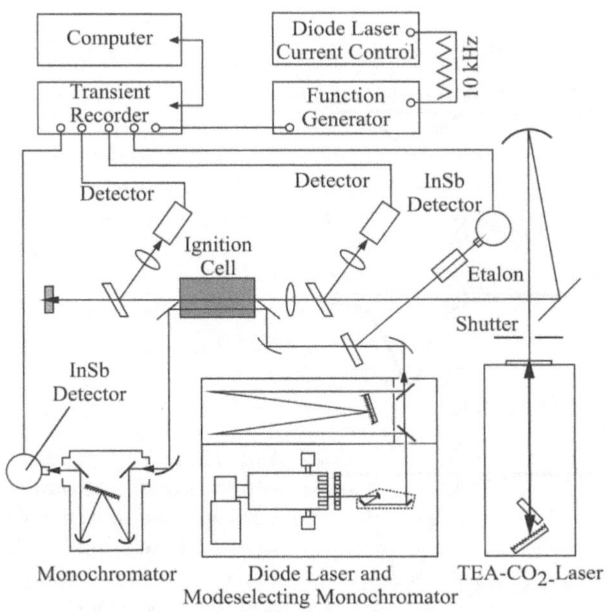

Fig. 13.10. Experimental set up for the investigation of CO_2 laser induced ignition by tunable diode laser spectroscopy

The ethanol–oxygen mixtures were premixed with a small amount of carbon monoxide added and stored in a reservoir from which the evacuated explosion cell could be filled up to the desired pressure, typically about 100 $mbar$. The ignition was initiated by a single pulse from a $TEA - CO_2$–laser, which was selected from the pulse train by an electronic shutter. The laser – operating at the $9P20$ line at $\lambda = 9.552\ \mu m$ – was focused along the ignition cell axis. With a rectangular intensity profile and a fairly constant photon flux over the profile pulse energies up to 0.7 J and a mean photon flux at the beam waist of 40 J/cm^2 was reached. The absorbed CO_2–laser radiation was determined from the intensity ratio between two beam splitter in front of and behind the cell. The diode laser was cooled with a closed cycle helium refrigerator to 28 K. The laser output was collimated, a single longitudinal mode was selected by a monochromator, and the laser beam was softly focused into the cell parallel to the CO_2–laser beam. The diode laser beam had a diameter below 1 mm inside the cell. A second monochromator behind the cell served as a narrow band filter to suppress the strong infrared emission from the combustion region. Finally the diode laser beam was focused on a liquid–nitrogen cooled $InSb$ detector. For a relative wavelength reference, a fraction of the diode laser beam was split off and the transmission

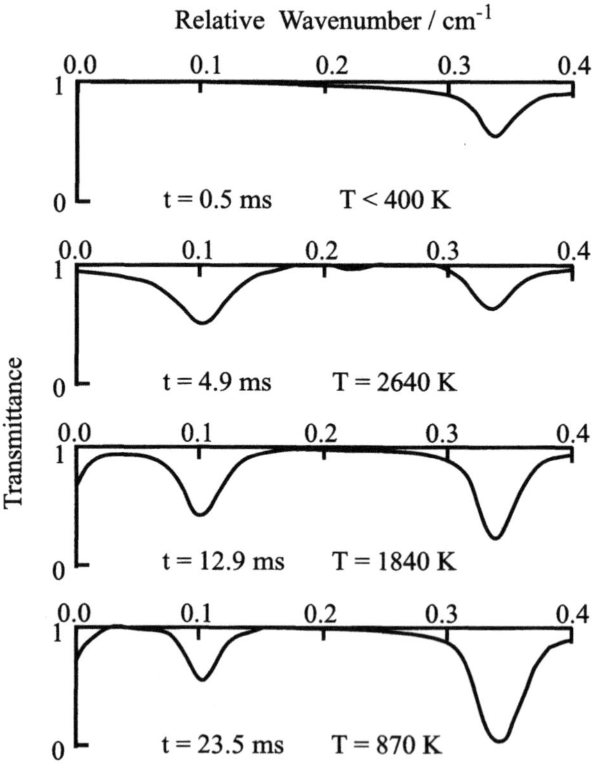

Fig. 13.11. Typical experimental transmission spectra of the CO lines $1R4$ (left) and $0P2$ (right) during ignition and flame propagation

spectrum of a solid germanium etalon was recorded with a second $InSb$ element. All detector signals were captured by a transient recorder which was started some milliseconds before the CO_2–laser pulse, so that the ignition time could be measured precisely.

The diode laser was tuned over a scan range of 0.4 cm^{-1} by a 10 kHz sawtooth shape injection current modulation, so that the two CO–absorption lines could be recorded every 50 μs. Examples of typical spectra of a single ignition event are shown in Fig. 13.11.

The first trace is the transmission spectrum of the gas mixture just before the arrival of the flame front. Only the absorption due to the $0P2$ line is visible. The next traces show the spectra of the hot gas at several times after the ignition event. The temperatures indicated in Fig. 13.12 were calculated from the ratio of the transmittances of the $1R4$ and $0P2$ lines as outlined above.

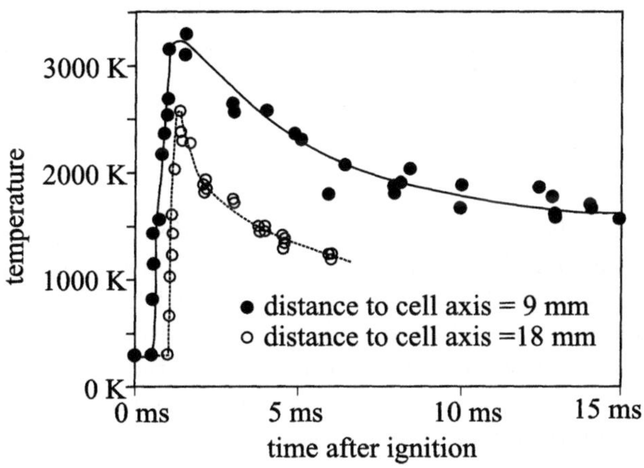

Fig. 13.12. Two typical temperature profiles during the combustion of CH_3OH / O_2–mixtures after CO_2–laser ignition

C. Harmonic detection techniques for the measurement of small absorptions

In recent years it has become clear that the increasing pollution of the atmosphere is a severe threat to our living. Even trace amounts of pollutants interfere with the atmospheric chemistry. Almost all of the pollution gases exhibit an infrared absorption spectrum which can be used for monitoring purposes. To detect the very small pollutant concentrations in the atmosphere an extremely sensitive technique is needed. Absorption techniques are one out of few which can provide absolute absorber concentrations, however, the technique suffers from the fact that the difference or ratio between two large quantities I and I_0 has to be evaluated very precisely, requiring very high resolution and dynamic range simultaneously. A drastic improvement in sensitivity is possible by combining fast tunable diode lasers, harmonic detection techniques and lock–in amplifiers to reduce the detection bandwidth and to remove unwanted signal components. This method is so successful that it has become state of the art in infrared absorption spectroscopy. With diode lasers modulation frequencies from the kilo–to the megahertz range can be used, taking advantage of the fact that the amplitude noise of these lasers has an $1/f$–dependence up to frequencies in the megahertz range. Due to the superior noise properties in the high frequency range the sensitivities could be improved down to the shot–noise–limit, i.e. fractional absorptions down to 10^{-7}.

The basic principle of harmonic detection is to modulate the signal source and to extract only the periodic part of the detector signal at exactly the modulation frequency or some integer multiple there of by a lock–in amplifier. This method monitors the change in light intensity due to absorption rather than the light intensity itself and profits further from the suppression of unwanted frequency components. Given that the integration time of the lock–in amplifier is long against the modulation period, the mathematical representation of this filtering process is given by

$$H_n = \frac{1}{\pi} \int_{-\pi}^{\pi} U(\theta) cos(n\theta)\, \partial\theta; \quad \theta = 2\pi f t \tag{13.38}$$

where t is time, f is the modulation frequency, $U(\theta)$ is the detector signal and H_n is the n–*th* harmonic signal, i.e. the n–*th* Fourier coefficient of the cosine transform of the detector signal. In harmonic detection with a tunable diode laser the laser wavelength or/and amplitude is modulated by direct injection current modulation. The wave number ν of the emitted light can then be expressed as: $\nu(\theta) = \nu + aF(\theta)$ where ν is the center frequency of the laser light, a the modulation amplitude, and $F(\theta)$ the normalized modulation waveform. For simplicity, we assume that the laser power does not change with modulation. Then the light intensity incident on the detector will only vary if a substance or optical element, whose transmission depends on the wavelength, is placed in the light path. Hence the detector signal will be proportional to:

$$U(\theta) \propto I(\theta) = I_0 T(\nu) = I_0 T(\nu + aF(\theta)) \tag{13.39}$$

In the case of sinusoidal modulation and isolated Lorentzian absorption lines analytical solutions are possible whereas for special modulation waveforms or complicated line shapes the Fourier coefficient has to be evaluated by numerical integration. If the modulation amplitude is sufficiently small the transmission function can be expanded as a Taylor series around ν_0 with n terms for the n–*th* harmonic signal:

$$T(\nu) = \sum_{k=0}^{\infty} \frac{1}{k!} \frac{\partial^k T(\nu_0)}{\partial \nu^k} (\nu - \nu_0)^k \tag{13.40}$$

$$T(\nu) \approx T(\nu_0) + \sum_{k=1}^{n} \frac{1}{k!} \frac{\partial^k T(\nu_0)}{\partial \nu^k} \; (a\ cos(\theta))^k \tag{13.41}$$

$$H_n \propto \frac{1}{\pi} \int_{-\pi}^{\pi} \left(T(\nu_0) + \sum_{k=1}^{n} \frac{1}{k!} \frac{\partial^k T(\nu_0)}{\partial \nu^k} \; (a\ cos(\theta))^k \right) cos(n\theta)\partial\theta \tag{13.42}$$

By transforming the cosine terms according to the formulas of De Moivre it is easily seen that all terms except the n–*th* term yield zero integrals. The harmonic signal is then given as:

$$H(\nu) = \frac{2^{1-n}}{n!} a^n \frac{\partial^n T(\nu_0)}{\partial \nu^n} \tag{13.43}$$

This small modulation amplitude limit is often referred to as derivative spectroscopy, as the output signals are proportional to the derivatives of the transmission function. With diode lasers the emitted laser power is usually proportional to the injection current and therefore varies linearly with the wavelength. It can be seen by the same arguments that this effect will produce only a constant offset in the first harmonic signal and vanishes in the second harmonic signal. In harmonic detection, one is only concerned with weak absorption lines. Consequently there is negligible loss in accuracy and generality if the transmission function of the absorption lines is approximated by:

$$I(\nu) \approx I_0(\nu)(1 - S \cdot \phi(\nu - \nu_0) \cdot c \cdot z) \quad \text{for } S \cdot \phi(\nu - \nu_0) \cdot c \cdot z \leq 0.05 \tag{13.44}$$

In the derivative spectroscopy limit the line shapes of the harmonic signals will be proportional to the derivatives of the line shape function ϕ. The Lorentzian line shape and its first and the second derivative are shown in Fig. 13.13 as an example. The height of the Lorentzian is normalized to unity at the line center and the abscissa is normalized to the half width at half maximum. One now has to find values of the modulation amplitude which optimize the signal–to–noise ratio. In this situation the Fourier integrals have to be solved individually for the specific shapes. For the Lorentzian line shape an analytical solution has been found [192,249].

Examples of the shape of the second harmonic signal are shown in Fig. 13.14 (left) in terms of the dimensionless modulation parameter m which is defined as the ratio of the modulation amplitude and the half width of the line. The second harmonic signal amplitude is maximum at the line center and is plotted in Fig. 13.14 (right) as function of m. The maximum harmonic signal is obtained for a value of $m = 2.2$. For other line shape functions the results are similar, e.g. the harmonic signal due to a Gaussian broadened line is maximum for $m = 2$.

As an example of the application of harmonic detection we describe measurements on the selective non–catalytic reduction of nitric oxide by ammonia

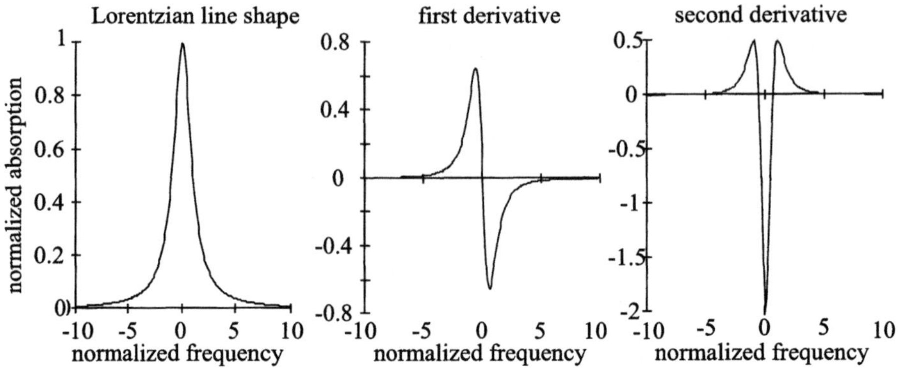

Fig. 13.13. The Lorentzian line shape and its derivatives

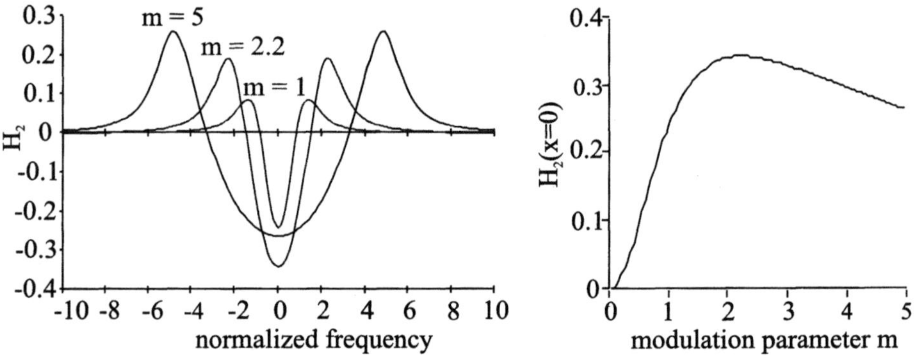

Fig. 13.14. Left: Second harmonic signal for different values of the modulation parameter m. Right: Second harmonic signal amplitude at line center for different modulation parameters m.

in a quartz reactor. This process proceeds only in a narrow temperature range around 1000 K, and the exact position of the window can be influenced by additives to the flue gas. The diode laser frequency is modulated with a triangle waveform from a signal generator and the detector signals are demodulated by lock–in amplifiers at the second harmonic as shown in Fig. 13.15. A second signal generator was used to superimpose an additional modulation with a frequency not harmonically related to the first one. Its main purpose is to average out high frequency interferences to the signal. A D/A–converter provided a 0.1 Hz ramp to scan the diode laser over the absorption lines. Product gases from the quartz reactor were sampled through a quartz capillary into a 40 cm long pyrex sample cell with 80 $mbar$ inner pressure to enhance spectral resolution hence selectivity. The cell and the capillary tube were kept at 200 $°C$ to minimize the adsorption on the walls. The laser beam was softly focused into the middle of the cell to avoid beam obstruction due to changes in the cell position as a consequence of the thermal expansion of the

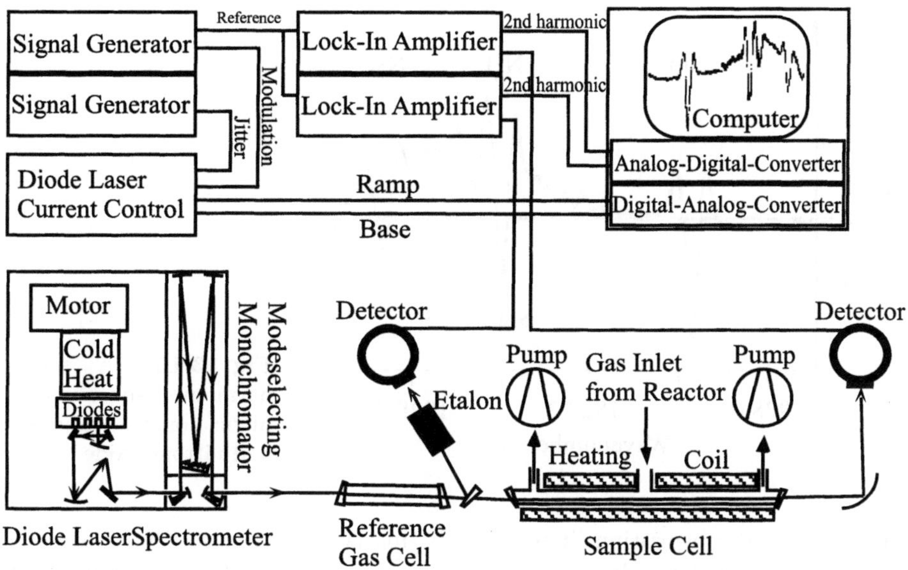

Fig. 13.15. Experimental set–up for absorption measurements using harmonic detection of the reaction products of the reduction of nitric oxide by ammonia

reactor between different temperatures. The light was collected behind the cell by a spherical mirror and recorded by a liquid–nitrogen cooled $HgCdTe$–detector. A reference cell placed in the beam path in front of the sample cell contained a known amount of a gas having an absorption line in close vicinity to the species under observation. Its harmonic signal amplitude served as a reference to ratio the signal amplitudes, as their absolute values showed a significant drift with time. An example for a second harmonic spectrum of three different species ($NO - 500\ ppmV$; $NH_3 - 750\ ppmV$ – an unresolved absorption doublet – and pure N_2O) is shown in Fig.13.16.

The signal amplitude is obtained as the peak–to–peak excursions of the lines after subtracting the background. Absolute concentrations can be obtained by calibrating the signal ratios with certified gas mixtures. To measure all species of interest four diodes were attached to the cold head simultaneously. These covered the specified spectral range from 10 μm to 5 μm. Switching from one diode to another could be done by shifting the collimation unit. With this experimental set–up the sensitivity limits were, for example, 10 $ppmV$ for HCN, 1 $ppmV$ for NO, and 0.1 $ppmV$ for NO_2. Great care had to be taken to reduce the so–called "optical noise" of the set up to a minimum level, which is caused by optical feedback of laser light into the laser diode and which strongly affects the tuning behavior. In contrast, if some of the laser light is reflected back and forth between any surfaces in the light path, a Fabry–Perot resonator i.e. etalon is formed, which superimposes a periodic structure on the baseline. This structure can completely

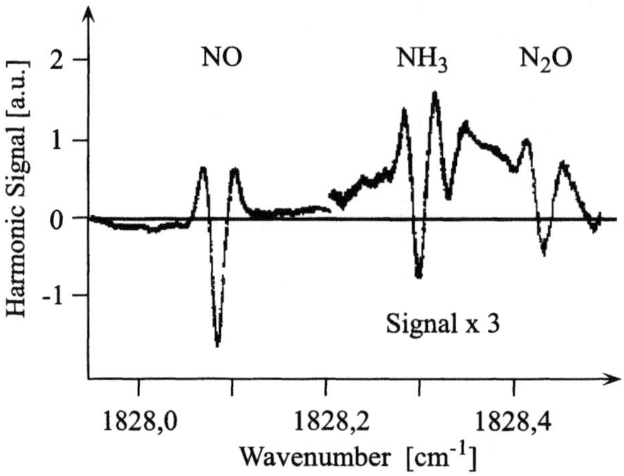

Fig. 13.16. A second harmonic spectrum of nitric oxide, ammonia and nitrous oxide

hide the weak absorption lines. Even if this is not the case, the phase of it is strongly time dependent since it changes with the relative positions of the optical elements. A time varying baseline structure can cause large additional errors on the extracted concentration values. Consequently all cell windows were wedged and arranged such that their surfaces were not perpendicular to the laser beam. Solely mirrors were used to redirect and focus the laser light. With these precautions the sensitivity was limited by the intrinsic noise of the laser diode itself.

D. Simultaneous *in–situ* detection of oxygen and water in a full scale waste incinerator with near infrared diode lasers

Combustion of municipal waste can be a valuable source of energy. Since fossil fuels are replaced, waste combustion has no negative effects on the CO_2 budget, whereas waste disposal in landfills, the most common disposal path in many countries, is responsible for long term emission of very efficient green house gases (e.g. CH_4) into the atmosphere. To take advantage of the thermal conversion process, however, it is compelling to maximize heat release and conversion efficiency of the overall combustion process by ensuring homogeneous combustion throughout the entire furnace. Since waste is known to show very strong variations in its fuel parameters, e.g. calorific value and waste humidity, fast, *in–situ* species and temperature sensors and efficient combustion control strategies, are needed, to allow an optimization of the primary process parameters.

For this purpose we started to develop on–line sensors for waste incinerators [224,240], aiming at fast *in–situ* detection of fuel quality fluctuations, and at an active control of the homogeneous combustion process above the solid fuel bed. Near infrared diode lasers (NIRDL) and high resolution absorption spectroscopy are employed for a simultaneous *in–situ* detection of O_2 and

H_2O along a single absorption path at the end of the secondary combustion zone of a full–scale municipal waste–to–energy plant (Fig.13.17).

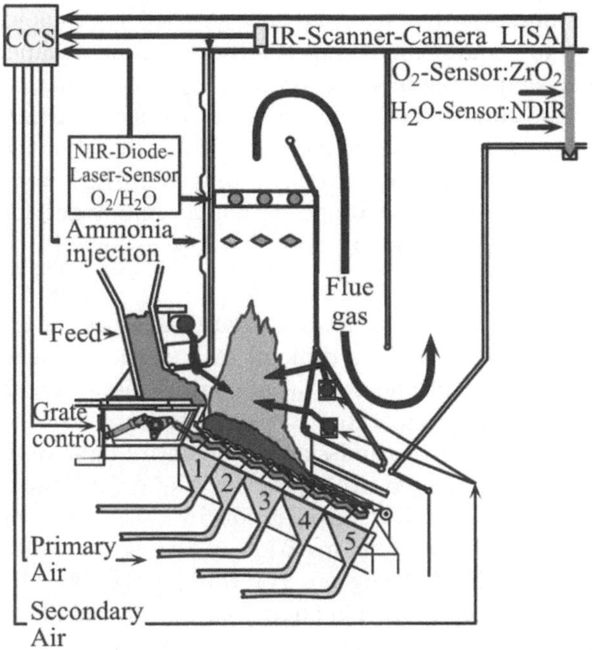

Fig. 13.17. Intersection through the waste incinerator at which the *in–situ* detection of O_2 and H_2O has been performed. A selection of important sensors (Diode laser *in–situ* sensor, Laser–In–Situ–Ammoniasensor, LISA and IR–Scanner–Camera, TACCOS) as well as the possible actuators, which are connected to the combustion control system (CCS) are indicated.

As proven in many laboratory applications, laser absorption techniques have a unique potential with regard to selectivity, resolution and speed. The possibility to measure over large distances avoids the numerous disadvantages of sensors based on local measurement principles especially in chemically aggressive, high temperature environments. Whether laser sensing can be transferred from a lab technique into an industrial control tool is closely connected to an adequate choice of the laser source. While for sensitivity and selectivity reasons the finger print region in the infrared ($2 - 20\ \mu m$) is most interesting for species sensors, the lack of infrared lasers suited for industrial environments prevented wide spread use. Regarding these issues NIRDL are the laser best suited for industrial sensor applications. Developed for communication technology together with high performance detectors and glass fibers, they are produced in enormous quantities. They are available with almost complete spectral coverage from 635 nm to 1650 nm and offer with regard to industrial applications a unique combination of positive properties like low cost, tiny size, room temperature operation, laser life time of 10^5 hours, single mode output power of up to several 100 mW, and excellent spectral resolution of 10^{-3} nm. Most important are their versatile tuning properties–slow and coarse tuning by a change of laser temperature (range: several nm; speed: some GHz/s), and extremely fast fine tuning by a modulation of the laser current (range: $30 \sim 100\ GHz$; speed: $10^5\ GHz/s$). This allows the application of very efficient noise reduction strategies such that sensitivities down to the ppm–level could be demonstrated under well–controlled laboratory conditions. Another main advantage–their continuous tuning–opens the possibility to acquire the complete absorption line shape, which allows either a rather simple signal interpretation or, by evaluation of the spectral line profile, the simultaneous extraction of physical boundary conditions (pressure, temperature, velocity, mass flux and others) in the measurement volume. Their major drawback, to cover only relatively weak transitions in the visible and near infrared range, is not as important for many process control applications since mostly majority species have to be monitored (CH_4, O_2, H_2O, CO_2).

However, the scale–up of these sensors to accommodate full–scale industrial processes in particular has proven to be a difficult task, since the weak molecular absorption signal is strongly affected by numerous disturbances [202,203]. For the application in the incinerator it was most important to correct for any changes in the effective transmission of the entire light path between laser and detector. These changes are caused by the quickly fluctuating dust load of the flue gases, window degradation by dust, heat, or chemical attack, vibrations or mechanical deformation of the boiler walls, and refractive index gradients or thermal lensing. For this purpose an all–analog electronic on–line transmission correction circuit was developed [201], which corrects each individual scan in real time. The idea to correct these fluctuations makes use of the diode lasers high speed tuning capabilities, and requires that the scan frequency is chosen well above the fastest transmission variations. Since this way all interferences are "frozen" and only the molecular signal component changes within the scan time, an electronic correction by division through the quasi–static background transmission signal possible. Additional noise reduction is achieved by averaging consecutive absorption profiles and extracting the absorption signal i.e. the area underneath the absorption line by a fast fitting algorithm.

Oxygen monitoring was established in the incinerator via the well known atmospheric A–band ($b^1\Sigma_g^+;\ v = 0 \leftarrow X^3\Sigma_g^-;\ v = 0$), a magnetic dipole transition in the $759 - 770\ nm$ region that shows line strengths in the order of $2 \cdot 10^{-4}\ cm^{-1}atm^{-1}$ [237]. By using a continuously tunable, distributed feedback diode laser we could access any line within the R–branch of the A–band, so that the temperature dependence of the line–strength could be minimized by choice of an adequate absorption line. Water vapor, which shows numerous strong, rotationally resolved, vibrational overtone and combination bands in the near infrared [190,191,219,244], was monitored via the relatively weak $(0,0,0) \rightarrow (1,1,2), (2,1,1)$ rovibrational transitions around $812\ nm$. This was possible since high power $30\ mW$– Fabry–Perot–type lasers are available for this spectral range. However, their discontinuous tuning behavior made a careful search for coincidences with H_2O absorption lines necessary. Theoretical absorption spectra based on the HITRAN 92 database [238] were calculated and verified by laboratory measurements in a high temperature absorption cell in the temperature range $300-1000°C$. The laboratory studies had to be performed because strong discrepancies between measurements and HITRAN 92 results were observed particularly at high temperatures [190,191,219].

The simultaneous single–path *in–situ* detection of O_2 and H_2O was performed in an incinerator with a boiler height of 18 m, a diameter of 4 m, and a thermal power of 20 MW. Combustion in such a waste incinerator takes place at atmospheric pressure with flue gas temperatures above $850°C$ and O_2 or H_2O concentrations around 8 $Vol.\%$ and 18 $Vol.\%$, respectively.

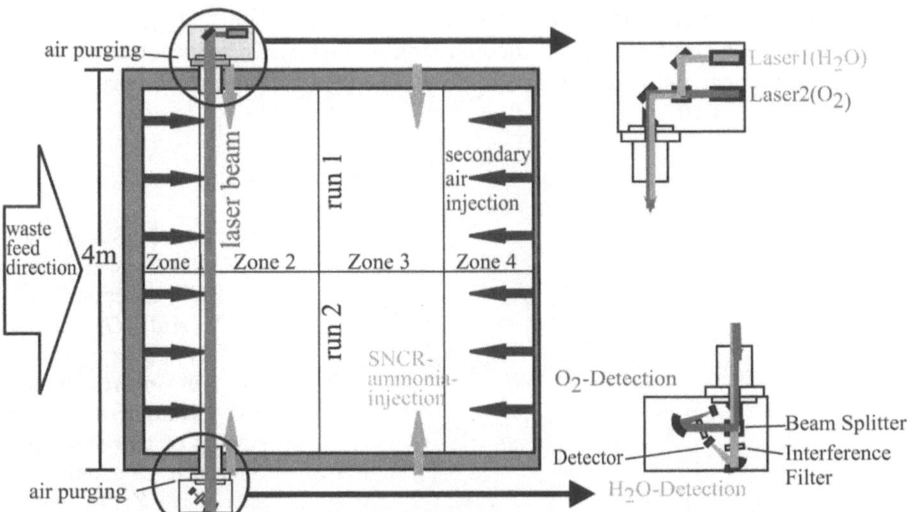

Fig. 13.18. Outline of the incinerator and experimental set–up of the diode–laser–based dual–species *in–situ* spectrometer

As depicted in Fig. 13.18 the absorption path was separated one meter from the furnace wall and located at the end of the secondary combustion zone (12.7 m above the fuel bed and 8.7 m above the secondary air injection nozzles). Lasers and detectors were mounted onto a flange on the furnace wall without water cooling or vibrational damping. Two opposing optical ports with 30×50 mm windows placed under the Brewster's–angle provided optical access to the furnace. Air purging of the windows was essential to prevent rapid condensation of opaque ammonia sulfates originating from the SNCR injection ports one meter below the windows.

Both lasers were scanned independently over the selected absorption lines by a triangular phase locked current modulation of each laser at 1 kHz. A dichroic beam splitter superimposed the two collimated laser beams, and a gold coated mirror directed them through the furnace. An identical dichroic beam splitter separated the two wavelength components of the transmitted light. The two following interference filters minimized the cross talk between the two wavelength channels and suppressed thermal background radiation. The resulting detector signals were pre–amplified, transmission corrected, filtered by analog correction electronics and digitized with 1.2 *Msample/s* by a dual channel 12 *Bit* A/D converter plugged into a computer. 200 consecutive scans were averaged for additional noise reduction. A fitting procedure extracted the concentration value assuming a Lorentzian line shape function and a quadratic background polynomial. Both laser wavelengths were actively stabilized to increase system stability. The dual–species spectrometer was exposed to several weeks of testing in an industrial environment (the aforementioned incinerator) under various load conditions. Typical results, which represent the first simultaneous, single–path *in–situ* measurements of oxygen and water using *AlGaAs*–diode–lasers in a full scale incinerator, are shown in Fig. 13.19.

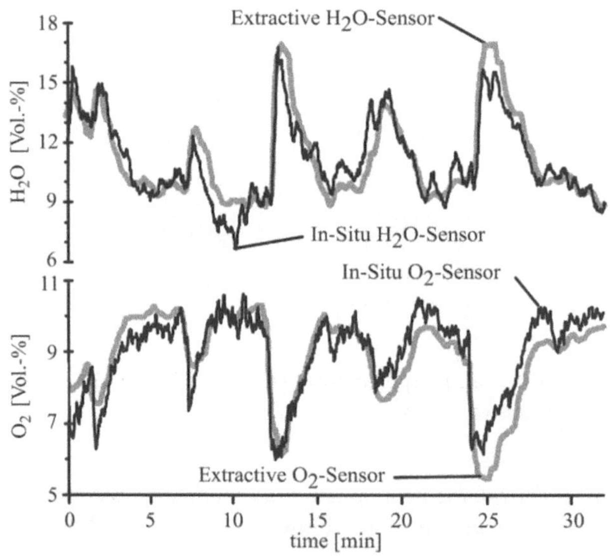

Fig. 13.19. Simultaneous diode–laser–based *in–situ* detection of H_2O and O_2 in a 20 *MW* waste incinerator. The *in–situ* signals are compared with signals of two standard reference sensors: an extractive ZrO_2–sensor for O_2 and an extractive non–dispersive infrared probe for H_2O, which probe the flue gas duct 45 *m* behind the diode laser sensor location (depicted in Fig. 13.17).

The steep concentration changes are caused by grate agitation and fuel feed events. Comparing the two *in–situ* signals, the expected close temporal coupling between O_2 consumption and H_2O production due to the overall reaction

$$C_xH_x + (x + y/4)O_2 \rightarrow y/2H_2O + xCO_2)$$

could be verified experimentally. Because alternative *in–situ* species sensors are unavailable thus far the laser signals were tested by comparison with standard gas sensors: an extractive ZrO_2–sensor for O_2 and an extractive non–dispersive infrared (NDIR) absorption sensor for H_2O. Both standard sensors analyze the gas at the relatively cold flue gas duct 45 m downstream of the laser measurement location (Fig. 13.17), and are expected to show severe artifacts like flow dependent time–delays due to the necessary gas transport to the sensor. However, after removal of these signal delays and low–pass filtering of the laser signals the *in–situ* and reference signals show good agreement for both species (Fig. 13.19).

Due to the lack of alternative *in–situ* signals the absolute calibration of the *in–situ* sensors is based on the reference devices (Fig. 13.20), which is done by direct comparison of reference and laser signal over a limited time period assuming homogeneous and stable conditions within the furnace. By this procedure we take into account the different response times and the different measurement locations of the sensors.

The line integrated laser signal is assumed to be a good representation of the average concentration in the furnace. The comparison is done by low pass filtering the laser signals with a moving average to achieve a time response comparable to the reference sensors. By comparing characteristic peaks in the time histories of the signals (s. Fig. 13.20a) the time delay between laser

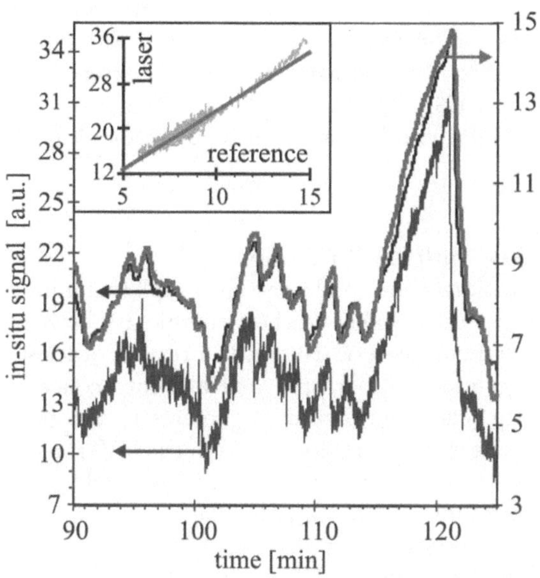

Fig. 13.20. a) Comparison of the laser *in–situ* signal and the reference signal for the case of oxygen. The bottom trace depicts the laser O_2–signal without further processing shifted to lower concentrations by 4 Vol.%. The top traces indicate the laser *in–situ* signal (averaged over 15 samples and time shifted by +20 *sec*; thin black line) and the O_2–reference sensor (gray thick line). The reference sensor is delayed by 20 *sec* due to its probing location 45 m further down the flue gas duct. **b)** Inset: Correlation of laser signal vs. reference signal over 30 *min* and a linear fit to the data. This fit is used for an absolute calibration of the laser signal.

and reference is removed. After this the offset and the differential sensitivity (Δ *laser signal*/Δ *reference signal*) are extracted by a linear fit to a scatter plot of the laser vs. the reference signal as depicted in Fig. 13.20b. With this procedure we demonstrate the good linearity of the laser signals and get an estimate of the concentration resolution of the sensor. By attributing all deviations from linearity to noise or errors we get for the depicted case a worst case estimate of 0.5 $Vol.\%$ of O_2. If we assume that these are real oxygen fluctuations the resolution is about 0.1 $Vol.\%$. Even though the calibration and the resolution estimate for the water transitions is more difficult, since the reference signal is more distorted due to the sampling process with the heated probe line (adsorption of water), they yield about the same results.

The major advantages of the *in–situ* measurement technique presented are the high time resolution of about 1 *sec*, and especially the fact that the *in–situ* signals are much faster than the delayed reference signals (with a delay of 20–40 *sec* for O_2 and $40 - 80$ *sec* for H_2O). This time gain can either be used to improve the bandwidth of a control loop significantly, or to set–up a feedback system despite relatively slow actuators. These properties together with the complete absence of corrosion problems make a laser–based single–species measurement already very attractive for active combustion control purposes. Even more interesting is the described possibility for simultaneous multi–species detection in a single–path, because all species are monitored *in–situ* with identical time response and within the same measurement volume, so that the local chemistry can be investigated directly without any errors due to different sensor response times and adsorption or chemical reaction effects in the probe lines. With such a device it should therefore be possible to monitor changes in the C/H–ratio of the waste (i.e. calorific value) on–line. On the other hand, if the C/H–ratio is constant, the combined *in–situ* O_2/H_2O signals allow a monitoring of non–combustion–related sources of water like changes in waste humidity. In the next development step we will extend our spectrometer to allow simultaneous O_2, H_2O and CO_2 detection, so that calorific values and the waste humidity can be determined simultaneously.

E. In situ determination of free radicals in flames

During the last few years investigations were made on the complete simulation of technical combustion processes. Because of their complexity, only a few idealized flames can be modeled completely. To test the results of these simulations it is necessary to determine temperature and absolute concentrations of important intermediate (chemical unstable) radicals in the test flame. These radicals usually show intense absorption bands in the UV–range. UV–wavelengths can be produced by dye lasers, the most widely used types of tunable lasers in the range from 300 nm to 1.2 μm. Their active media are organic dye molecules solved in liquids, which display strong broadband fluorescence spectra under excitation by visible or UV–light. Below the determination of absolute $OH-$ and CH_3 radical concentrations in two different test flames will be described.

In the first example absolute OH radical concentrations have been detected via UV–laser absorption measurements in a laminar counter–flow diffusion flame, burning at atmospheric pressure [241]. As shown in Fig. 13.21 (left), a pulsed, frequency doubled dye laser pumped by a $XeCl$ excimer laser was used to measure absorption spectra in the $A^2\Sigma \leftarrow X^2\Pi, (v'' = 0, v' = 0)$ band of the OH radical (Fig. 13.21 (right)).

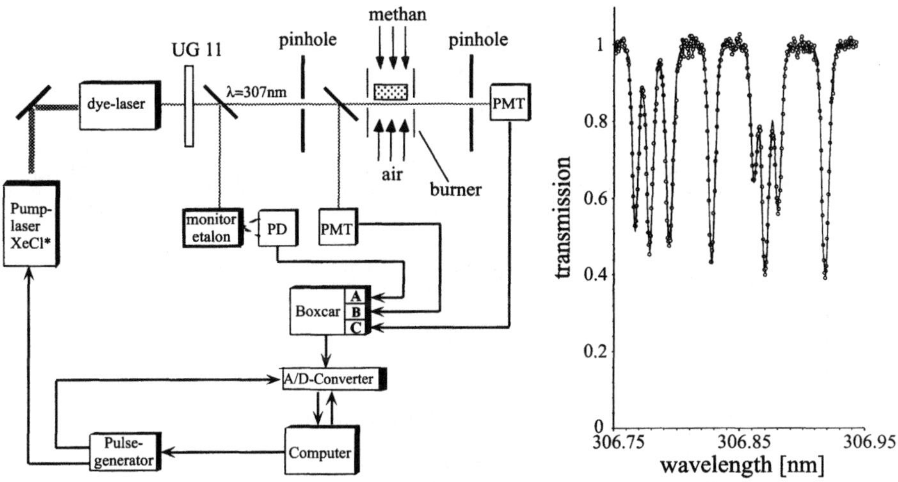

Fig. 13.21. Left: Experimental set up for the determination of absolute OH radical concentrations in a laminar counterflow diffusion flame at atmospheric pressure. Right: Measured (•) and synthetic (–) spectrum of OH, P–branch, $A^2\Sigma \leftarrow X^2\Pi$, (0,0).

To resolve the structure within the approximately 1 mm thick reaction zone of this flame type a high spatial resolution was needed. This was achieved by two 50 μm pinholes in front of the burner and behind it. The second pinhole also served to block the diffraction pattern produced by the first one. Stray light and flame emission were suppressed by Schott filters (UG 11) and short gate times of the boxcar integrator. Lines of the R–bandhead near 307 nm were recorded and synthetic spectra using published molecular data were fitted to these spectra to obtain rotational temperatures and absolute concentrations of OH radicals. The OH line shapes were well described by a Voigt profile, since the Voigt parameter depends on the unknown gas composition the Lorentzian contribution to the profile was evaluated by the fitting routine. Changes of ± 15 % in the Voigt parameter resulted in measured temperature and concentration changes smaller than 2 %. The laser profile was determined with a high resolution Fabry–Perot interferometer and was taken into account in the calculation. Figure 13.21 (right) shows the measured and the fitted spectra of OH near 307 nm. In the reaction zone of the diffusion flame a maximum concentration of $3 \cdot 10^{-8}$ mol/cm^{-1} was measured.

In the second example absolute CH_3–concentrations in a laminar stoichiometric premixed low pressure CH_4/O_2 flame have been determined [205,206]. CH_3 radicals exhibit an absorption band ($^2A_1' \leftarrow {}^2A_2''$) next to 216 nm. Because of the strong predissociation of the $^2A_1'$–state, the CH_3–spectrum exhibits no vibrational rotational structure. Only the R–and the $P+Q$–branch can be resolved. The experimental set–up is shown in Fig. 13.22.

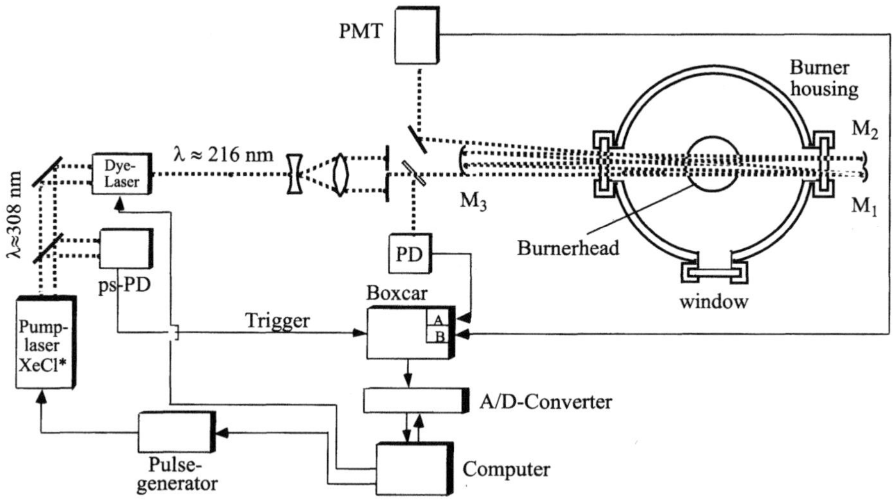

Fig. 13.22. Determination of absolute CH_3–radical concentrations in a laminar premixed CH_4/O_2 flame, 10 $Torr$: Experimental set–up

The burner head consists of a circular sinter plate 6 cm in diameter. Laminar flow of the premixed fuel through the sinter plate was ensured. Because of this geometry, the flame can be treated as one–dimensional, e.g. temperature and species concentrations will be constant in a plane parallel to the burner surface. The wavelength needed for absorption measurements in the far UV was produced by a frequency doubled, excimer pumped dye laser. The laser beam was expanded by a Galilean telescope and then apertured by a pinhole to reduce the intensity. A beam splitter was used to monitor a part of the light with a photodiode (PD). Because of the low concentration of CH_3–radicals in the flame, a White type multi–pass absorption set up was used. It consists of three spherical, concave mirrors having the same radius of curvature. Two mirrors, M_1, M_2, have to be positioned close together at one end of the absorption cell, the third mirror, M_3 on the other side. The centers of curvature of M_1 and M_2 are on the front surface of M_3 and the center of curvature of M_3 is halfway between M_1 and M_2. The laser beam enters through the pinhole close to one end of M_3, travels to M_1, from there to M_3, then to M_2, back to M_3, to M_1, and so on. The mirrors M_1 and M_2 form several images of the pinhole on the surface of M_3, until another image is formed beside M_3. To achieve a total absorption of 50 % in the low pressure flame 24 passes (total absorption length: 111.30 *cm*) were necessary. After passing the absorption set–up, the beam was detected by a photomultiplier (PMT). It was important that the beam was adjusted parallel to the burner surface to access regions in the flame with identical physical conditions.

The dependence of the absorption on the relative height to the burner head has been determined. The signals from the photodiode and the photo–multiplier were sampled by a boxcar integrator and then fed into an A/D–converter. The ratio of the photo–multiplier and the monitoring diode signals was used to eliminate pulse–to–pulse–variations of the dye–laser.

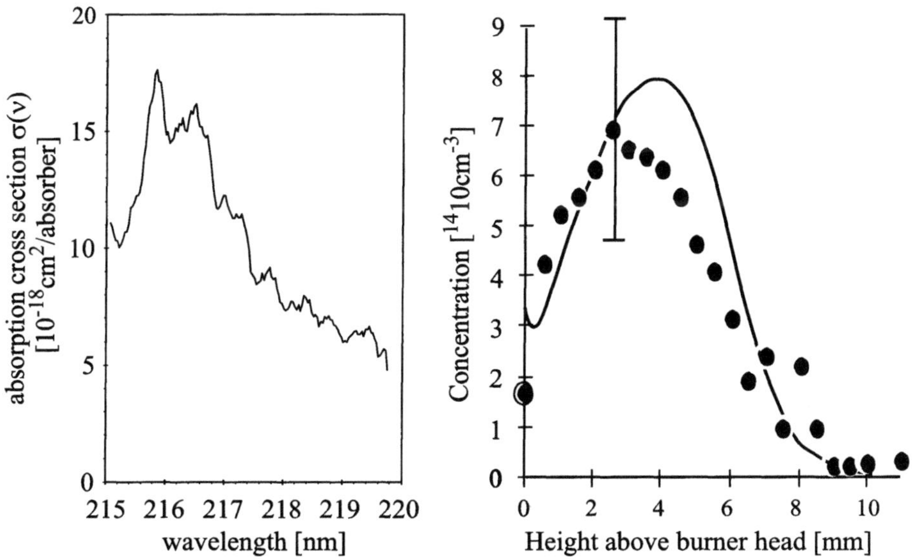

Fig. 13.23. Left: Absorption cross section of the CH_3–radical, ${}^2A_1' \leftarrow {}^2A_2''$–band, measured in 0.5 *mm* height above the burner head, $T = 660\ K$. Right: CH_3–radical concentration as a function of the height above the burner head. measurement (•); model (–)

The relative absorption coefficient has been measured as a function of the wavelength at 0.5 *mm* from the burner surface ($T = 600\ K$) and could be calibrated using the known temperature dependence of the absorption coefficient [225] (Fig. 13.23 (left)). The CH_3–concentrations were calculated using the Lambert–Beer law. To calculate the ratio I/I_0 the transmitted light in each height was divided by the transmitted light in a reference height where no CH_3 is present.

Fig. 13.23 (right) shows the measured and modeled mole fraction of the CH_3–radicals as a function of the height above the burner head. The experimental error results mainly from the uncertainty of the temperature as a function of the distance from the burner plate, which causes an error in the (temperature dependent) absorption coefficient.

14 Pyrometry and Thermography

Udo L. Glückert, Robert Schmidt

A Critical Survey
General Basics and Application Examples

14.1 Introduction

Herschel[1] discovered the infrared spectral region of optical rays about the year 1800. Although the first applications for thermography devices date back to 1900, the field of pyrometrics took a different course of development. An exact definition of the two measurement methods shall not be made at this point. The following characteristics will serve to illustrate the differences between the two methods. Pyrometry is a method which allows measurements of temperature to be recorded from surfaces emitting temperature radiation also known as heat radiation or thermal radiation. These temperature radiations are recorded by a pyrometer, a "radiation thermometer" resulting in point like information. In contrast, when the information obtained is optically recorded as an entire scene ie. buildings, operational equipment etc. and displayed as a "temperature picture" this is known as thermography. A firm grasp of the difference between infrared photography and similar methods – which record only the **reflected** spectral portion of the optical spectrum (only in the near region of the infrared spectrum) and which generally do not permit any temperature intepretations of the recorded scene to be made – is essential to understanding this phenomena.

To arrive at a very general statement concerning temperature measurement without direct contact, Fig. 14.1 presents a schematic diagram with three individual sectors: the source, the method of transmission and the receiver (detector and the hole signal generating process). We will study these three sectors individually, this method appearing to be the most appropriate of the measurement methods employed. Due to their historical background, pyrometrics and thermographics are often treated separately and studied

[1] *See the appendix for additional information.*

individually in the literature. This is because the development of thermography was conducted primarily in the military sector. In the early 1990's this situation began to change, as civilian fields of study began to develop a larger number and wider variety of applications. Though the high costs of thermography equipment do not allow for widespread application increased development has led to a decrease in prices in recent years. This has led, for example, to the use of these systems in the building construction sector as a service offered by surveyors on a rental basis. Pyrometry has also been employed as a standard method in the field of measurement technologies.

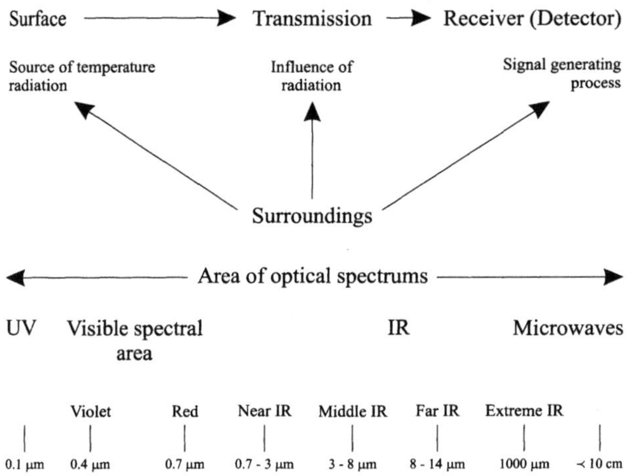

Fig. 14.1. Selected systematization of the topic and subdivision of the optical spectral region

14.2 Temperature Radiation

In this article, only surface temperature radiation will be treated, the field of volume radiation, (e.g. gases, flames, etc.) will not be examined. Every surface, due to its temperature, radiates electromagnetic waves (optical radiation) which we define as temperature radiation (heat radiation). It should be noted without going deeper into the subject matter, that optical radiation can be studied in a two–fold manner; the wave aspect or the quantum approach (light quant aspect), depending on which is more appropriate in solving the problem at hand. Figure 14.2 illustrates the relationship between the radiated power of a surface over the individual wavelength λ and the corresponding temperature T. It was Max Planck's accomplishment that he was able to explain these well–known relationships in his definition of the radiation laws (1900). Figure 14.2 shows the spectral distribution of a physical ideal, a black body whose definition will be examined later in the paper. The spectral emissive power, also known as radiant emittance M, is viewed over the half–space ($2{\cdot}\pi$) and is presented in the usual isothermal representation,

$$M_{(BB,\lambda,T)} = 2 \cdot \pi \cdot c_1 \cdot \lambda^{-5} \cdot (\exp\,(c_2/(\lambda \cdot T)) - 1)^{-1} \tag{14.1}$$

with

$$c_1 = h \cdot c_0{}^2 \quad \text{and} \quad c_2 = h \cdot c_0/k \tag{14.2}$$

The indices in 14.1 are:

- BB – black body,
- λ – spectral quantity (in relationship to the wavelength),
- T – relates to an isotherm,
- c_0 – speed of electromagnetic waves in vacuum,
- c_1 , c_2 – the first and the second Planck–constant.

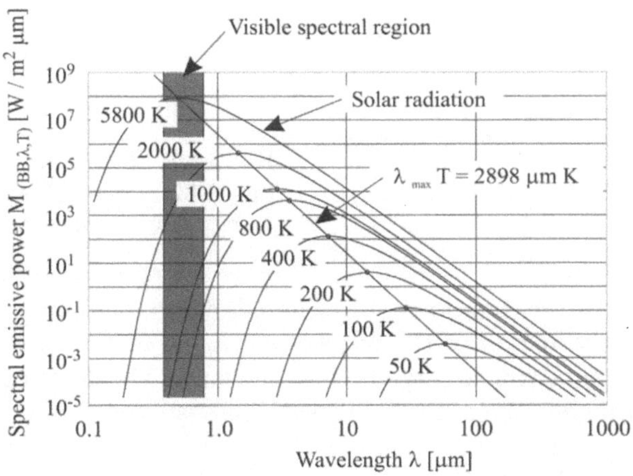

Fig. 14.2. $lg(M_{(\lambda)})$ as a function from λ and T

In the double–logarithmic representation of Fig. 14.2 the individual maximas of the isotherms are all connected by a straight line, the Wien's displacement line. Wien's displacement law

$$\lambda_{\max} \cdot T = 2897.8\ \mu\text{m} \cdot \text{K} \tag{14.3}$$

or

$$\lambda_{\max} \cdot T = c_3 \tag{14.4}$$

states that the product of the absolute temperature T and $\lambda_{\max}$ (wavelength of maximum energy radiation) for an ideal black body is a constant. Wien's displacement law is derived from Planck's radiation law. These thermodynamic considerations were, however, based a few years previously on those of Wien himself and can thus be classically derived. The practitioner applies Wien's displacement law to determine the wavelength region in which his measurement problem lies. If the wavelength in Fig. 14.2 is represented as linear, we obtain Fig. 14.3. Here we can clearly see that in the wavelength area of the maximum energy radiation, the optical data of all components influencing the measurement must be well known, otherwise an error in the temperature evaluation can occur.

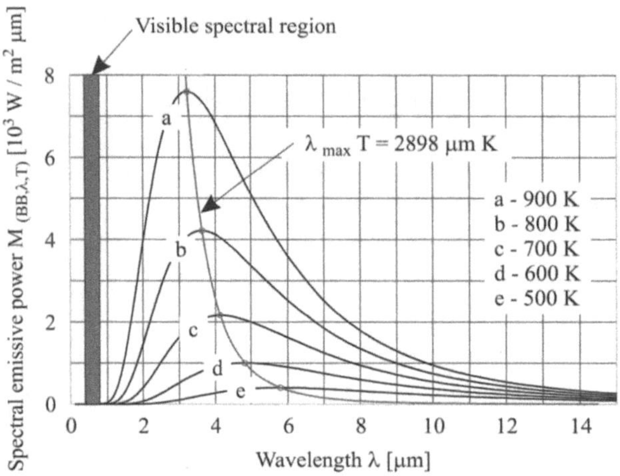

Fig. 14.3. Identical to Fig. 14.2. Drawn however in linear scale for λ

The relationship of the practical majority of radiated energy or temperature radiation, located in the region of the wavelength $\lambda_{\max}$, is clearly illustrated in Fig. 14.4. The function in Fig. 14.4 is known as the fraction function $F_{0-\lambda \cdot T}$ and is probably the most universal demonstration of the Planck radiation laws. $F_{0-\lambda \cdot T}$ is the fraction of total black body intensity or emission potential lying in the spectral region between the wavelength 0 and the wavelength λ. It applies to all temperatures and wavelengths (black body) and is important for all practical calculations. The wavelength examined is calculated from 0 to λ, the fraction function of a wavelength span is then determined as the difference of 2 values of fraction functions. If we integrate the Planck radiation equation for all wavelengths, we arrive at the Stephan–Boltzmann law. For black body radiation the radiated power, known as radiant emittance M, increases with the fourth power of temperature.

$$M = \sigma \cdot T^4 \quad \text{exact} \quad M_{(BB)} = \sigma \cdot T^4 \tag{14.5}$$

It should be noted that the units for M (radiant emittance) and $M_{(\lambda)}$ (spectral radiant emittance – a spectral property) are not identical. M is defined as the unit power per area (W/m^2) while $M_{(\lambda)}$ is the unit power per area and wavelength (W/(m$^2 \cdot \mu$m))! For real bodies and surfaces there are modifications which will be discussed later. The use of metals better fulfils the real

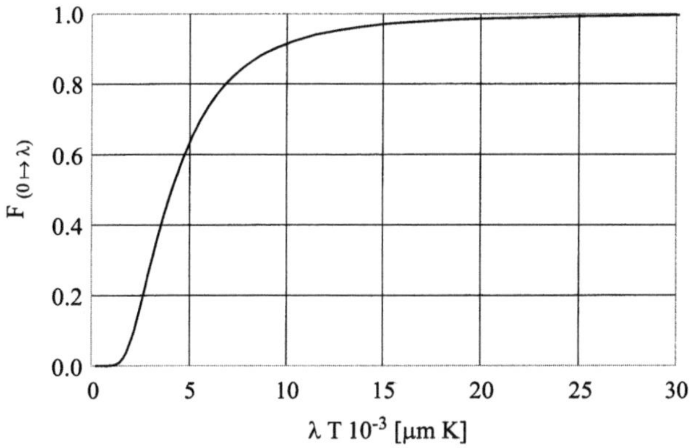

Fig. 14.4. Part of the radiant emittance M as a function of the product of λ times T, called as fraction function $F_{0-\lambda \cdot T}$

relationship $M \sim T^5$, but for dielectric materials the Stephan–Boltzmann law is correct. It is clear however that as thermography and pyrometry become more simplified, the object temperature (equipment price) will become higher. In order to describe a surface radiator properly, we also need the Kirchhoff radiation law. In a state of thermal equilibrium: the entire radiated energy absorbed by a body is again emitted as follows:

$$\text{emissivity } \varepsilon = \text{absorptivity } \alpha \tag{14.6}$$

Despite the fact that the radiant emittance M has the unit of a power, it is practical to use the term energy, as no measuring system can react so fast and is thus always a product of power and time as presented in a practical reading. This law (14.6) is constantly being confirmed in our surroundings. If (14.6) were not true ($\varepsilon \neq \alpha$), we should constantly be observing that via radiation exchange, bodies or objects in a relatively isothermal room would cool down or heat up relative to the surroundings. In addition to ε and α, one can also define the reflected radiation fraction ϱ as well as the fraction which passes unimpeded through the body (degree of transmission τ) and the fraction of light scattered by the body or degree of extinction e. In the literature this is also the sum $(\alpha + e)$ understood as extinction:

$$\alpha + \varrho + \tau + e = 1 \tag{14.7}$$

In a general case one must consider the spectral quantity:

$$\alpha_{(\lambda)} + \varrho_{(\lambda)} + \tau_{(\lambda)} + e_{(\lambda)} = 1 \tag{14.8}$$

This is clearly shown in Fig. 14.5. In the literature there are a very few adequate definitions that distinguish between the transmission factor (transmittance) and pure–transmission factor, whereby the transmisssion factor would then, in our case, be written as the sum of $e + \varrho + \tau$.

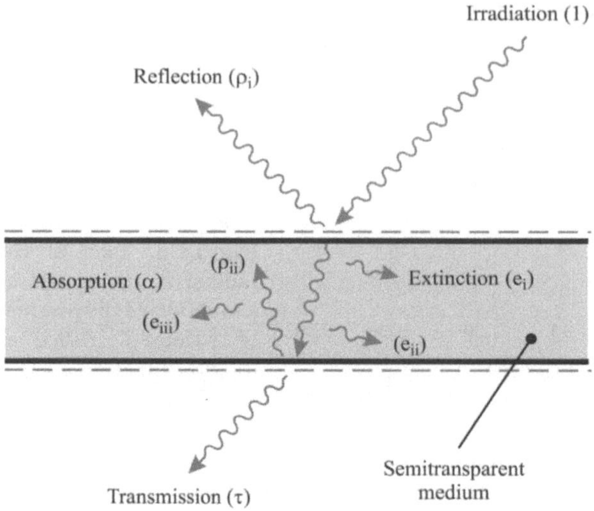

Fig. 14.5. An illustration of (14.8)

As we are investigating surface radiation only, and not volume radiation we have:

$$\alpha_{(\lambda)} + \varrho_{(\lambda)} = 1 \quad , \quad \varepsilon_{(\lambda)} + \varrho_{(\lambda)} = 1 \tag{14.9}$$

In (14.7) α can not actually be written in this manner since α was defined for "non–radiation permeable" surfaces. We understand α or $\alpha_{(\lambda)}$ as absorptivity in the surface; bulk effects are not in our discussion. We are now in a position to define the black body as:

$$\varepsilon_{(\lambda)} = \alpha_{(\lambda)} = 1 \tag{14.10}$$

for all values of the wavelength. Thus we can also write $\varrho_{(BB)} = 0$. This is true of radiation of all wavelengths for all the angles to the surface. A black body is physically ideal, it is a perfect absorber for optically incident radiation. In comparison to real temperature radiators, a black body radiates the maximum energy possible for a given temperature. This is clearly illustrated in Fig. 14.6. Grey radiators or grey bodies are defined as bodies which radiate a constant percentage of the energy of the black body over all wavelengths ($\varepsilon < 1$ and $\varepsilon_{(\lambda)} \approx const.$). This amount of energy is naturally smaller than that of a black body. The definition of a selective radiator is given in Fig. 14.6. Now we can write (14.5) in a more practical form as:

$$M_{(T)} = \sigma \cdot T^4 \cdot \varepsilon \tag{14.11}$$

Generally speaking, the temperature (T_{amb}) of the the radiators in the surroundings can not be neglected. In this case, the effective specific radiation is:

$$M_{(T)} = \sigma \cdot (T^4 - T^4_{\text{amb}}) \cdot \varepsilon \tag{14.12}$$

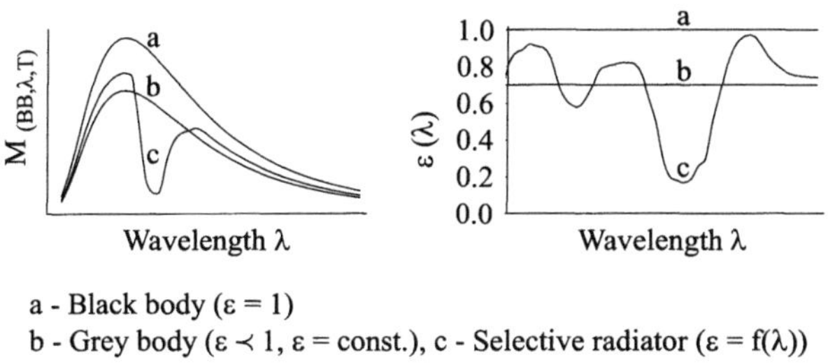

Fig. 14.6. Definition of grey and selective temperature radiators

with T_{amb} as the surrounding temperature. In (14.12) $\varepsilon_{\text{amb}} = 1$ is assumed as is usual in practice. When investigating a surface radiator, we are not measuring the entire specific radiation emanating from the object. As a consequence, we must define a quantity which takes into account the actual solid angle observed as already applied in Fig. 14.2.

$$L_{(T)} = \frac{M_{(T)}}{\Omega \cdot \cos\phi} \tag{14.13}$$

The radiance L reflects what portion of M is radiated into the solid angle Ω with the angle ϕ to the surface normal.

The difference between the temperature radiators and the non Planck radiators is important. The examples given are fluorescent tubes, lasers, LEDs etc. which are not temperature radiators. As a result, the above mentioned characteristics do not apply. They can, in specific wavelength regions, emit higher energy than their appropriate temperature! The physics of temperature radiators is extensively presented in introductory physics texts and literature so that they do not need to be listed in detail here. An additional quantity however has to be treated in more detail, the previously introduced degree of emittance in relation to a black body, the emissivity ε.

$$\varepsilon_{(T)} = \frac{M_{(T)}}{M_{(BB,T)}} \quad , \quad (\varepsilon_{(BB)} = 1) \tag{14.14}$$

ε can also of course be defined by spectral properties

$$\varepsilon_{(\lambda)} = \frac{M_{(\lambda)}}{M_{(BB,\lambda)}} \tag{14.15}$$

In addition, we define the emissivity via radiance to:

$$\varepsilon_{(T)} = \frac{L_{(T)}}{L_{(BB,T)}} \tag{14.16}$$

Equation (14.16) forms the condition as to how we can measure the emissivity in a test section in relation to a black body in the practical field. Fig. 14.7 illustrates how complex the practical application can be.

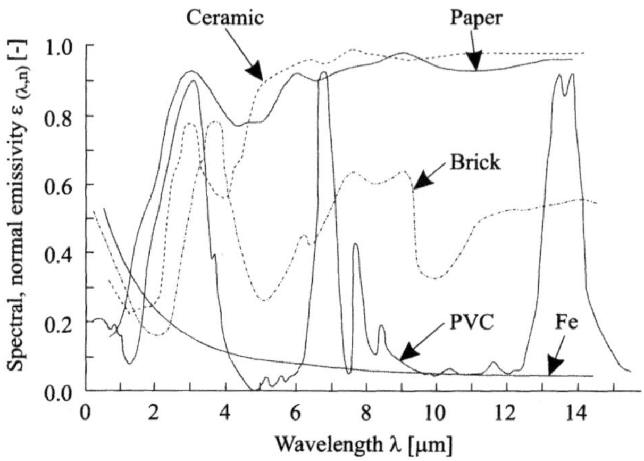

Fig. 14.7. Spectral emissivity values of various technical surfaces

Generally speaking, the emissivity is dependent on many factors.

$$\varepsilon \;=\; f(T, \phi, \psi, \lambda, SQ) \tag{14.17}$$

where SQ represents the surface quality. Furthermore T is the absolute temperature, λ the wavelength and ϕ and ψ the angles to the surface (acimuthal and zenithal). As a result, ε is a very difficult quantity to plot. A technical measuring evaluation of the emissivity is mostly eliminated in practical applications. For example, one can often apply an additional coating like a special varnish or soot to the surface that is to be measured in order to achieve an energetic evaluation via the pyrometer or thermographic equipment and record a valid temperature reading. Fig. 14.8 shows the angle dependency of the emissivity, which often leads to disturbing factors when conducting readings for metal surfaces. In the appendix are listed all the known definitions of ε. $\varepsilon_{(T)}$ and $\varepsilon_{(\lambda)}$ are of interest for practical purposes. Usually $\varepsilon_n^{;}$, the hemispherical emissivity measured normal to the surface is tabulated as a function of temperature.

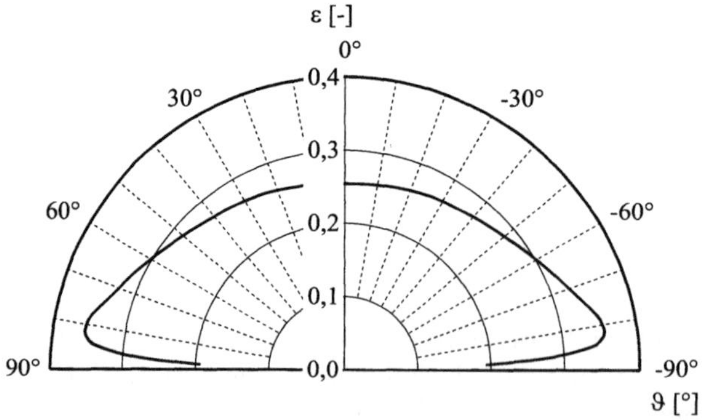

Fig. 14.8. Emissivity as a function of angle for a polished aluminium surface

It is important for a practical reading that the spectral emissivity usually increases with an increase in wavelength for non–metallic matter (s. also Fig. 14.16). The uncertainty concerning the value of ε is, as a result, smaller in the far infrared region.

Thus, we have represented the important characteristics of the surface to be measured via pyrometer or by thermographics. The essence is: the higher the temperature of an object at the surface, the lower the technical measurement effort required to arrive at a reliable statement of the surface temperature. The closer ε is to unity, the simpler is the evaluation reading. When the specific radiant emittance M is higher the reflected portion of the background radiation (surrounding area) becomes compulsorily smaller ($\varepsilon \uparrow \;\rightarrow\; \varrho \downarrow$). Polished metals are generally more difficult to measure with radiation thermometers (pyrometer) than dielectrics. This is due to their small $\varepsilon(0.05 - 0.4)$ and to the fact that metals exhibit a strong angular dependency of the emissivity arising from polarization effects. The temporal change of the emissivity value, eg. weather, ageing, oxidation, is also of importance in practical readings due to its low value. An example of our recordings is shown by stainless steel plates which when new $\varepsilon = 0.15$ and after being exposed to climate conditions (on a slope of a busy road for two months) $\varepsilon = 0.51$.

Additional information for the emissivity and some values of important materials are listed in a table in the appendix (Chap. 14.8.2).

14.3 Method of Transmission

Ideal conditions for transmission where the thermal radiation is not influenced by disturbances between the transmitter of the temperature radiation and the receiver detector for the optical radiation sensor can best be studied under laboratory conditions. Even in the laboratory the actual sensor must be protected against environmental influences with proper window material. As a result the detector window as well as the entry optics of the pyrometer or the thermographics system is included in the method of transmission. The temperature radiation is reduced by absorption, scattering and reflection. This reduction is naturally a function of the wavelength. For short transmission lengths (optical paths of several meters), air is viewed as an ideal medium under normal conditions of the atmospheric values e.g. with not too high content of water vapor. For terrestrial thermographics (infrared image processing) from satellites or airplanes, the atmospheric dampening of air is important in the selection of a detector (region of spectral responsivity). Fig. 14.9 shows the transmission of a 10 km layer of air as an example.

Three important atmospheric windows, the main regions of transmission, have been recognized whereby the proper detectors need to be selected according to the object (surrounding temperature 20°C $\rightarrow \lambda_{\max} \approx 10\ \mu m$ or jet exhaust of an airplane 800°C $\rightarrow \lambda_{\max} \approx 2.7\ \mu m$). For short range measurements we can imagine for example, measuring the temperature of a

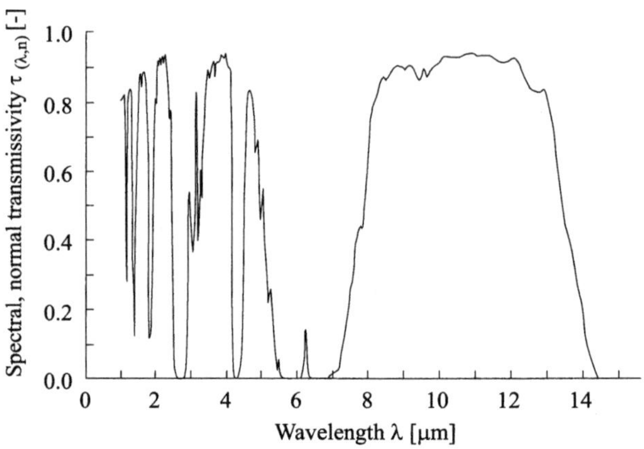

Fig. 14.9. Schematized transmission of a 10 km large air layer

probe through an oven window with a pyrometer. Under these circumstances, the glass plane is of extreme significance if the interaction temperature radiator or oven window detector device allows a proper reading to be taken. As stated previously this problem will be discussed later. Fig. 14.10 shows the permeability of various window materials.

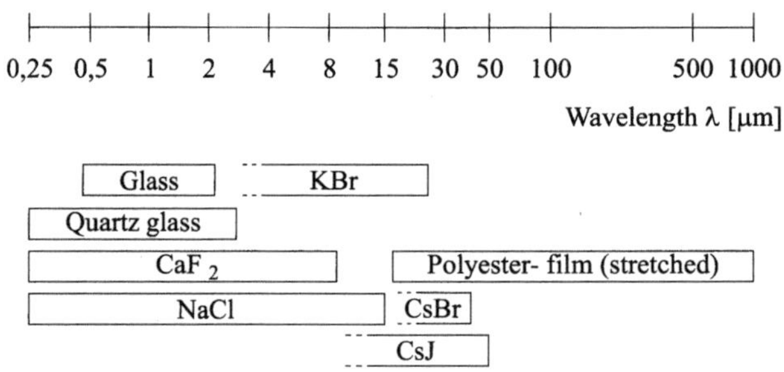

Fig. 14.10. Permeability of various window materials for sensors in the infrared spectral region

Germanium, for example, can be used in the wavelength region where the object to be measured emits radiation to the surrounding area. Glass, in this distant wavelength region, is not permeable. Thus I would like to mention here the greenhouse effect: short wave radiation from the sun passes through the glass roof and is absorbed by the ground. The ground becomes warmer, emits long wave radiation for which the greenhouse is "radiation sealed". The result is a rise in temperature into the greenhouse. It is clear that extreme conditions of the transmission method (blast furnaces, glass factory, etc.) as well as vapors and dusts can also impede the measurement reading.

Many window materials have much higher refractive indices in the infrared region than in the preferred visible region of glass, for example (Ge $\rightarrow n \approx 4$). The high refractive index causes a drastic increase in reflection making anti–reflex coatings of the individual optical components extremely important.

Through "proper" filtering of the optics, one can strongly influence the performance of the pyrometer or the thermographic system. Spectral significant disturbing radiation can also be minimized by using filters to effectively reduce the unwanted background percentage of the measured signal (as in sun filters infrared systems).

14.4 Radiation Receiver (Detector)

Of the operation and the actual physical mechanisms which actually enable thermal radiation to be recorded, the difference in **quant** and **nonquant** detectors is proposed and is the best way to a good understanding of physical background. With the exception of one pyrometric type, the filament pyrometers, all sensors and detectors can be catalogued under this scheme. For non–quantum detectors radiation is detected by heating up the entire sensor element; here we have a classical bulk effect. The advantage of these sensors is that by proper blackening (high value of ε) of the absorbing surface it is possible to obtain a uniform responsivity for a wide wavelength region. Thus one can define the spectral responsivity region by selecting the proper window materials. Two major disadvantages of these detectors are the very low responding speed (edge frequency several Hz) and the relatively low responsivity.

Table 14.1. Non–quantum detectors

Effect	Operation	Type of Detector
Seebeck – Effect	Thermoelectric effect between two different metals.	Thermocouple Thermopile
Bolometer – Effect	$R = f(T)$. Electrical conduction is a function of temperature.	Bolometer (Metal) Thermistor (Semiconductor) Superconductor–Bolometer
Thermo – Pneumatic Effect	$p = F(T)_{\text{gas}}$, for a definit gas volume.	Golay – Detector
Thermo – Optical Effect	Change of optical properties of materials.	Thermo – and Luminicence detectors.

Tables 14.1 and 14.2 list several quantum detectors whereby differences are shown between the internal and external photo effect. Several infrared sensitive films can also classified as "quantum detectors" but cannot be clearly categorized into external and internal photo effect. In an external photo effect, an incident light quant whose energy is characterized by $h \cdot \nu$ (a light quant with a shorter wavelength ($\lambda \downarrow \;\; \rightarrow \;\; \nu \uparrow$) has more energy than a light quant with a greater wavelength) releases an electron out of the solid

Table 14.2. Quantum detectors

Effect	Operation	Type of Detector
a) Quantum detectors → Photo effect		
Photocathode	Electron (in vacuum) is accelerated by an electric field. Target is a fluorescent screen.	Vidicon tube TV camera Photomultiplier
b) Quantum detectors → Photoconductive effect		
Semiconductor detector	Light quant changes the conductivity (Photo conductor). Light quant generates an elctron in the "frozen electrical field" (p–n junction of a diode), as a result → voltage.	PbS – Detector CdHgTe – Detector InSb – Detector (Solarcell)

state bond. It holds that $h \cdot \nu \geq W_p$ whereby W_p is the electron work function. As a result, the electron can be influenced by electrical fields, for example in a vacuum. The electron can be accelerated, takes on energy in an electrical field and appears on the screen where it releases kinetic energy and thus activated states of the illuminating phosphor which leads eventually a to recognizable screen illumination with an image amplifier. With the internal photo effect the absorbed light quant leads to a change of conductivity either by forming an electron–defect electron or a electron is carried up from the valence band to the conduction band. With defined semiconductors only a light quant whose

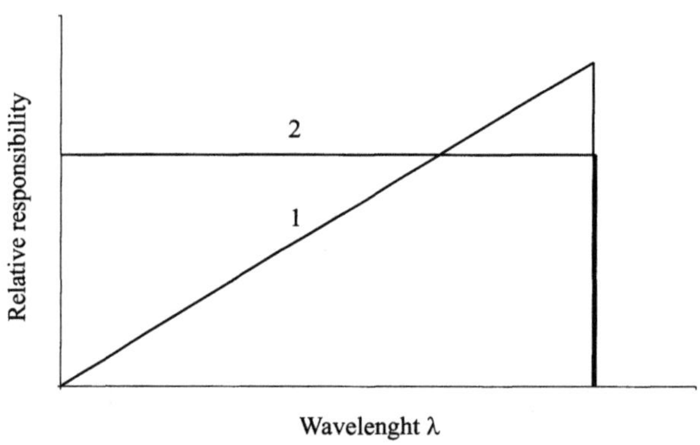

Fig. 14.11. Qualitative responsivity progression of quantum detectors– in relationship to the same number of light quants (2), and in relationship to the irradiated energy (1)

energy $(h \cdot \nu)$ is larger than the band distance in the semiconductor can lead to absorption (fundamental absorption). An example of this effect is shown by the following.

In the shortwave wavelength region, germanium can be used as a semiconductor material for photo detectors (absorption). Wavelengths whose light quants ($h \cdot \nu <$ band gap) do not have enough energy to "lift" an electron from the valence band into the conduction band cannot be absorbed. Germanium is also suitable as a window material for infrared detectors beyond this cut–off wavelength (no absorption, only reflection on the surface). This leads, as a result, to the following responsivity characteristics (s. also Fig. 14.11). In Fig. 14.11 the spectral responsivity is represented in relationship to the irradiated energy (curve 1) and in relationship to the constant number of light quants (curve 2). Fig. 14.12 shows several spectral responsivities of photo detectors used today.

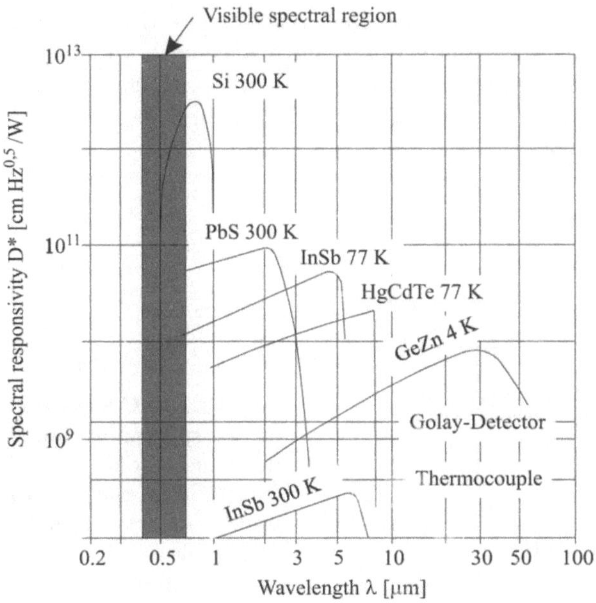

Fig. 14.12. Commonly used detectors (quantum and non–quantum detectors) and their spectral responsivity (schematized)

If detectors are employed in the far wavelengths (far IR) or with lower signal levels, a more favorable signal to noise relationship detectivity can be obtained by cooling the detector element. When quantum detectors are employed in the far infrared region cooling is important. These detectors, for example CdHgTe (CMT–material), have, when used in far infrared regions, such a narrow band gap that at room temperature the conduction band becomes "flooded" with charged carriers via thermal activation (electrons) and the material looses its semiconductor properties in this temperature region. CMT–material as a ternary semiconductor, from a technological standpoint, is difficult to work with. Binary semiconductors, such as InSb are easier to employ. Ternary semiconductor materials have the definite advantage based on their relatively "free" composition in band gap. Band gap determines the wavelength of maximum responsivity so it is possible to compose a detector which gives a good definition of the measurement problem. The important characteristics for practical application of photodetectors e.g. detectivity and responsivity are listed in the appendix.

It is important to consider the region of the spectral responsiveness as an optimizing criterion when using quantum detectors. In addition, the differences between the two spectrally limited detectors are listed as an example. Maximum radiation power on the detector itself is required. Figure 14.13 illustrates this optimization problem. One recognizes that for up to 500 K the long wave detector is better, temperatures above 500 K are sensed better by the detector in the 3 – 5 μm region. This detector "sees" more of the object–radiation. To summarize, we can make the following qualifying points about quantum detectors: the advantages are high reaction speed and higher responsivity, as a disadvantage there is a maximum wavelength which limits application. Quants, whose energy is smaller than the band gap and/or electron work function, lead to no technical measurement in the interaction between radiation and detector.

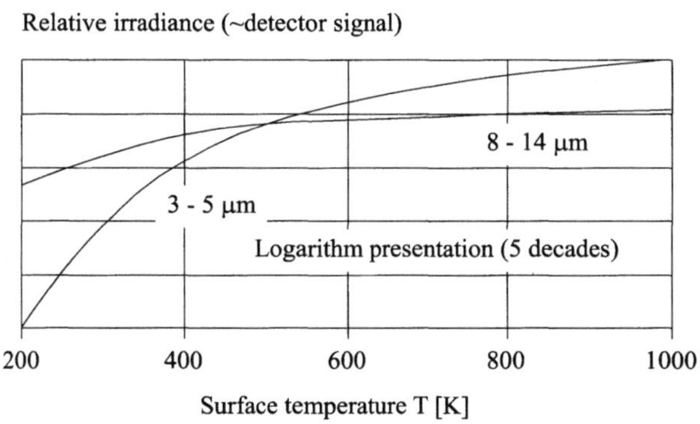

Fig. 14.13. Comparison of two spectrally limited detectors, from 500 K ($\approx$ 227 °C), the "short wave" detector is more favourable

14.5 Thermal Cameras – Thermography Image Systems

We have now described the conditions and prerequisites necessary to understand how thermographic equipment operates. This, of course, also holds true for the pyrometer described in Chap. 14.6. Theoretically, it is possible to generate a thermal photograph with one single photo sensitive sensor, thus creating a temperature distribution of an object to be investigated. This is frequently employed in practice, but would lead to relatively long recording times (typical image generating time $\approx$ 20 Hz). In practice today, technically advanced and expensive systems use an array–configuration which allows parallel recording of several "image lines" or "interlacing" simultaneously. One can recognize two major directions of development which are currently being followed. Array detector configurations are usually selected where fast, contrasting image information is more important than quantitative temperature. For example, infrared sight flying (forward looking infra red) FLIR–systems at night or under poor visibility–primarily for military use. With technical measuring applications, a faster single detector is preferred for a "temperature photograph" since such problems as calibration, ageing and spectral responsivity which influence the recording results are technologically more easily solved for a single detector element.

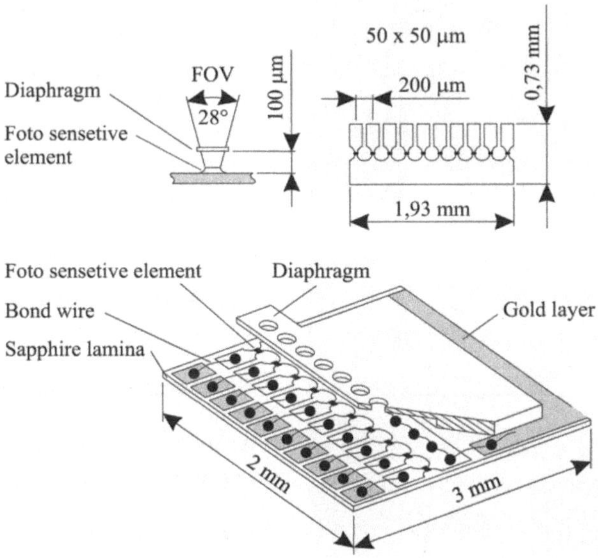

Fig. 14.14. Cross–section view of detector array (cooled) for thermographic equipment. The diaphragm (FOV) is located in the cooled region which leads to higher values for the signal to noise ratio

In this special case the distance between the individual photo sensitive elements is (50 μm × 50 μm) which means an entire full screen display with an interlacing factor of 4 is achieved. A proper colour TV–picture needs about 400 lines. We must also have 100 individual elements (in Fig. 14.14 only 10 elements are shown). This may be realized with two synchronized oscillating–mirror systems: one mirror scans the image area while, a second mirror realizes the line spacing (interlacing). The trend is currently towards thermoelectrically cooled detectors (InSb) which achieve satisfactory results in practical applications. The system resolution or decomposition of the optical system, is determined by the detector area and diaphragm and does not achieve the resolution quality that one obtains with electro–optical systems in the visible region due to physical reasons e.g. wavelength.

In the last years we looked back to two important developments for thermography systems. On the one hand Focal Plane Array (FPA), array detectors came successfully on the market. On the other hand, microbolometer were realized as FPA detectors with good results. The advantage of microbolometer–FPA that you need no high performance in cooling equipment. In this direction we have a lot of new developments in future.

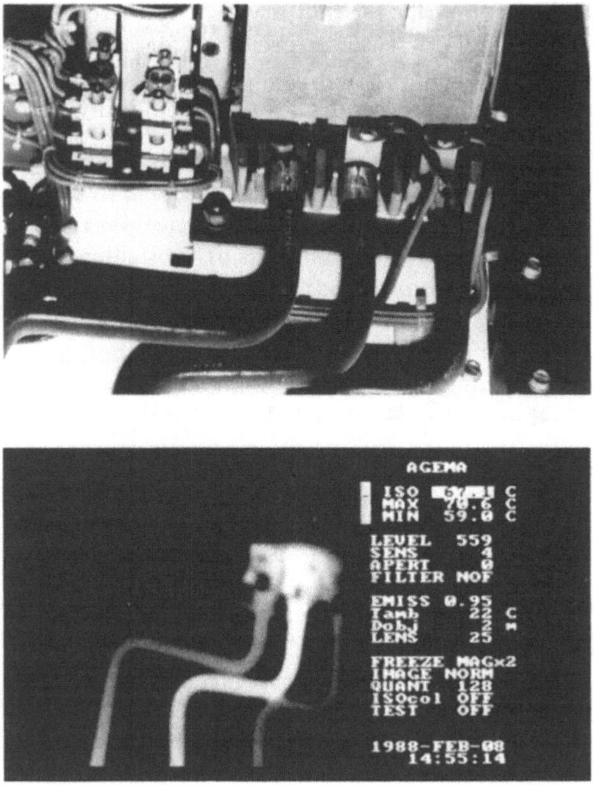

Fig. 14.15. Above–Picture in the visible area, below–thermographic photograph (segment) with easily recognizable heated areas. (Pictures, AGEMA IR–Systems GmbH, Germany / today a company of FLIR Systems, Portland).

Fig. 14.15 was provided by AGEMA Company, Germany. Here one recognizes the "hot" areas of an electrical connecting block. This picture is also a good example of how IR–systems are gaining importance in material and system investigations. The temperature information is merely seen as evidence for additional area of complexity. For example: hot spots, weak insulation points, cavity in glassfibre equipped polymer structures, etc. The temperature resolution as a result is more important than the accuracy of the absolute value of temperature.

14.6 Pyrometers

Radiation thermometry – pyrometry provides accurate and efficient measurement and control of temperatures because there is no physical contact with the surface from the object. Pyrometry is the preferred method for temperature measuring in the following cases:

- Small object and or high velocity
- Fragile goods
- Thin films or continuous sheets
- Very high differences in temperatures to surroundings
- Non–ideal conditions in the location (vacuum recipient)
- Possibility of direct contact with thermometers

When these or similar conditions exist, then a pyrometer is an extremely valuable instrument for controlling and measurement. In many cases the temperature is reached faster and more accurately than with other devices. It is very easy to develop an understanding of measurement, sensing and how an electrical signal is generated in relation to the irradiation energy.

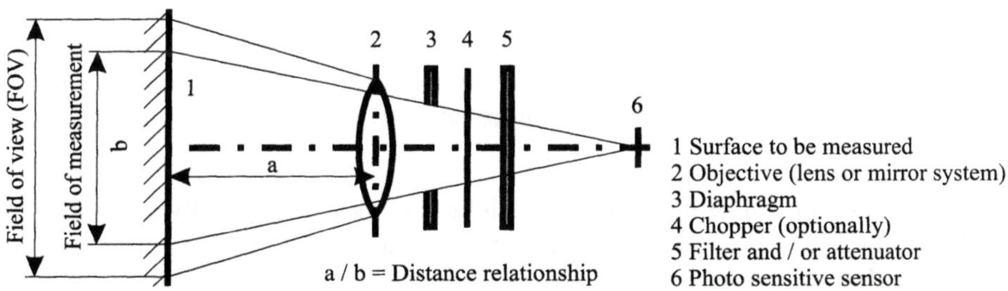

Fig. 14.16. The diagram illustrates the basic definitions the concept of a radiation pyrometer

The radiation receiver (sensor, detector) evaluates the integrally obtained temperature radiation from field *b*. The spurious radiation result emanates from all the radiating surfaces from the area in the remaining angle and the reflected radiation from the surface coming from the background. A pyrometer that is used to measure very low temperatures encounters difficulty here because the solid angle region from which the spurious and disturbing radiation of the construction falls on the pyrometer housing detector, is usually much larger than the solid angle of the radiation to be measured. One possibility would be to insulate the unit body; yet this can involve a great deal of effort. A more practical solution is to modulate the radiation from the measurement object via the chopper. Only the modulated signal segment is then electronically processed, i.e. the spurious radiation, the disturbing radiation of the unit body is greatly reduced and no longer disturbs the measured signal. The AC portion of the signal, produced via chopper, forms the signal. The DC aspect will be greatly reduced. Fig. 14.17 shows a schematic view of this type of construction. The modulated fraction of the radiation is conducted to the pre–amplifier via an electrical capacity. The constant part of radiation or unchopped radiation–thermal background radiation from the pyrometer housing body–no longer carries a signal after the modulator or Chopper. The electric capacitor creates an AC signal and blocks DC fractions. From a construction viewpoint, the modulator (impeller, diaphragm with free segments, oscillator mirror or similar unit) must be built in such a way that the emission potential on the sensor side is kept to a minimum. The object side from the chopper should also be free of reflecting construction parts. Fig. 14.17 shows the mirror optics used with Cassegrain optic, especially in pyrometers that are used to record low object temperatures. It is more cost effective to work with mirrors, as oppsed to more expensive lens optics.

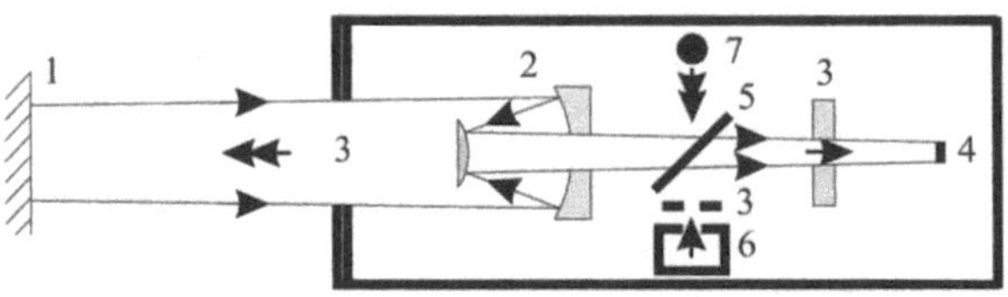

1 Surface
2 Cassegrain-Objective Pyrometer
3 Diaphragm
4 Sensor
5 Chopper (mirror)
6 Reference (black body)
7 Spot lamp

Fig. 14.17. Schematic view of radiation pyrometers

The question of which recording field the receiver actually evaluates is rather difficult to answer. To eliminate errors here, the manufacturer recommends either a certain distance relationship (Fig. 14.16) or offers instrumental aids to avoid errors. Pyrometers used for low object temperatures can, for example, mark the measurement region with a light spot – an illuminating beam from the pyrometer. During the actual measuring operation the beam is then switched off. These aids should not be employed if the object temperatures are high enough that the radiation percentage in the visible region leads to noticeable self–illumination in the visible region. Many pyrometers have a view finder. A circular ring in the viewing optics, for example, helps in selecting the surface to be measured.

The calibration of the pyrometer is generally related to the emission of the black body. As a result, the emissivity ε of real radiators must be known in order to obtain a relevant recording value from the signal. The manufacturer either supplies a correction table with the device or electronically modifies the recorded value. This means that the pyrometer has an adjusting mechanism with which one can take into account the emissivity value of the recording surface. Let us return to (14.9), as a result with

$$\varrho_{(\lambda)} = 1 - \varepsilon_{(\lambda)} \tag{14.18}$$

The radiance of a surface radiator can be written as follows:

$$L_{(\lambda)} = \varepsilon_{(\lambda)} \cdot L_{(\lambda,T)} + \varrho_{(\lambda)} \cdot L_{(\lambda,T_{\mathrm{amb}})} \tag{14.19}$$

whereby $\varepsilon_{(\lambda)} \cdot L_{(\lambda,T)}$ is the emitted and $\varrho_{(\lambda)} \cdot L_{(\lambda,T_{\mathrm{amb}})}$ is the reflected fraction of the radiation of a surface (special case $\varepsilon_{(\lambda)} = 1 \rightarrow \varrho_{(\lambda)} = 0$). If the transmission is ideal, the subsequent irradance E is the resultant radiation density on the receiver, written as

$$E_{(\lambda)} = A \cdot d^{-2} \cdot L_{(\lambda)} \tag{14.20}$$

Applying equation varions (14.19) and (14.20), we obtain

$$E_{(\lambda)} = A \cdot d^{-2} \cdot (\varepsilon_{(\lambda)} \cdot L_{(\lambda,T)} + \varrho_{(\lambda)} \cdot L_{(\lambda,T_{\mathrm{amb}})}) \tag{14.21}$$

For the black body radiator, (14.21) can be simplified to

$$E = A \cdot d^{-2} \cdot L_{(T)} \tag{14.22}$$

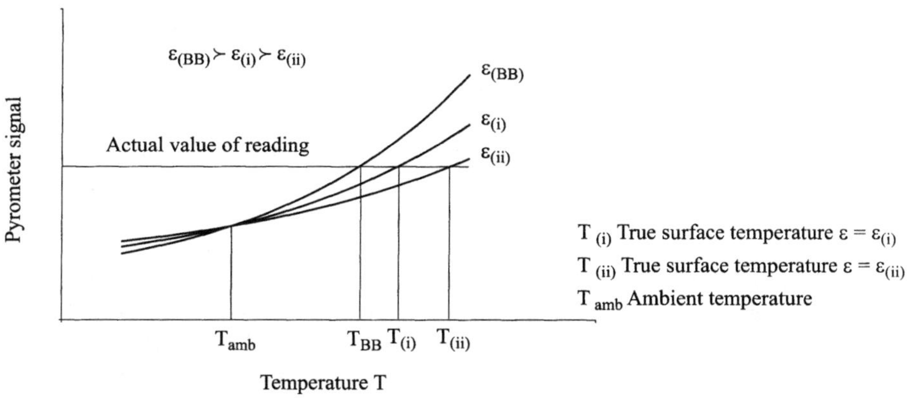

Fig. 14.18. Signal function of a pyrometer

Fig. 14.18 provides another opportunity to understand this procedure. Is it also important that the pyrometer signal is **independent** from the distance when a homogeneous temperated surface is examined (E decreases quadratically, A increases quadratically for a circular measuring field). The temperature T can be determined from the pyrometer reading only if ε is adequately known. The functions in Fig. 14.18 must, strictly speaking, be known for various practical surrounding temperatures T_{amb} (ambient temperatures).

14.6.1 Classification According to Construction Types

Pyrometers can be grouped into various categories. Radiation percentages within the visible spectral area are sensed by optical pyrometers though this definition can be very misleading because the region of optical radiation is larger than the visible area. The second group can be categorized as thermoelectrical pyrometers. There is also a third distinguishing factor. Pyrometers which assesses the "entire" emitted spectral region are identified as "total–radiation pyrometers". However a pyrometer which assesses 90 % of the entire radiation is still considered as a "total–radiation pyrometer" as opposed to spectral– and partial–radiation pyrometers which are sometimes also classified as band–radiation pyrometers. A spectral pyrometer is usually a very narrow band sensitive pyrometer. We have learned from the previous chapters that total–radiation pyrometers are employed in areas with very low object temperatures, where detectors whose responsivity in the far infrared is determined by the internal photo effect fail due to extensive cooling requirements and resulting high costs. Pyrometers are commonly named after the type of recording process by the various manufacturers, e.g. PbS–Chopper pyrometer, etc.

14.6.2 Filament and Quotient Pyrometers

The quotient pyrometer employs the characteristic that E is usually constant for smaller wavelengths. In two almost neighboring wavelength areas a reading is taken which is based on the relationship between the two. As a result, E is shortened from the evaluation and we achieve a very stable reading without prior knowledge of the emission value. External influences which effect the optical path and specific damping are disregarded, in so far as they have the "same" effect on both wavelength areas.

The filament pyrometer is used extensively today. However due to its physical operation it is not included in outline mentioned above. Up to the present, employed recording units have been employed which are based on a specific irradiant of radiation density on the pyrometer. The filament pyrometer directly evaluates the radiation density on the surface with color intensity comparisons. Filament pyrometers are not adversely influenced by the recording distance from the object if the optical paths are not considerably long. The only limitation is that the glowing filament must adequately cover the field to be measured. The operation is as follows: A self–illuminating object is focused with the filament pyrometer. The pyrometer contains a calibrated lamp with a tungsten filament. The amount of energy which flows through the lamp is regulated in such a manner that the filament disappears in front of the background, in this case the object which will be measured. The radiant density of the object to be measured and the lamp are identical. The ampere meter which indicates the lamp current can be directly calibrated by the temperature values. The emissivity must also be known. Errors may occur during the procedure if the operator attempts to regulate comparisions of the brightness colour.

14.6.3 IR Recording Heads

For monitoring and automatization tasks more simplified pyrometers are utilized which are not strictly designed to measure temperatures. Under certain circumstances IR recording heads are equipped with only one adjustable event electronic which indicates only conditions such as an open or closed oven door, automatically transported goods adequately heated yes/no, and other similar temperature problems. These simple pyrometers are becoming more important in automatic production and in control procedures.

14.7 Error Potential

14.7.1 Error Sources During Recording

Here is a brief list of error sources that can occur during measurement readings:

- Emissivity not adequately known
- Lack of homogeneous temperature in measuring field
- Receiver "integrated" above an area of the measured surface which does not correspond to the recording objective i.e. cold or hot neighbouring areas – portions of slag of a foundary
- The actual temperature radiation is disturbed by a reflecting background or disturbing radiation that cannot be corrected
- The transmission path between the object and the pyrometer is underestimated by the spectral influences (not properly read due to influences from the spectral field).

This last item will be discussed further at a later point in this work thus illustrating further the difficulties of influences on the spectral field. When utilizing a total–radiation pyrometer the object is to be recorded through a glass oven window.

Fig. 14.19 illustrates that a reading may be discussed in a proper sense only at very high temperatures. At lower temperatures the signal value of the pyrometer cannot be correlated to the actual temperature curve. One is immediately confronted with an array of difficulties when trying to obtain a useful reading if the idea that the signal value can be corrected with the transmission of the glass plate is followed. The transmission of a glass plate of various thickness is strictly nonlinear. The reflection on both surfaces requires a transmission lower than 100 % also without $\alpha = 0$! But the degree of transmission that can be reached is 100 % ($\tau = 1$), see (14.7), (14.8) and

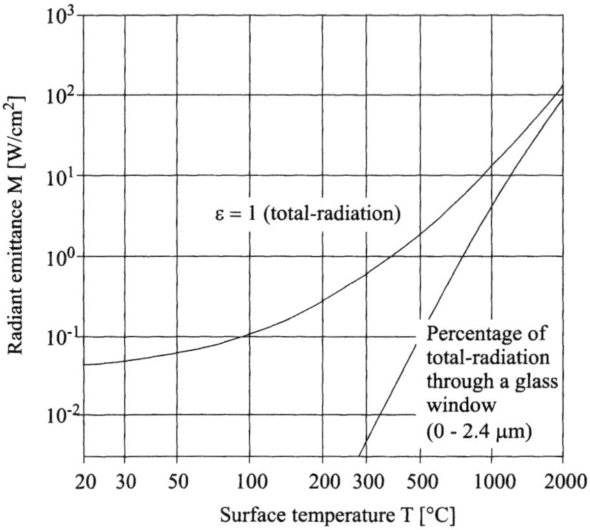

Fig. 14.19. Coordinate relationship during a reading with a total radiation pyrometer through a glass plate

the following remark. It should also be noted that glass as a technical material, according to the additional substances it contains, can have a strong diverging influence on the transmission spectrum. Traces of metal have a very large influence on the transmission–particularly iron. This example again illustrates the extreme error potential that can exist during temperature readings in practice and how important it is to recognize the physical factors of pyrometrics involved during a recording. As a rule, one would work with a spectral pyrometer when confronted with these recording problems, and the transmission of a glass plate should also be considered.

An additional example serves to give an idea of error sources when using a spectral pyrometer. Fig. 14.20 assumes on two narrow band sensitive pyrometers. Pyrometer 1 is sensitive at λ_1 and pyrometer 2 is sensitive at λ_2. The relative change in signal between temperatures T_1 and T_2 is dependent on the form of the function of the Planck radiation law, with the wavelength larger at wavelength λ_1 than the λ_2. One recognizes that when using a spectral pyrometer, one is tempted to select the shortest wavelengths of spectral responsivity. It is important to note that with decreasing wavelengths where the spectral pyrometer is more responsive, the lower limit of the minimal object temperature is reached. In the first case a minuscule amount of temperature radiation can be detected (T_3, Fig. 14.20); the second case occurs when the temperature radiation reaching the receiver creates a signal which cannot be properly pinpointed due to detector and amplifier noise.

14.7.2 Equipment Error

- Due to ageing, the radiation receiver has a decrease in its absolute and spectral responsivity

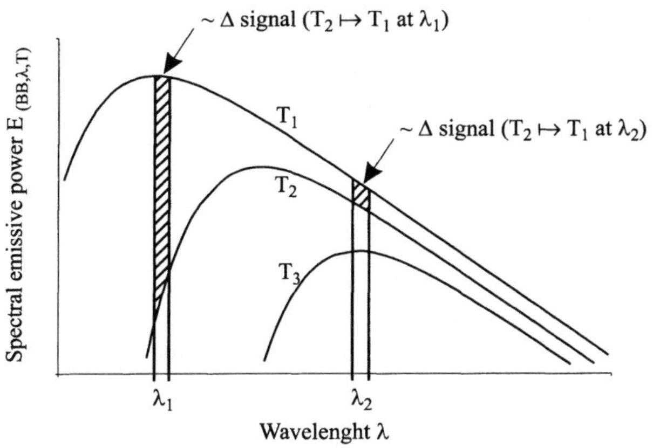

Fig. 14.20. The relative temperature sensitivity of a pyrometer increases towards the shorter wavelengths

- The spectral region which is "seen" by the receiver, corresponds no longer to the recording problem and
- Disturbing radiation (housing/body temperature) is unadequately compensated for

One recognizes that numerous errors can occur during a recording and that a precise reading requires enough knowledge about the problem in hand. Nevertheless, we list common error values that can occur during any recording operation to serve as a guide. The resolution of the pyrometer is generally extremely high. Given this limitation the following difficulties may occur during transforming the signal to temperature readings:

- Accuracy of the emissivity at defined surfaces 5–10 % (table values)
- "Estimation" of the surface quality (rough–smooth etc.) 5–10 %, in addition to the irradiance
- System inaccuracies of the pyrometer
  1. total–radiation pyrometers ± 1 to 5 K
  2. spectral pyrometers ± 1 to 3 K
  3. the same holds true for very high temperatures ($>$ 1500 °C $>$ 10 K).

This information is orientational in character and the given values can be minimized with an increase in operating expenses as seen in the following example. The temperature of a flexible plastic film is to be evaluated with a radiation pyrometer. It is known from laboratory experiments that the plastic material between the wavelengths λ_1 to λ_2 has a strong absorbing band, and in this region it can be selectively viewed as a "black body". If we use a spectral pyrometer, whose spectral responsivity is between λ_1 and λ_2, the measurement task can be simplified by assuming $\varepsilon = 1$, subsequently we have nearly the ideal conditions for measurement.

14.7.3 Problems with Thermograph Readings

In general, the majority of thermograph manufacturers claim to record temperature fields with a very high accuracy. The problems and error sources that occur with radiation thermometers also apply when using thermal cameras. Here we can recognize that the temperature resolution of a thermal camera ($\Delta T < 0.1$ K) should not be confused with the evaluation of an absolute temperature–which cannot be better than comparable pyrometers. In practice, the question to analyze is not to determine what the surface temperatures of an object are, but the relative differences. For example, we can, without a doubt, determine the blood circulation problems of a patient who smokes too much; the actual absolute skin surface temperature plays a secondary role and is deterred by the temperature radiation in the room and many other factors. The advantages of thermographics lies in locating the "weak spots", be they hot spots in microchips, or inadequate insulation of high–voltage powerlines or insulation defects in a building. The following example illustrates how limited the recording possibilities of thermal cameras can be. A building facade is to be analyzed via thermographics. A reading is only then effective when there is a temperature difference of at least 10 to 15 K between the room and the outside temperature. Since the sun can negatively influence the reading, it should be conducted at night so that the memory effect of the sun's influence on the facade construction is not detected as an apparent "leak". A relatively long period of time must lie between sunset and measurement to achieve a proper reading. If the building facade has an extra, rear–aired isolation layer, a recording of the outer side with a thermal camera is relatively useless. Which demonstrates the definite limitations of thermographics applied in pratice. Nevertheless, thermographic and pyrometric applications, when properly employed, are very effective and sometimes indispensable recording methods. See appendix for additional information.

14.8 Appendix

14.8.1 Important Constants

Table 14.3. List of important constants

constant	value	unit	name
c_0	$2.998 \cdot 10^{8}$	$\mathrm{m \cdot s^{-1}}$	speed of electromagnetic waves in vacuum
c_1	$5.955 \cdot 10^{-17}$	$\mathrm{W \cdot m^{2}}$	1.constant in Plancks energy distribution
c_2	$1.439 \cdot 10^{-2}$	$\mathrm{m \cdot K}$	2. constant in Plancks energy distribution
c_3	$2.898 \cdot 10^{-3}$	$\mathrm{m \cdot K}$	constant in Wien's displacement law
e_0	$1.602 \cdot 10^{-19}$	$\mathrm{A \cdot s}$	electrical elemental charge
h	$6.626 \cdot 10^{-34}$	$\mathrm{J \cdot s}$	Planck's constant
k	$1.3807 \cdot 10^{-23}$	$\mathrm{N \cdot m \cdot K^{-1}}$	Boltzmann constant
σ	$5.670 \cdot 10^{-8}$	$\mathrm{W \cdot m^{-2} \cdot K^{-4}}$	Stefan–Boltzmann constant

14.8.2 Further Information and Tables

Herschel, Friedrich Wilhelm

Born on 15th of November 1758 in Hannover, Germany. Herschel moved as child to London and worked later as director of a music corps. In this time he has also studied mathematics and astronomy. On 13 March 1781 he discovered Uranus, naming the planet in honour of the King of England (Georgium sidus). To acknowledge his accomplishment Herschel was appointed astronomer to the royal court of George III.

In the year 1800 Herschel discovered that a blackened liquid thermometer in the spectrally broken sunlight, according to spectral colour, and under uniform conditions, displayed different temperatures. He was fully aware of the importance of his findings in that area which followed after the red end of the sun's spectrum, the highest temperatures were found. Herschel called this area "dark heat". Later, this spectral region was known as the "infrared spectral region", shortened IR to the region. Today we generally speak of a near IR, a middle IR, and a far infrared (IR) region. Since the transition to microwaves is continuous according to the application employed, a region of "extreme infrared" up to $\lambda = 1000\ \mu m$ is given in the literature.

Planck, Max

German physicist, born on 23 April 1858, awarded the Nobel Prize in 1918. Based on his discovery of a proportionality factor (h), the "Planck constant" he found the equation form which describes the radiated energy distribution of a black radiator, the black body (1899/1900). He assumed for a long period of time the mathematical "inaccuracy" of his equation, knowing that energy can only exist in quants (multiple values of h) which became the basis of the quant theory. The Planck radiation law was confirmed by experiments conducted by Rubens and Michel in 1921 which finally ended the discussions concerning the validity of the radiation laws.

Definitions of Emissivity

Emissivity is the quantity which describes the extent a real body (surface) radiates energy compared to the ideal black body. We need further definitions and normalies as we do have not enough information about the value. Mostly, we need an averaging over all wavelengths and over all directions (completely solid angle). This is the hemispherical total emissivity or the averaged emissivity:

$$\varepsilon = f(T) \tag{14.23}$$

It is an emissivity averaged over all wavelengths and the total solid angle evaluated.

For a concrete wavelength λ, the emissivity is a spectral quantity

$$\varepsilon_{(\lambda)} = f(\lambda, T) \tag{14.24}$$

the hemispherical spectral emissivity. It is averaged over all the solid angle (Ω) evaluated quantity at the wavelength λ.

The directional total emissivity

$$\varepsilon' = f(\phi, \psi, T) \tag{14.25}$$

is integrated over all wavelengths and over the angle ϕ and ψ. Most frequently, this quantity is given as $\varepsilon' = \varepsilon'_n$. The emissivity measured normal to the surface is denoted as ε_n (without dash). The dash symbolizes a directional quantity.

$$\varepsilon'_{(\lambda)} = f(\lambda, \phi, \psi, T) \tag{14.26}$$

is the directional spectral emissivity. In contrast to ε' , $\varepsilon'_{(\lambda)}$ is a spectral quantity.

Additional Remark

The ideal black body is characterized by the fact that all the definitions from above are identical:

$$\varepsilon_{(BB)} = \varepsilon_{(\lambda,BB)} = \varepsilon'_{(BB)} = \varepsilon'_{(\lambda,BB)} = 1 \tag{14.27}$$

Here we recognize that this idealization is physically impossible to achieve. A special construction of a cavity (a hollowed form of a metal cylinder with a small opening and high emissivity on the inner cavity surfaces), with small angle deviations in direction of the surface normal can reach

$$\varepsilon_{\text{CAVITY}} > 0.99 \qquad \text{or} \qquad \varepsilon_{\text{LABORATORY}-BB} \approx 1 \tag{14.28}$$

This is reached by multiple reflections in the cavity.

Important Differences

Concerning the previously defined quantity ε, it should be noted that for the **volume radiators** (for example flames, glass enamels) an emissivity coefficient $\tilde{\varepsilon}$ analogous to the emissivity ε is defined. In contrast to the emissivity (unit 1), the emissivity coefficient $\tilde{\varepsilon}$ has the unit, for example, $\mathrm{W} \cdot \mathrm{m}^{-4} \cdot \mathrm{sr}^{-1}$, since the layer thickness and the volume as well as the solid angele which is emitted by the radiation, are included here.

Figure of Merit from which Detectors

The responsivity of radiation sensors (detectors) is related to the receiver area:

$$R_S = \frac{U_S}{E \cdot A} \tag{14.29}$$

An additional feature that allows the comparison of different detectors is the detectivity D^* (spelling D star).

$$D^* = \frac{\sqrt{A} \cdot \sqrt{\Delta f}}{NEI \cdot A} \tag{14.30}$$

Here, Δf is the electronic measuring bandwidth, and NEI the noise equivalent irradiance, where a signal–noise relationship of 1 (U_S/U_R) is reached. The product $NEI \cdot A$ represents a noise equivalent radiation irradiant on the receiver area. The quantity D^* represents recording conditions such as modulation frequency, reference radiators etc. The manufacturers usually attempt to specify the most favorable measuring parameters for their detectors. This is important to note when comparing and purchasing detectors.

Table 14.4. Black body function–fraction function

$\lambda \cdot T$ in $\mu m \cdot K$	Fraction function $F_{0-\lambda \cdot T}$	$\lambda \cdot T$ in $\mu m \cdot K$	Fraction function $F_{0-\lambda \cdot T}$
1000	0.000321	7000	0.80807
1500	0.01285	8000	0.85625
2000	0.06673	9000	0.88999
2250	0.11031	10000	0.91416
2500	0.16135	15000	0.96893
3000	0.27322	20000	0.98555
3500	0.38290	30000	0.99529
4000	0.48086	40000	0.99792
5000	0.63372	50000	0.99890
6000	0.73779	60000	0.99935

Table 14.5. An example of the effect $\varepsilon = f(\lambda)$

Wavelength region in μm	1.8 ... 2.7	3 ... 5	8 ... 14	
Vigorous leaves	0.67	0.90	0.92	Values for the
Sand	0.56	0.82	0.93	emissivity $\varepsilon_{(\lambda)}$

Table 14.6. Normal total emissivity ε'_n tabled as ε

Material	Temperature in K	ε
Metals (electrica)		
Aluminum		
– highly polished	480	0.038
– heavily oxidized	370	$0.2 \div 0.35$
– thick oxid layer (Eloxal)	360	$0.7 \div 0.9$
Brass		
– polished	370	0.09
– dull	370	0.2
– oxidized	370	> 0.6
Copper		
– highly polished	310	0.02
– black oxidized	310	0.78
Gold polished	370	0.018
Nickel		
– polished	370	0.1
– oxidized	370	> 0.6
Steel		
– polished sheet	310	0.1
– rough oxide layer on sheet	310	0.8
– stainless steel (1.4145 ≈ type 301)		
brush polished	300	0.15
	370	0.17
Tungsten	310	0.03
– filament	3500	< 0.4

For most metals, the emissivity increases with increasing temperature. An angle dependence in emissivity for highly polished materials also exists.

Material	Temperature in K	ε
Dielectrics ("isolaters")		
black body	--	1.00
Brick red (rough)	310	0.93
Candle soot	310	0.95
Clay smooth	310	0.91
Concrete, rough	310	0.94
Glass	340	0.93
Glimmer	310	0.75
Marble	310	0.94
Paint (practically all qualities and colours)	310	0.8 ÷ 0.95
Paper (roofing)	310	0.93
Paper, white	310	0.96
Plaster	310	0.9
Porcelain	310	0.91
Skin (human)	305	0.95
Sand	300	0.94
Vigorous leaves	290	0.9
Water		
– thick layer	273 ÷ 350	0.96
– ice smooth	270	0.97
– ice rough	270	0.98
– snow	270	(0.82)
Wood	310	0.85 ÷ 0.95

For dielectrics, the emissivity normally decreases at higher temperatures.

15 Tomographic Measurement and Reconstruction Techniques

Mathias Buchmann and Dieter Mewes

15.1 Introduction

Tomographic measurement techniques which allow the measurement of three dimensional concentration, temperature and velocity fields within an investigated volume without influencing physical processes have been developed in the past fifty years within the fields of medicine, electron microscopy and radio astronomy. Tomographic measurement techniques are especially well suited to the analysis of unsteady phenomena and their application to the analysis of materials and to the field of chemical engineering has risen markedly in recent years. This chapter describes the different measurement techniques used in tomography. The mathematical methods implemented in the reconstruction of the measured physical properties are reviewed and the quality of the reconstruction is critically evaluated. Some applications of tomographic techniques are also discussed.

In the broad field of measurement tasks in chemical engineering, data must be acquired simultaneously at all points within the volume of interest. The analysis of unsteady three–dimensional velocity, temperature and concentration fields is of importance in a variety of engineering applications. The use of miniature sensors allows high accuracy three–dimensional local measurements of velocity, temperature and concentration in fluids at rest or in steady flows. For this purpose, several sensors are distributed in the investigated volume or a single sensor is translated through it. When using such sensors in unsteady flow fields, only information on the temporal variation of the field parameter at a certain point in the volume can be detected. In unsteady flows, one is faced with the difficulty of relating sensor readings taken at different instants of time and varying spatial locations.

The aim of space–resolved investigations is to determine the three–dimensional distribution of the physical parameter of interest as time history at discrete time instants. In such applications, tomographic measurement techniques are recommended. Reconstruction of three–dimensional data from line–of–sight measurements from a single projection image is not possible, except in such special cases as axially symmetric flows. By using tomographic techniques, the physical property of interest is recovered from observations integrated along the different paths of probe radiation for each cross–sectional plane of the measurement volume. The three–dimensional information is then reconstructed from measurement data related to several layers of the volume.

Radiation of different natures, wavelengths and energies are suitable for tomographic measurements. The intensity and the phase of radiation is affected by different physical properties, such as the absorption coefficient or the refractive index of the irradiated fluid or solid. These physical properties are termed field parameters. They carry information on the variation of the investigated physical property in the measurement volume. The variation is originally caused by the variations of the temperature, concentration, density and velocity fields under consideration. The variation of the field parameter along the ray path is described by the field function.

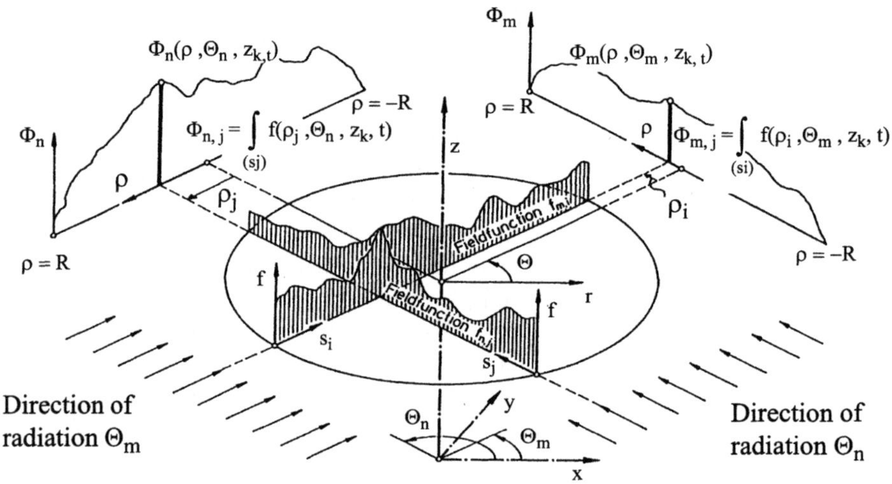

Fig. 15.1. Line integrals of the field function along two different paths of irradiation across the measurement volume

In Fig. 15.1, the variation of a field parameter is presented. It is related to the physical property of interest, along two different ray paths determined by the directions m and n in the selected cross–sectional plane of the investigated volume presented. The integrated value of the field function Φ is projected onto the image plane. The field function f is variable in the cross–sectional plane. The projection corresponds to the line integral of the field function along the path defined by the ray traversing the measurement volume, and it is detected by appropriate sensors located outside the volume. If a sufficient number of integrated projection values originating from various transillumination angles Θ are recorded simultaneously, the local values of the field function f can be reconstructed with specific algorithms.

By using image reconstruction techniques, the dimension of the image is increased by one. The field function for a two–dimensional cross section is recovered from a set of line projections. Three–dimensional data are reconstructed from sets of planar images. The number of projections necessary for reconstruction of desired accuracy depends on the selected irradiation and reconstruction technique.

The mathematical relationship between the value of the field parameter and its projection along different paths in the investigated cross–sectional plane is described by the equation:

$$\int_{s_i} f(x, y, z, t)\, \mathrm{d}s = \Phi_i(\rho_i, \Theta, z, t). \tag{15.1}$$

In(15.1), Φ_i denotes the measured value of the projection, f is the field function, s_i the path through the measurement volume and ρ_i is the coordinate of the projection. The coordinate z corresponds to the height of the investigated cross–sectional plane in the measurement volume and t denotes time.

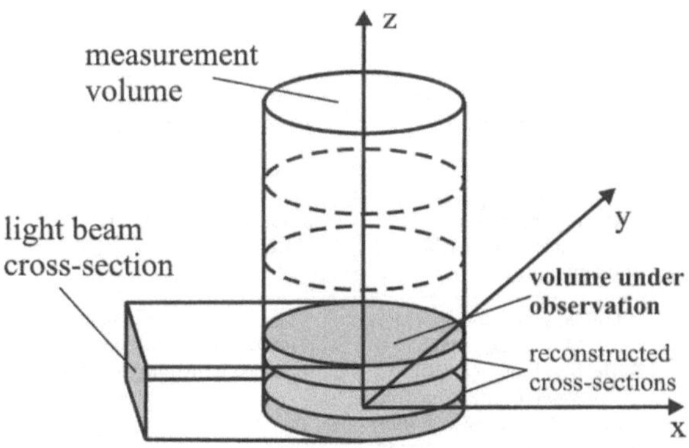

Fig. 15.2. Measurement planes dividing the investigated volume at various heights

The measurement volume is analyzed by subdivision into several two–dimensional cross–sectional planes which correspond to constant value of the height coordinate z, as illustrated in Fig. 15.2. If the field parameter is considered in the selected cross–sectional plane at a specific time instant only, (15.1) can be rewritten as:

$$\int_{s_i} f(x, y, z_0, t_0)\, \mathrm{d}s = \Phi_i(\rho_i, \Theta, z_0, t_0). \tag{15.2}$$

For the cross–sectional plane corresponding to the height coordinate z_0, (15.2) can be expressed in simplified form for the projection angle Θ at the time instant t_0 as follows:

$$\int_{s_i} f(x, y)\, \mathrm{d}s = \Phi_i(\rho_i, \Theta). \tag{15.3}$$

The nature of the radiation used and the field parameter of interest influence the choice of the techniques used to obtain the measurement of the value of the projection The most commonly used measurement techniques will be presented in the following section.

15.2 Integral Measurement Methods

To obtain the integral measurements Φ different techniques can be used. Reinecke et al. [367] give in Fig. 15.3 a summary of the known tomographic techniques. They are sorted according to the physical interaction between the sensor and the material inside.

physical measurement property	**physical interaction**		
	transmission	**reflection, refraction, defraction, scattering**	**emission**
corpuscular radiation (α, β+, β-)	linear accelerator, radioisotopes [2,3], neutron radiation [4]		'Positron Emission Tomography' (PET) [5,6]
highenergetic electromagnetic waves (γ and x-ray radiation)	x-ray [CT], γ-ray [7-9]	photon scattering [10] , x-ray scattering, 'Compton-Backscatter' [12]	'γ-Emission Computed Tomography (ECT)' [13], 'Single Photon Emission Computed Tomography (SPECT)' [14]
ultraviolet and infrared radiation, lightwaves	Holography [15, 17], Interferometry [16] Spectroscopy [18]	scattering techniques [19], optical refraction [21,22], defraction [20]	Infrared-Thermography [23,26], light emission [24,25]
low energetic electromagnetic waves (micro- and radio waves)	amplitude attenuation [28], phase shift [27], transmission time, 'Multi-Incidence Transmission Imaging' (MITI) [29]	'Modulated Scattering' (MST) (30), 'Multi Frequency Reflection Imaging' (MFRI)[31], 'Microwave Diffraction Tomography' (MDT) [32], 'Radio-frequency Inverse Scattering Techniques' (RIST) [30]	'Nuclear Magnetic Resonance' (NMR) [34,35], 'Magnetic Resonance Imaging' (MRI) [33]
quasi stationary electromagnetic fields	'Electrical Capacitance Tomography' (ECT) [37,38], 'Electrical Impedance Tomography' (EIT) [36,39,40], 'Electromagnetic Tomography' (EMT) [41]		'Electrostatic Charge Tomography' (ESCT) [42], 'Electrodynamic Tomography' (EDCT) [43]
sonic waves (pressure or ultrasonic)	ultrasonic transmission [44,46], 'Time Of Flight' (TOF) [45], attenuation techniques, Doppler-techniques [47]	ultrasonic reflection techniques [48,49]	'Radiation Induced Acoustic Imaging' (RIAI) [50,51]

Fig. 15.3. Known tomographic techniques [367]

In Fig. 15.4 various physical interactions are depicted. During transmission, the physical property of the electromagnetic wave or particle radiation emitted by the source, is influenced by the material inside the measurement plane. The change of the physical property is measured by the receiver, which is mounted on the opposite side of the sensor. The change of the physical property can be the intensity, the polarization or change of the run time of the beam. For emission–techniques, the intensity and the spatial orientation of the radiation which is emitted from inside the measurement plane is measured. The material in the measurement plane itself is the source of the radiation. The source of the radiation can be natural or due to external excitation.In reflection techniques, the measured value is the position and the change of the physical property of a wave or a particle, which has been reflected on an interface. Refraction and diffraction techniques are similar to reflection techniques. The change of the direction or the physical property of a wave or corpuscular radiation is measured at discrete or continuous interface in the measurement plane. When molecules or atoms are hit, corpuscular radiation or high energy electromagnetic waves cause scatter effects. These scatter effects can also be used for tomographic measurements. In this case the measured values are the statistical distribution of the property vector as well as its orientation.

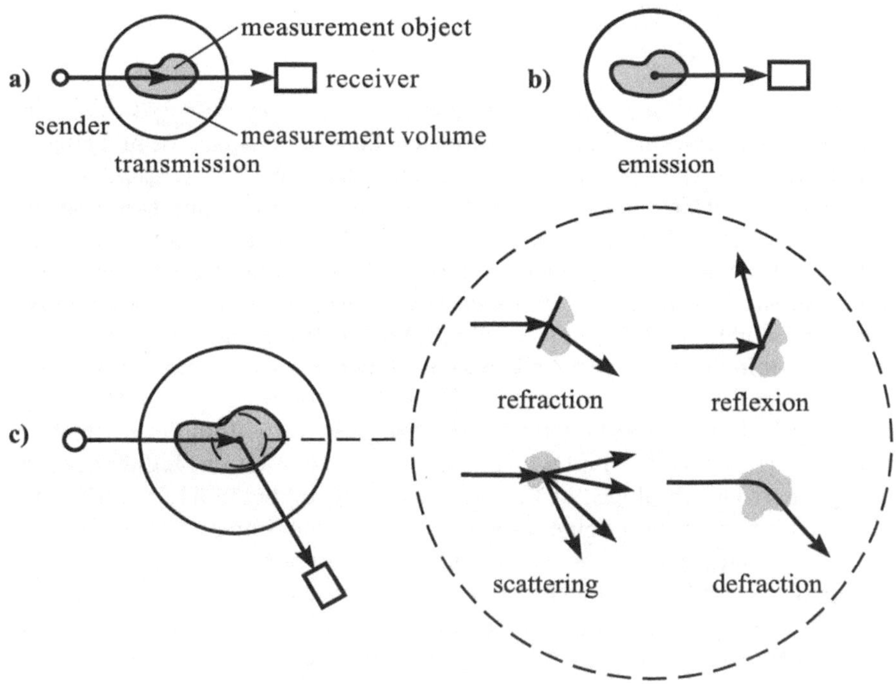

Fig. 15.4. Physical interactions used for tomographic measurements [367]

The following explanations focus on optical transmission techniques namely absorption and interferometric methods.

15.2.1 Absorption Methods

In absorption methods, the attenuation of the intensity of radiation is measured, which traverses a transparent fluid or solid. The investigated medium has to be transparent to the type of radiation used. This does not necessarily imply optical transparency. Here the field parameter corresponds to the spatially varying absorption coefficient μ. The generally applicable Lambert–Beer absorption law can be written as:

$$I = I_0 \exp(-\mu s). \tag{15.4}$$

In (15.4), I_0 denotes the initial intensity of the radiation at the entrance into the investigated volume, I is the intensity leaving the volume, μ is the absorption coefficient and s the irradiation pathlength. Consequently, in (15.1) the measured projection is the natural logarithm of the ratio of the intensity of the radiation entering the test volume to the intensity leaving it:

$$\ln\left(\frac{I}{I_0}\right) = -\int_{s_i} \mu(x, y, z, t)\, \mathrm{d}s \tag{15.5}$$

The intensity of the radiation is measured with sensors selected to match the nature of radiation. The absorption coefficient is, apart from being a function of the physical properties of the substance in the investigated volume, also a function of the wavelength of the radiation. For this reason, monochromatic waves should be used in absorption measurements. Only the use of monochromatic waves relates the reconstructed absorption coefficients unambiguously to the investigated physical property describing, for example concentration field. From the total spectrum of electromagnetic radiation, X–rays and visible light are primarily employed in tomographic measurements. X–rays are used to measure density differences in solids or opaque fluids. In special applications, for example to verify the presence of bubbles in steel or concrete, γ–rays may be used as well [385]. In this situation, higher energy radiation has to be employed. A further interesting application of γ–radiation is the emission–computer tomography of barrels containing radioactive materials [300] which enables the localization of the ingredients emittingγ–rays. Visible light is employed to measure concentration differences in transparent fluids. The concentration differences can be visualized by introducing dye solutions into the liquid or by initiating a color change reaction [316].

15.2.2 Interferometric Methods

In addition to the change of the intensity of the wave leaving the measurement volume, the relative shift of its phase compared with a reference wave can also be measured. The phase shift corresponds to the speed of the radiation inside the measurement volume, which is governed by the field parameter along the irradiation path. The general relation between the field function f and the measured parameter Φ defined by (15.1), can be rewritten as:

$$\Phi_i(\rho_i, \Theta, z, t) = \int_{s_i} \Delta f(x, y, z, t)\, \mathrm{d}s, \tag{15.6}$$

for interferometric techniques.

In this equation, the change of the field parameter related to the reference state is denoted by Δf. The advantage of interferometric methods over absorption techniques is that the measured projection values are *equal* to the average change of the field function. Conversely, in absorption techniques, the measured projection is only *proportional* to the average change of the field parameter. Thus, in interferometric techniques there is no need for calibration. Conversely, the techniques for recording the projection values are usually more complex in interferometric measurements. The recording of interference phenomena is possible using coherent, monochromatic light waves. Lasers are powerful monochromatic light sources providing sufficient coherence length for measurement purposes. The coherence length of the light emitted by an argon–ion laser, which is frequently used in tomographic measurements, is over 1 m and its power is in the range of several Watts.

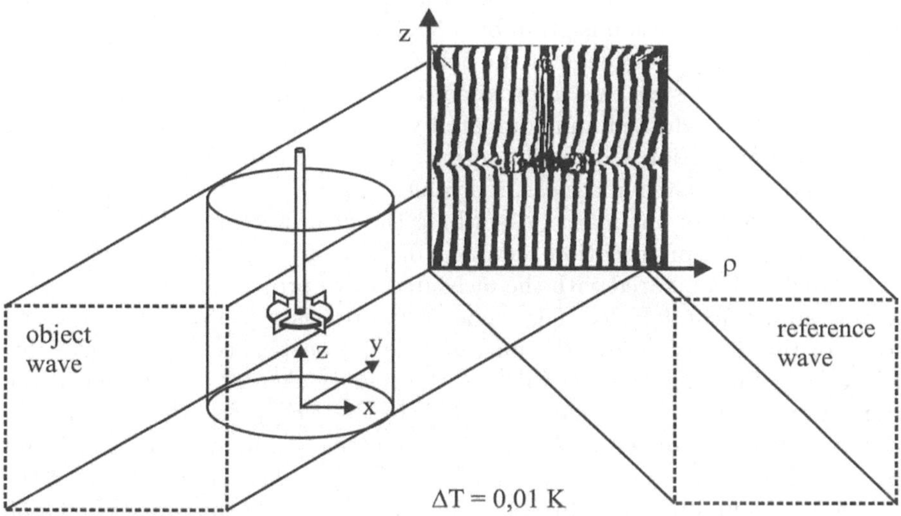

Fig. 15.5. Recording an interferometric projection by superimposing the reference beam and the object beam

In Fig. 15.5 the transillumination of a mixing vessel from one direction is shown. Here the field parameter to be recovered is the refractive index. Change of the refractive index causes change of the optical pathlength of the light wave propagating in the measurement volume. This wave is called the *object wave*. By superimposing the *reference wave* which propagates outside the measurement volume to the object wave, the effect of interference can be observed. The resulting interference pattern is recorded on the interferogram. The deviation of the interference fringes from their initial shape is related to the change of the refractive index in the volume during the measurement. Knowledge of the dependence of the refractive index on the temperature and the density of the fluid in the measurement volume, enables the evaluation of the spatial temperature and density distribution. The variety of recording and evaluation methods implemented in interferometric measurements is treated in detail in numerous publications [305,338,340,353,354,389].

15.3 Mathematical Reconstruction Methods

The integral measurement techniques described in the preceding section enable the recording of projections of the field function in the irradiated measurement volume. Before introducing more sophisticated methods commonly used in the reconstruction of field parameters and before analyzing the quality of reconstruction, a simple iterative reconstruction procedure will be discussed.

The procedure is illustrated in Fig. 15.6. In the top left corner of the figure, the original field function is given by a 3×3 matrix with four of its measured projections. The initial guess for the field function is obtained from the values of projection 1. Using this approximation, the values of the projection in direction 2 are evaluated and compared with the actual – measured – value of the same projection. Correction terms can be evaluated from the difference between the calculated and measured values. By using these correction terms, the initial guess for the matrix describing the field function can be adjusted. From the new approximation obtained in this way, the third projection is calculated and compared with the measured third projection. After this step, a new adjustment follows. The same procedure is repeated for projection four. In the first iteration, the maximal deviation between the exact value of the field function and the value of the matrix elements evaluated in the iterative procedure was found to be 50%. In the second and third iterations, the same value is reduced to 6% and 2%, respectively. The system of equations in this illustration is overdetermined, resulting in fast convergence of the algorithm. In practice, the dimension of the matrix is in the range of 50×50 or more elements. For matrices of this size the described simple algorithm which compares the measured and calculated values of the projections is neither efficient nor sophisticated enough.

The various reconstruction algorithms typically implemented can be divided into algebraic methods and explicit solution techniques. In Fig. 15.7, the commonly used reconstruction procedures are summarized. Combined techniques have been implemented, in numerous technical applications. their

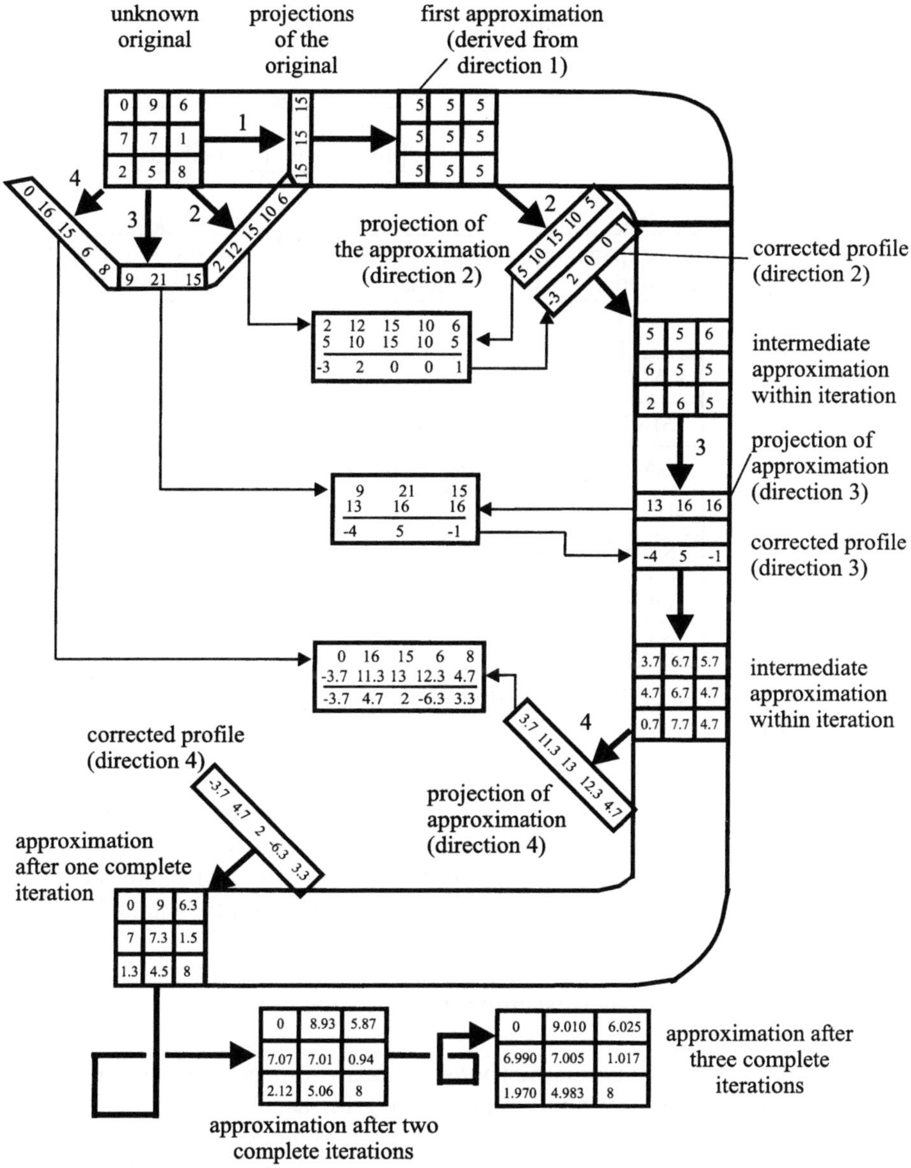

Fig. 15.6. Reconstruction steps in a simple, iterative procedure for recovering the field parameter

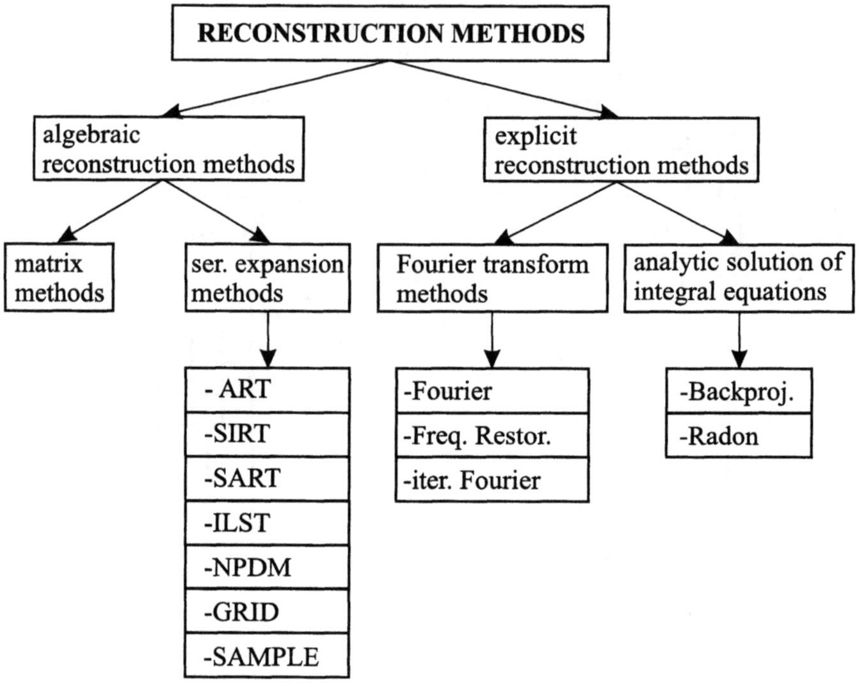

Fig. 15.7. Overview of the mathematical reconstruction techniques for tomographic measurements

allocation to a single group would be ambiguous. For example, the Fourier transform method is frequently combined with series expansion. In this chapter, the mathematical bases of commonly implemented evaluation techniques are discussed. Further reviews can be found in the literature [323,332,344].

15.3.1 Algebraic Reconstruction Methods

The algebraic reconstruction methods are based on the discretization of the cross–sectional plane by superimposing a grid, as illustrated in Fig. 15.8. Initially, a constant value of the field function is assumed for each grid element of the mesh. The algebraic reconstruction methods can be realized either as direct matrix methods or as methods using series expansion.

Matrix Methods

In matrix methods, the system of linear equations formed by the elements of the field function for the selected grid and the measured projection values is solved using direct matrix calculations. The solution algorithm can be presented in simplified form as:

$$
\begin{array}{cc|c}
f_1 & f_2 & \Phi_1 \\
f_3 & f_4 & \Phi_2 \\
\hline
\Phi_3 & \Phi_4 &
\end{array}
$$

In the system of linear equations, f denotes the elements of the field function and Φ is the value of the projection. The corresponding system of linear equations results from the following matrix equation:

$$\mathbf{W} \bullet \mathbf{F} = \mathbf{P}, \tag{15.7}$$

with the matrices:

$$\mathbf{W} = \begin{bmatrix} w_{11} & w_{12} & w_{13} & w_{14} \\ w_{21} & w_{22} & w_{23} & w_{24} \\ w_{31} & w_{32} & w_{33} & w_{34} \\ w_{41} & w_{42} & w_{43} & w_{44} \end{bmatrix}, \quad \mathbf{F} = \begin{bmatrix} f_1 \\ f_2 \\ f_3 \\ f_4 \end{bmatrix}, \quad \mathbf{P} = \begin{bmatrix} \Phi_1 \\ \Phi_2 \\ \Phi_3 \\ \Phi_4 \end{bmatrix} \tag{15.8}$$

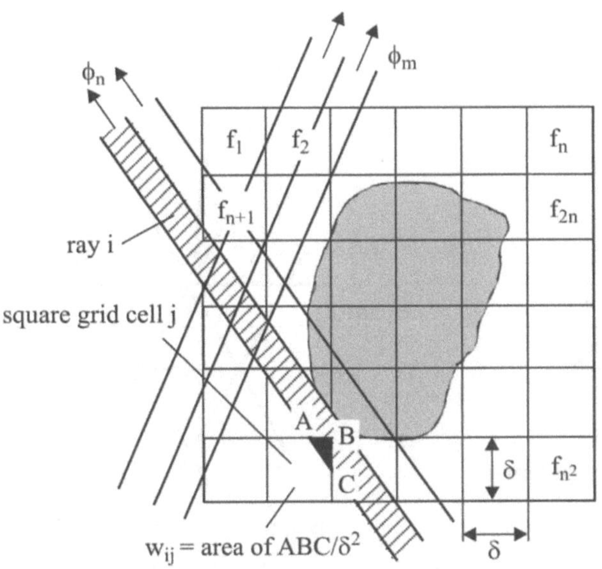

Fig. 15.8. Discretization of the investigated cross–sectional plane by superimposing a square grid and the paths of the rays traversing the measurement volume parallel to the cross–sectional plane

Matrix **W** is termed the allocation matrix and is dependent on the shape of the investigated cross–sectional plane. The coefficients of this matrix are weighting factors, indicating the contribution of the jth grid element to the ith ray integral. In Fig. 15.8, the propagation of the ray of finite width in measurement plane subdivided into square grid elements is presented schematically. The coefficients w_{ij} correspond to the fractional area of each grid element j which is being intercepted by ray i. The matrices (15.7) describing the system of linear equations can be written as summation:

$$\sum_{j=1}^{N^2} w_{ij}\, f_j \;=\; \Phi_i\,, \quad i = 1, 2, \ldots, M. \tag{15.9}$$

From (15.9), the following system of equations is obtained in explicit form:

$$\begin{aligned} w_{11}f_1 + w_{12}f_2 + w_{13}f_3 + \cdots + w_{1N^2}f_{N^2} &= \Phi_1 \\ w_{21}f_1 + w_{22}f_2 + w_{23}f_3 + \cdots + w_{2N^2}f_{N^2} &= \Phi_2 \\ &\;\;\vdots \\ w_{M1}f_1 + w_{M2}f_2 + w_{M3}f_3 + \cdots + w_{MN^2}f_{N^2} &= \Phi_M . \end{aligned} \tag{15.10}$$

It should be noted here that for the situation illustrated in Fig. 15.8 many of the coefficients w_{ij} (s. (15.8)) are zero, due to the fact that only small number of grid elements is intercepted by the ray, thus contributing to particular projection.

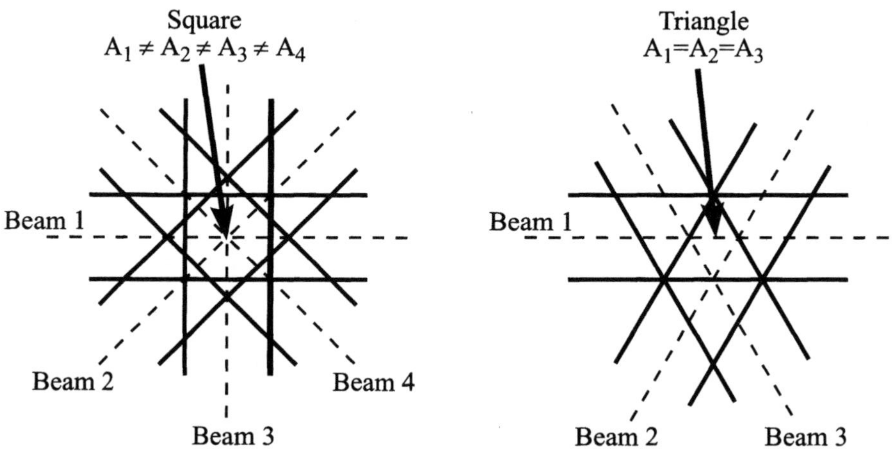

Fig. 15.9. Discretization with quadratic and with triangular elements [366]

Reinecke et al. [366] propose discretization of the investigated cross–sectional plane with an adapted element scheme. In Fig. 15.9 non adapted quadratic elements in a four projections set up are compared with adapted triangular elements in a three projections set up. In the latter case the weighting coefficients w_{ij} are either one or zero, because an element belongs either completely to a specific beam or not at all. Therefore the structure of the allocation matrix **W** is simpler which yields faster reconstruction. In Sect. 15.3.3 the accuracy of reconstruction is compared for the two different element schemes.

In order to reconstruct the field function in a cross–section with $N \times N$ grid elements, a system of N^2 equations has to be solved. The solution is given for particular configurations of matrix **W** only. For this reason, the direct matrix inversion procedures are limited to a few specific applications [327]. In practical applications, underdetermined equation systems are usually obtained. This is due to the limited number of available projection values. Using the series expansion procedures are described in the following sections, an approximate solution is found iteratively.

Iterative Series Expansion

Iterative methods are characterized by good reconstruction quality in applications with a small total viewing angle and a small number of irradiation directions. The unknown local values of the field function in the underdetermined equation system are first approximated by using the projection values in the individual viewing directions and then evaluated iteratively. These procedures are described by Herman [321] and censor [293]. Munshi et al. [380] describe several iterative algorithms and evaluate their performance.

Algebraic Reconstruction Technique (ART)

The algebraic reconstruction technique is a frequently used iterative technique. It was introduced by Gordon et al. in 1970 [312]. In later publications [313,320,311], numerous modifications of the ART procedure have been described. The procedure is based on the comparison of the calculated projection values with measurement data after each iteration step. From the results of the comparison, correction term for the field function is obtained. The procedure starts by setting all values of the image matrix either to zero or another nonzero initial value. The value of the elements is then improved in the iteration steps that follow. The calculated projection value $\Phi_i^{(k)}$ after the kth iteration step is obtained using (15.9), it usually differs from the measurement data Φ_i. The deviation $\Delta\Phi_i^k$ can be evaluated as:

$$\Delta\Phi_i^{(k)} = \Phi_i - \Phi_i^{(k)} = \Phi_i - \sum_{j=1}^{N^2} w_{ij} \cdot f_j^{(k)} . \tag{15.11}$$

This value is then used in the correction term to determine the elements of the field function in the iteration $k + 1$. Mathematically the correction algorithm is given as:

$$f_j^{(k+1)} = f_j^{(k)} + \frac{w_{ij} \cdot \Delta\Phi_i^{(k)}}{\alpha_i}. \tag{15.12}$$

The correction coefficient α_i in (15.12) is evaluated as:

$$\alpha_i = \sum_{j=1}^{N^2} (w_{ij})^2. \tag{15.13}$$

The correction method using (15.12) is called the *additive ART method.* The second method is called the *multiplicative ART method.* In this procedure, the initial guess for the field function is $f_j^{(1)} = 1$, the correction term is given by:

$$f_j^{k+1} = f_j^{(k)} \left(\frac{\Phi_i}{\Phi_i^{(k)}} \right). \tag{15.14}$$

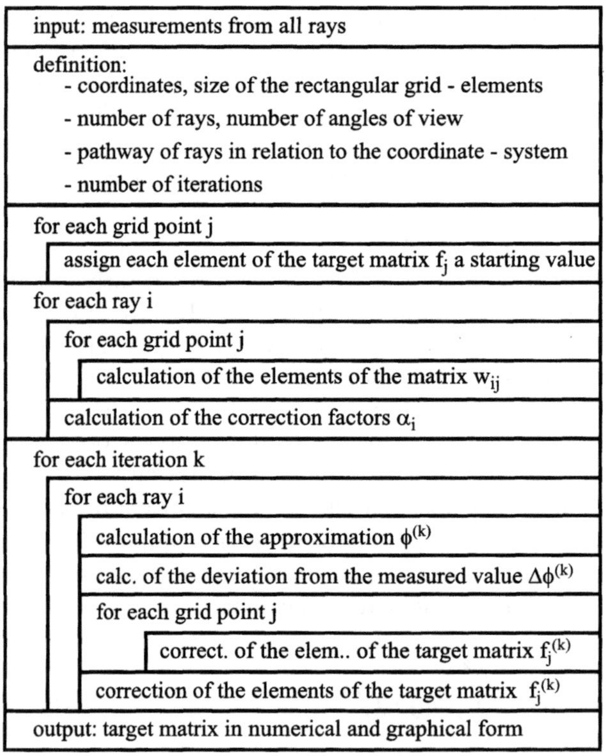

Fig. 15.10. Structure of the ART algorithm

The ART procedure is illustrated by the algorithm presented in Fig. 15.10. First, the measured projection data for all rays are entered and the listed program parameters are defined. In the following step, initial values are assigned to the image matrix. Next, the coefficient matrix w_{ij} is determined and the correction coefficient α_i is evaluated for each ray using (15.13). Then the iteration given by (15.11) and (15.12) is carried out. In addition, weighted smoothing of the field elements may be performed, to suppress the discontinuities of the field function. In each iteration, the number of the corrections performed on the image matrix corresponds to the number of rays in the measurement. This means that all elements of the image matrix are affected by each comparison with the projection data. Therefore it is not possible to define unique termination criterion.

In practice, two different termination criteria are applied. One of them is based on the sum of squares of all residuals given by (15.11). The second criterion specifies the number of iteration steps, as indicated in Fig. 15.10. As the field function is unknown, the required number of iterations is determined by introducing test functions. These are analytical three–dimensional functions similar to the possible distributions of the field parameter in the investigated volume. The test functions are integrated along different ray paths. The calculated mean values of the field function simulate the measured projection data, which is then used to recover the field function. The deviation of the tomographic reconstruction from the original test function depends on the number of iteration steps. The mean deviation is a direct measure of the fidelity of reconstruction. The required number of iteration steps has been achieved when no significant improvement of the accuracy of reconstruction is observed in successive iterations. The advantage of this termination criterion is that it is based on the deviation of the field function and not the projection data, which is the case when residuals are used. The least number of iteration steps necessary has to be determined for a particular application by using the described test functions.

Simultaneous Iterative Reconstruction Technique–SIRT
In addition to the most frequently used ART method, other iterative reconstruction procedures, such as the simultaneous iterative reconstruction technique (SIRT), have been developed as well. The method has been introduced by Gilbert [307]. In SIRT, the elements of the field function are modified after all the projection values corresponding to individual rays have been considered. As in the ART method, additive or multiplicative corrections are possible in the iteration steps.

Simultaneous Algebraic Reconstruction Technique–SART
The SART method combines the ART and SIRT algorithms described above. This technique was introduced by Anderson [281], and is characterized by high reconstruction fidelity and fast convergence of the iterative procedure [332]. The main feature of the technique is that all rays of an irradiation direction are considered simultaneously in the evaluation of the correction term. This is different from the ART and SIRT methods for which the correction is performed for single ray or for all rays of an irradiation direction, respectively. Additionally, the accuracy of reconstruction is increased by applying a weighting factor related to the ray path.

Iterative Least Square Technique (ILST)

This method is based on the minimization of the deviation between the measured and calculated projection data using the least squares method. The data originally described by (15.7) are given in matrix notation by:

$$\mathbf{W} \bullet \mathbf{F} = \mathbf{P}\,.$$

The above equation may be rewritten as:

$$\mathbf{W} \bullet \mathbf{F} - \mathbf{P} = \mathbf{\Delta P}. \tag{15.15}$$

In (15.15), the allocation matrix is denoted by $\mathbf{W}$, $\mathbf{F}$ is the matrix of field function elements to be determined and $\mathbf{P}$ corresponds to the measured projection data. The difference vector $\mathbf{\Delta P}$ is to be minimized by suitable algorithms [327].

Natural Picture Decomposition Method–NPDM

In the method, the grid superimposed to the cross–sectional plane consists of non–rectangular elements, their size and position being determined by the ray paths. In all the methods described in the preceding chapters, the grid is rectangular. For this reason in NPDM the grid elements no longer fit regularly; instead they overlap each other. The system of linear equations to be solved is significantly less underdetermined than for the methods described above, especially in applications with limited viewing angle. NPDM was introduced by Buonocore [287] and later implemented by other authors [294,306].

Grid methods

In order to calculate the elements of the field function using grid methods, the grid shown in Fig. 15.11 is superimposed to the investigated cross–sectional plane of the measurement volume. At the beginning of the reconstruction procedure, constant value of the field function is assigned to each element of the grid. The change of the physical property to be determined causes a change of the property of the ray traversing the measurement volume. This change can be related to the sum of the individual grid elements intercepting the ray path, the path length along each grid element being considered in the integration process. From the summation equation:

$$\Phi_i(\rho, \Theta_i, z_k) = \sum_{j=1}^{N^2} f_j\, l_{ij}. \tag{15.16}$$

the projection value Φ_i corresponding to beam i is obtained, with the distance from the origin of the coordinate system ρ and the viewing angle Θ. In (15.16), l_{ij} is the pathlength along a grid element and f_j is the corresponding field function value. When the number of irradiation angles, i.e. measured projections, is equal to the number of grid elements, the solution of the resulting system of linear equations is obtained by elementary techniques.

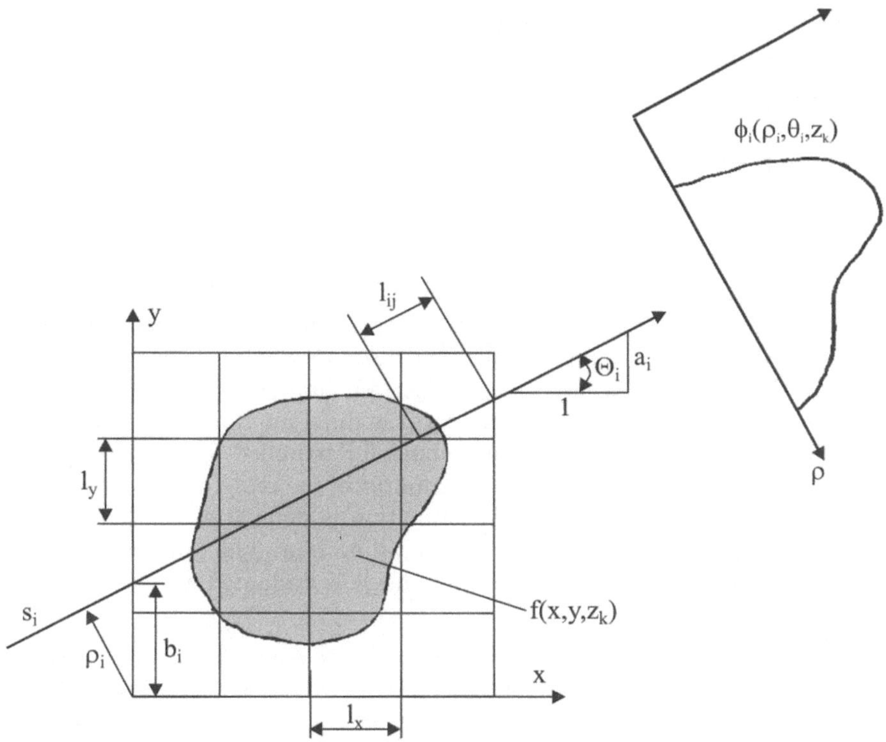

Fig. 15.11. Subdivision of the reconstruction plane into a grid for use in the grid methods

A very fine grid is required for high accuracy reconstruction. The number of equations to be solved increases quadratically with decreasing grid size. The number of measurements to be taken also increases simultaneously with the increase of the number of equations. Therefore, if high fidelity of reconstruction is needed, the requirements concerning the experimental equipment and computer capacity also increase. According to Sweeney [381], redundant projection value measurements must be taken. Otherwise, suitable iterative methods have to be used [338].

Sample method
The sample method belongs to the class of solution algorithms based on series expansion techniques. It combines the advantages of the integral transform methods (to be discussed later) and the grid methods. It is based on the discrete Fourier transform, which is used to calculate the values of the field function at discrete grid points of fine grid. The algorithm has been described and used by Sweeney for the reconstruction of unsteady temperature fields [381].

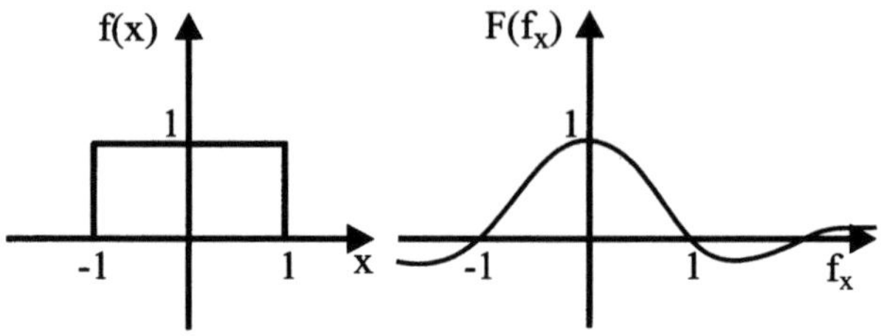

Fig. 15.12. The rect–function $f(x)$ and its Fourier transform, the sinc–function

The discrete Fourier transform method is based on the Whittaker–Shannon sampling theorem [310]. According to this theorem, an unknown function can be described by a sum of unknown discrete values, when the interval between the values does not exceed a certain limit. A function is referred to as "bounded" if its Fourier transform is different from zero only in the bounded region. However, the requirement for bounded function is not fulfilled for real field functions. A simple example of the rect–function, presented in Fig. 15.12, will be discussed. The function is defined as $\text{rect}(x) = 1$ for $-1 \leq x \leq 1$ and $\text{rect}(x) = 0$ for all other values of x. The Fourier transform of this function is termed sinc–function and it is evaluated as:

$$\int_{-\infty}^{\infty} \text{rect}(x)\, \exp(-2\pi i x \tau_x)\, \mathrm{d}x \;=\; \text{sinc}\,(\tau_x). \tag{15.17}$$

The sinc–function is defined as follows:

$$\text{sinc}\,(x) \;=\; \frac{\sin(\pi x)}{\pi x}, \tag{15.18}$$

and is plotted in Fig. 15.12. Its amplitude decreases rapidly with increasing frequency. However, even for infinite frequencies, it remains different from zero; i.e. the original function is not bounded. By applying filtering algorithms, higher frequencies are eliminated in the sample method. The deviation of the reconstruction from the exact solution is small, as, due to small amplitudes at higher frequencies, their impact is not significant.

From the approximate calculation according to the Whittaker–Shannon theorem, the desired function $f(x,y)$ is obtained as the sum of discrete values [381]

$$f(x,y) \;=\; \sum_m \sum_n f(l_x\, m, l_y\, n)\, \text{sinc}\left(\frac{x - l_x\, m}{l_x}\right)\, \text{sinc}\left(\frac{y - l_y\, n}{l_y}\right). \tag{15.19}$$

In Fig. 15.11, the grid with the value $f(l_x m, l_y n)$ of the field function $f(x,y)$ and grid sizes l_x and l_y for the coordinate axes x and y respectively is presented. By transformation and substitution of (15.19) into (15.9), the following general solution is obtained

$$\sum_m \sum_n w(a_i, b_i) \cdot f(l_x\, m, l_y\, n) \;=\; \Phi_i(\rho_i, \Theta). \tag{15.20}$$

In (15.20), $w(a_i, b_i)$ is a function of the ray with slope a_i:

$$w(a_i,b_i) = \begin{cases} \left(1+a_i^2\right)^{0,5} \cdot |l_x| \left(\dfrac{b_i + a_i l_x m - l_y n}{l_y}\right) \\ \text{for} \quad 0 \le |a_i| \le \dfrac{l_y}{l_x} \\ \left(1+{a_i}^2\right)^{0,5} \dfrac{|l_y|}{|a_i|} \left(\dfrac{b_i + a_i l_x m - l_y n}{a_i l_x}\right) \\ \text{for} \quad \dfrac{|l_y|}{|l_x|} < |a_i| < \infty \\ |l_y| \left(\dfrac{\rho_i + l_x m}{l_x}\right) \\ \text{for} \quad |a_i| = \infty \end{cases} \tag{15.21}$$

The coefficients $w(a_i, b_i)$ are determined for each ray i with the slope a_i and the intercept with y denoted by b_j. They correspond to the distance of the ray from the origin of the coordinate system ρ_i and the grid point j, denoted here as $f(l_x m, l_y n)$.

The system of (15.20) has $m \times n$ unknowns which is equal to the number of grid points in each measurement plane. An elementary solution of the system exists, if the same number of linearly independent equations is available.

Sweeney [381] even recommends that the system should be overdetermined in order to improve the accuracy of reconstruction. This can be achieved by providing more measured projection values in an irradiation direction than the number of unknowns to be calculated. If these values are not available, an approximate solution of the system of equations can be found using iterative techniques. The ART, SIRT and SART methods discussed in the preceding chapters may be considered in that case.

Converse to the conventional grid methods, the sample methods permit greater grid sizes l_x and l_y for the same reconstruction accuracy. Thus, the number of unknowns in the system can be reduced. In this combined technique, smaller number of projection values for the same irradiation angle is required than for the plain Fourier transform method.

15.3.2 Explicit Reconstruction Methods

The continuous field function of interest can be approximated using explicit solution techniques. In this case, the value of the field function is evaluated not only at the grid points but at all points of the cross–sectional plane. The different reconstruction methods can, as illustrated in Fig. 15.7, be subdivided into methods using the Fourier transform and those providing the analytical solution of the integral equation. Further reviews and discussions of these methods have been published by Lewitt [335] and Smith et al. [376].

Fourier Transform Method

By using the Fourier transform, any function of position and time can be transformed into a spectral representation in the frequency domain, as any signal can be represented as a sum of sine waves [288]. In Fig. 15.13, the application of the Fourier transform is illustrated for the mathematical reconstruction of the field function from measured projection data. The measurement data from the spatial domain are transformed into the frequency domain and plotted as function of frequency. The original function can be recovered by the inverse transform. Integral transforms are frequently implemented to recover the field function in medical applications. Many mathematical operations, such as filtering and smoothing, are easier to carry out in the frequency domain than in the spatial domain. Also, superimposed disturbances can be eliminated more easily. For this reason, Fourier transform methods are often used in digital signal processing [278].

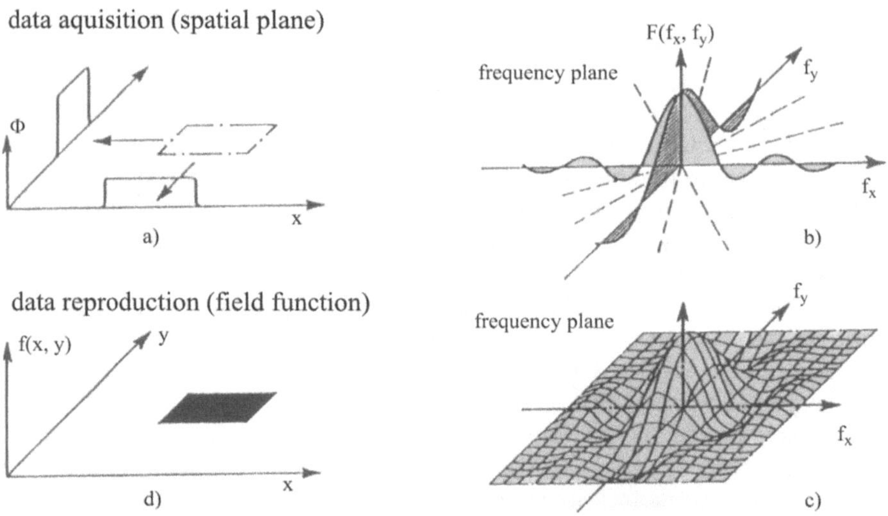

Fig. 15.13. Reconstruction of the field function using the Fourier transform

When applying the Fourier transform in tomography, it is convenient to introduce a coordinate system oriented along a ray path. The coordinate transform is carried out using equations

$$\rho = x\cos\Theta + y\sin\Theta \quad \text{and} \qquad s = -x\sin\Theta + y\cos\Theta. \tag{15.22}$$

The ρ, s coordinate system is rotated over an angle Θ relative to the initial coordinate system x, y. The s coordinate is parallel to the ray path, as indicated in Fig. 15.1. Thus the integral equation to be solved, obtained by transforming (15.3), becomes:

$$\Phi(\rho, \Theta) = \int f(\rho, s)\, \mathrm{d}s. \tag{15.23}$$

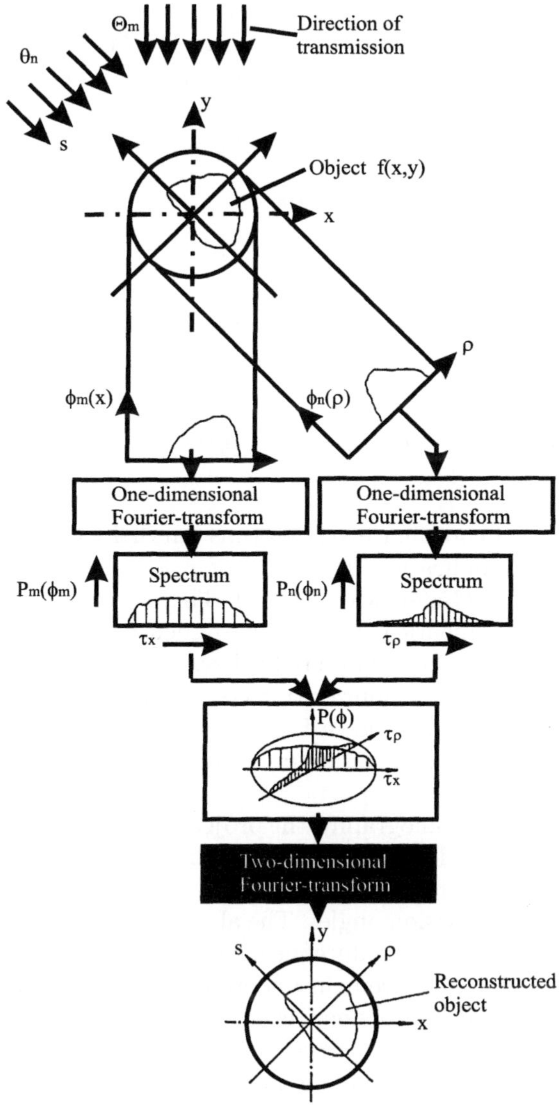

Fig. 15.14. Tomographic reconstruction of the field function using the Fourier transform method

The tomographic reconstruction of the field function using the integral transform is illustrated in Fig. 15.14. The measured projection values $\Phi(\rho, \Theta)$

are transformed into the frequency domain by means of the Fourier transform as:

$$P(\tau_\rho, \Theta) = \int_{-\infty}^{\infty} \Phi(\rho, \Theta) \exp(-2\pi i \rho \tau_\rho) \, \mathrm{d}\rho. \qquad (15.24)$$

In (15.24), $P(\tau_i, \Theta)$ is the Fourier transform, τ_ρ is the coordinate in the frequency domain and i the imaginary number $\sqrt{-1}$. If in (15.24) the measured projection data Φ are replaced by the field function f, the following equation is obtained:

$$P(\tau_\rho, \Theta) = \int_{-\infty}^{\infty} \int_{-\infty}^{\infty} f(\rho, s) \exp(-2\pi i \rho \tau_\rho) \, \mathrm{d}\rho \, \mathrm{d}s. \qquad (15.25)$$

The Fourier transform can also be applied to functions of two independent variables ρ and s, such as the field function. Thus, the two–dimensional Fourier transform can be written as:

$$F(\tau_\rho, \tau_s) = \int_{-\infty}^{\infty} \int_{-\infty}^{\infty} f(\rho, s) \exp\left(-2\pi i(\tau_\rho \rho + \tau_s s)\right) \, \mathrm{d}\rho \, \mathrm{d}s. \qquad (15.26)$$

When the axis of the rotated coordinate system runs parallel to the ray path, which is the case in this application, the value of τ_s is zero and the relations obtained by (15.25) and (15.26) are identical. Therefore, the one–dimensional Fourier transform is identical to the two–dimensional Fourier transform of the field function. The field function to be determined is obtained by applying the two–dimensional inverse Fourier transform to the previously transformed measurement data, as indicated in Fig. 15.14. If all the projections over a range of 180° are available, all values of the field function can be recovered.

In medical applications of X–ray tomography, the projections are recorded using angle subdivisions of 1° [383]. However, in most applications larger angle subdivision intervals are used, as the complexity of the experimental hardware increases with decreasing subdivision angles. The absence of projections at angles not covered in the recording procedure means that the corresponding amplitudes in the frequency domain remain undetermined. This causes the so–called "smearing" or "aliasing distorsion" effects in the inverse transform. As an example, in Fig. 15.15 the reconstruction of an ellipse is shown [332] for different numbers of projections and different numbers of rays in each projection direction. The increase of reconstruction accuracy with increasing number of projection directions and greater ray density is evident.

The reconstruction accuracy can be improved by filtering operations. Nevertheless, the required number of projections for good reconstruction quality

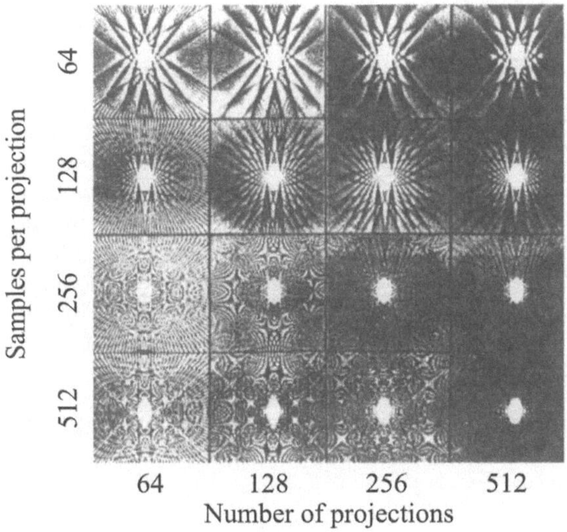

Fig. 15.15. Reconstruction of an ellipse for different number of projections and rays in the projections (Reproduced with permission of the authors) [332]

remains high. Tomographic methods based on the Fourier transform are primarily suitable for the investigation of steady phenomena. The reconstruction of unsteady phenomena requires complex and sophisticated experimental equipment, as all projections have to be recorded simultaneously or at least in a rapid sequence.

In chemical engineering applications, the Fourier transform technique was first implemented by Rowley [370] for the reconstruction of the refractive index. The originally laborious inverse Fourier transform was simplified by Junginger [331]. Sweeney [381] has introduced a modified version of the inverse transform called "frequency domain restoration". In this procedure, the missing measurements are added in the frequency domain, as indicated in Fig. 15.16. In the spatial domain, only projections over viewing angle $\pm\beta$ are measured. The Fourier transform provides data in the frequency domain over 2β. The objective is to calculate the frequencies that lie outside the known range. This is possible due to the fact that the Fourier transform of spatially bounded function $f(x, y)$ in the domain of interest can be expressed analytically. By using series expansion, approximate data for the missing frequencies can be evaluated.

Another modification of the method, the Fourier transform with iterative testing, has been introduced by Sato et al. [372,373]. It is recommended for applications where the number of projections or the projection angle is small [327].

Analytical Solution of Integral Equations

In this approach, the relation between the field function and its projections is described by a set of integral equations, which are solved analytically. For

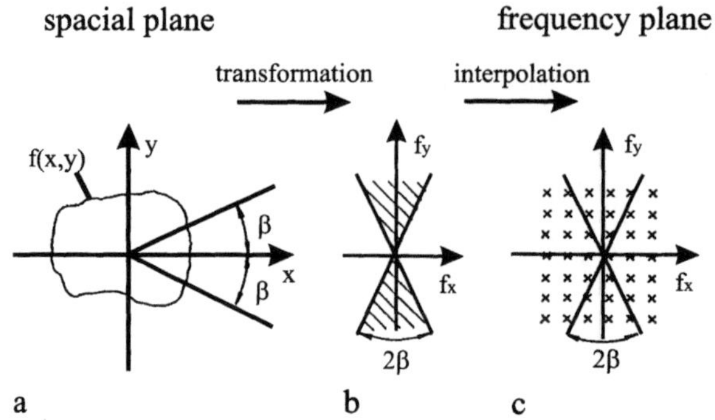

Fig. 15.16. Illustration of the restoration in the frequency domain. **a** field function in the cross–sectional plane. **b** frequency domain. **c** added values in the frequency domain

this purpose, either the back projection method or the Radon inversion are applied.

Back projection
The back projection method is based on a very simple algorithm [323]. A point of the field function is back projected by addition of all the projection values which correspond to the rays passing through the point of interest. The mathematical formulation is given by:

$$f(r,\varphi) = \int_0^\pi \Phi[r\cos(\varphi-\Theta),\Theta]\,\mathrm{d}\Theta. \tag{15.27}$$

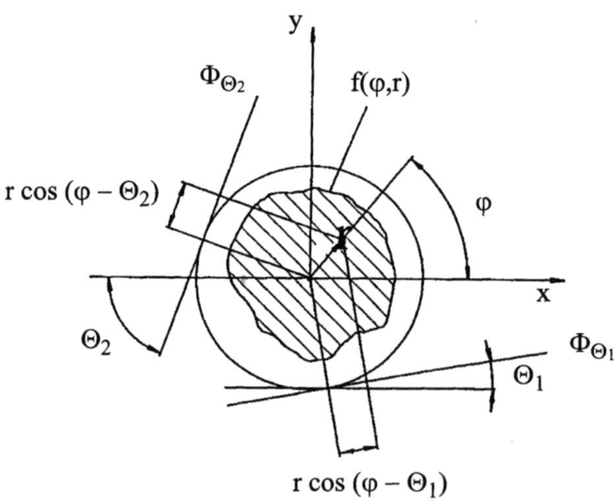

Fig. 15.17. Coordinate system for the back projection method

In (15.27), the field function of interest is denoted by f, and the measured projection data Φ are given in the polar coordinate system r, φ for the projection angle Θ. The method is illustrated in Fig. 15.17. In practical applications the number of available projections and the number of rays in each projection is limited. The approximate individual value of the field function f_j' is thus obtained from the sum of all measured projection values for each ray in direction k as:

$$f_j' = \sum_k \Phi_i(\Theta_k) \,. \tag{15.28}$$

Due to the missing projection values, the effect of smearing is observed, i.e. the reconstructed field function is blurred. Further improvements of this method are based on the filtered back projection and the convolution back projection. In these methods, the Fourier transform and the various filter functions described previously, are integrated. More details on this problem are available in the reviews papers [323,324] and in [301–303,371,377,378] numerous applications are described.

Direct analytical inversion methods
In 1917, Radon [363] has found an explicit formula for the solution of the linear integral equation. He has shown that the following inversion formula applies for the evaluation of the field function:

$$f(x,y) = \frac{1}{2\pi} \int_0^{\pi} \int_{-\infty}^{\infty} \frac{1}{x\cos\Theta + y\sin\Theta - \rho} \, \frac{\partial\Phi(\rho,\Theta)}{\partial l} \, \mathrm{d}\rho \, \mathrm{d}\Theta \,. \tag{15.29}$$

The investigated cross–sectional plane is presented in Fig. 15.11. More detailed proof of the discussed inversion technique has been given by Herman [323]. Modifications of the Radon formula for discrete grid and applications with incomplete projection values are implemented in various inversion methods. A common trait of all these methods is that no Fourier transform is performed. The availability of a large number of projection values is a prerequisite for good quality reconstruction. An interesting application of this technique is the measurement of velocity fields using ultrasound [318].

15.3.3 Accuracy of Reconstruction and Comparison of Reconstruction Methods

The selection of the most convenient reconstruction method for a particular application is a difficult task, as numerous tomographic algorithms based on different mathematical methods as well as modifications of the basic techniques have been described in the literature. In addition, the effectiveness of the algorithm and the reconstruction accuracy strongly depend on the quality of the measurement data and on the ray geometry. Very often, no direct correspondence between the reconstruction results of different authors can be established.

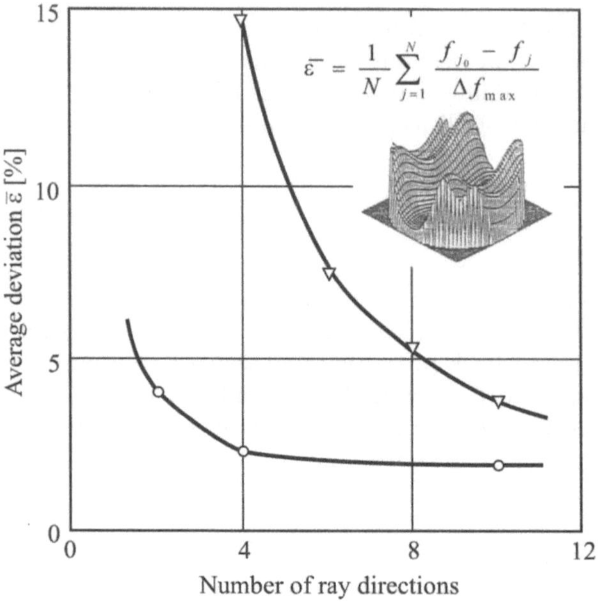

Fig. 15.18. Mean deviation of the reconstruction of the test function for the ART and the Fourier transform method versus the number of irradiation directions [338]

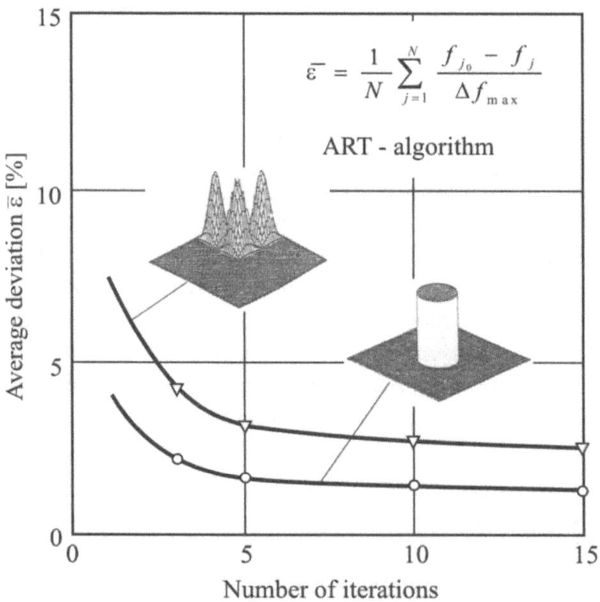

Fig. 15.19. Mean deviation of the reconstruction of the test function using the ART algorithm versus the number of iteration steps [316]

In order to be able to analyze the accuracy of the reconstruction, precisely defined fields in the investigated volume – so called test objects or test functions – are reconstructed. The reconstructed three–dimensional profile should be a close approximation of the test object. In the case of test functions, the projection value is evaluated as an integral along the ray path. The deviation of the reconstructed function from the original can be calculated when test functions are available. Lübbe [338] has compared two reconstruction methods using such test functions: (i) the ART algorithm combined with the sample method and (ii) the Fourier transform method with discretization [331]. In Fig. 15.18, the mean deviation of the reconstruction of the test function versus the number of ray directions is plotted. The better performance of the ART method is of particular advantage in applications involving few beam directions. When using iterative methods, the reconstruction of test functions is helpful for choosing the number of iterations to be carried out in the reconstruction procedure. In Fig. 15.19, the mean deviation of the reconstruction of the test function versus the number of iteration steps for the ART method is presented [316]. It should be noted that symmetrical field functions are easier to reconstruct accurately, and that the reconstruction accuracy can be improved only up to finite residual deviation.

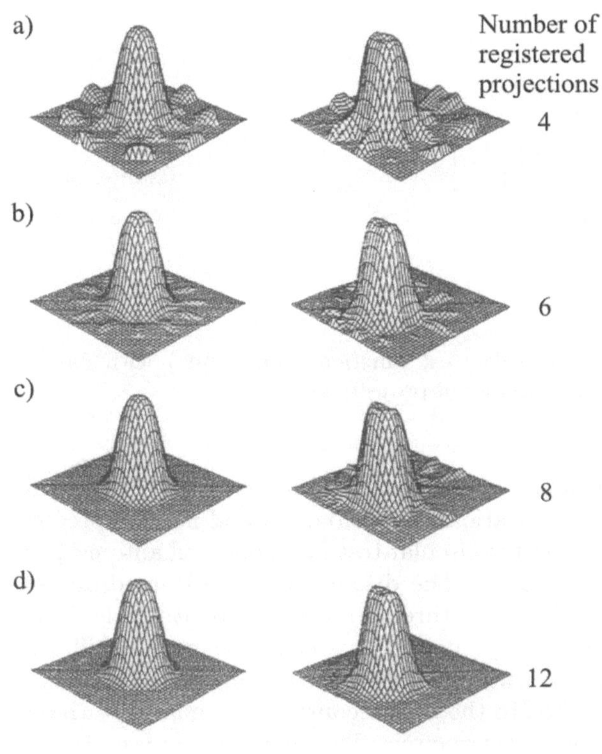

Fig. 15.20. Reconstructions in a cross–sectional plane of a gas jet for 4, 6, 8 and 12 projections [338]

In chemical engineering applications which are frequently concerned with the analysis of unsteady phenomena, the small number of possible viewing angles is of particular importance. Merzkirch et al. [337] have tested the convolution back projection method (cf. Sect. 15.3.2). The results of the analysis for 4, 6, 8 and 12 projections are presented in Fig. 15.20. The maximal appearing in the edge regions for small number of projections are called aliasing artifacts. They appear symmetrically in the transition zones between the individual projection directions. Their influence becomes negligibly small when the number of available projections increases. Aliasing artifacts are caused by missing projection values and they appear in all reconstruction algorithms based on the Fourier transform. If such reconstruction techniques are used with only four projection directions, the negative influence of non-symmetrical field functions on the quality of reconstruction is pronounced.

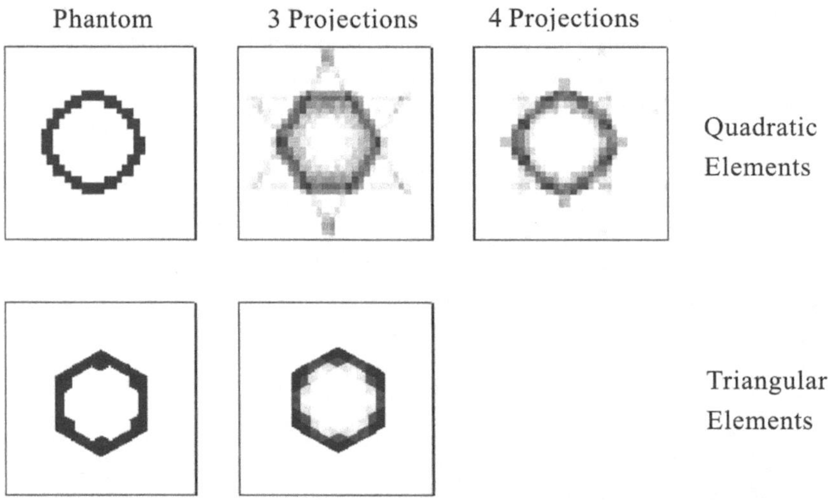

Fig. 15.21. Reconstruction of circular test function ("phantom") with different element schemes and a different number of projections [366]

Reinecke et al. [366] compare the reconstruction accuracy of the ART method for two different discretizations, with adapted and non adapted elements (Fig. 15.13). In Fig. 15.21 the circular test functions ("phantoms") and the reconstructed results are shown. The discretization with quadratic elements leads to strong artifacts if only three projections are available. If four projections are available the reconstruction quality is higher, but still some artifacts are visible. The discretization with adapted triangular elements and three available projections leads to the best reconstruction quality with only very small artifacts in the inside of the circle. The deviation between the test function and the reconstructed results for these discretizations is shown in Fig. 15.22 versus the number of iterations.

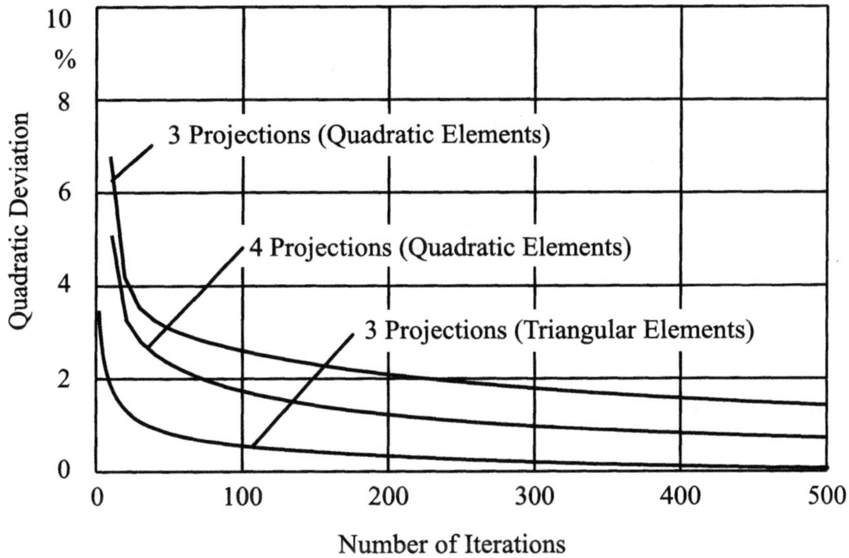

Fig. 15.22. Quadratic deviation for the circular test function (Fig. 15.21) versus the number of iterations[366]

In addition to the requirements on the reconstruction accuracy, important criteria for the selection of a suitable mathematical method are computer time, memory requirements, simplicity of instrumentation and easy realization. In several of the reviews referred to in the preceding chapters, the discussion and evaluation of the different reconstruction techniques can lead to confusing contradictions. The objects under investigation and the reconstruction methods vary from author to author, yet it is important to note that reconstruction methods are selected to match the particular problem considered. When using special instrumentation and filters, the selected method appears to be more efficient than others. For example, the SIRT method is rated by Gilbert [307] to be better than the ART method in all analyzed cases. Conversely, Herman and Lent [321] claim the ART to be the better algorithm for all situations when the projection data is not corrupted by strong noise.

In situations when measurements for numerous projections over the total viewing angle of 180° are possible, the back projection method or the Fourier transform is recommended. These methods are usually faster to compute and require less computer memory than those using series expansion.

The most suitable methods for reconstruction with a limited number of projection values are the ART and SIRT algorithms. In these techniques, the reconstruction accuracy may be significantly improved by the suitable choice of the initial guess. Owing to the increasing interest in the application of tomographic techniques in natural sciences, the algorithms used in the reconstruction of the field function have been improved. This is of particular importance when a limited number of projection values is available. Hence, for the so called "limited view conditions", the convolution back projection [295], and the Fourier transform [328] method are frequently employed.

Finally, the chapter on reconstruction methods will be closed with the conclusion of Herman and Rowland [319] stating that it is not possible, even on the basis of a great deal of experience, to predict the performance of a known solution method for entirely new applications.

15.4 Implementations

In the preceding sections, the tomographic measurement techniques and the mathematic methods used in the reconstruction of the three–dimensional field function have been discussed. At this stage, the different implementations of tomographic techniques in engineering measurements will be discussed.

In the field of medicine, computerized tomography has already been established as an important diagnostic tool. The most commonly used techniques, apart from X–ray tomography, are ultrasound and nuclear magnetic resonance. The required number of projections recorded from different irradiation angles is relatively high. This implies that prevalently steady or nearly steady objects can be examined in this way. The reconstruction is usually performed using the Fourier transform method.

Tomographic methods are frequently applied in nondestructive material testing [365,388]. They enable the detection of localization and size of material and manufacturing defects in the investigated measurement plane. The computerized tomographic equipment originally developed for medical applications can often be applied for the investigation of low density materials, such as plastics or aluminum. However, special techniques must be applied for the investigation of higher density materials or large samples. For example, computerized tomography has been developed for the testing of generator shafts with a diameter up to 1.60 m [384]. The measurement technique is based on the reflection of ultrasonic waves. The penetration depth of X–rays and γ–rays is insufficient for such measurements. The measurement of the annual rings of trees [351] and the investigation of oil distribution in rocks [387] are special applications of X–ray tomography. γ–radiation is used to analyze polymers. In this way, density differences of interlaced polymers in cured caoutchouc are detected [357].

Recently, tomographic measurements have been increasingly implemented in the field of chemical engineering. Some of these applications and the results of the investigations will be described in the following sections. As opposed to the applications discussed above, the objects of investigation are unsteady processes. By using computerized tomography, these three–dimensional, time dependent phenomena can be analyzed.

Apart from the review publications referred to in the first sections [345], numerous applications are described in conference papers [360–362].

15.4.1 Measurement of temperature fields in stirred vessels

Mayinger and Lübbe [338,341] as well as Ostendorf [353] measure three–dimensional temperature fields using optical tomography. The unsteady mixing of a liquid in a stirred vessel is a function of the spatial coordinates and time. In the investigations, two liquids of different temperature and identical concentration are used. The temperature distribution, due to the different initial temperatures of the liquids, are recorded simultaneously from four directions using the technique of holographic interferometry.

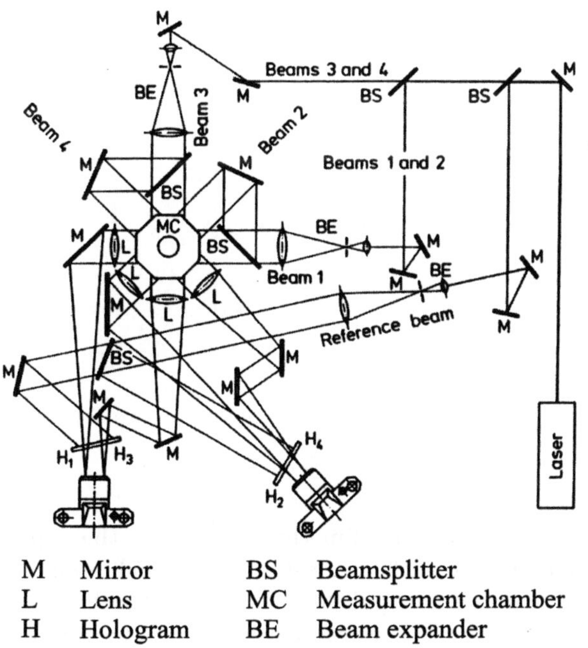

Fig. 15.23. Optical arrangement for tomographic measurements of unsteady temperature fields in a stirred vessel [353]

The optical arrangement developed for these investigations is presented in Fig. 15.23. In the measurements, a 2 W argon–ion laser is used. The first beam splitter divides the laser beam into an object beam and reference beam. The object beam is then divided into two portions by the second beam splitter. These two beams are divided again after leaving the beam expanders, and they transilluminate the measurement volume simultaneously with separation angle of 45°. Two object waves are then projected onto the same holographic plate by mirrors. The reference beam is also expanded and divided into two beams by a beam splitter. The two reference beams hit the two holographic plates without passing through the measurement volume. The interferometric images developing behind the holographic plates are recorded by two simultaneously triggered cameras. The diameter of the stirred vessel is 100 mm.

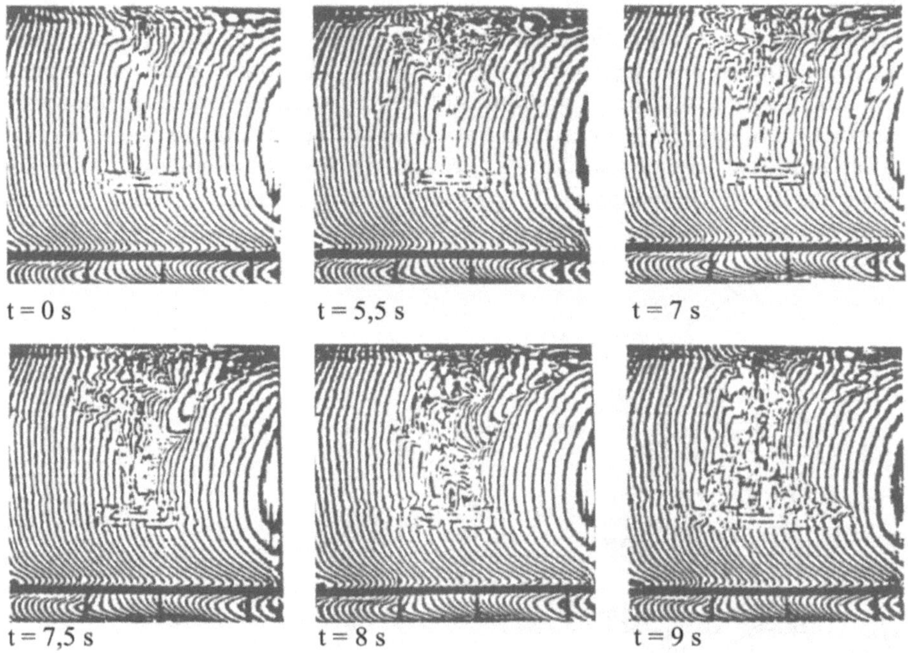

Fig. 15.24. Interferometric images recorded during the mixing of two liquids at different temperatures [353]

Lübbe [338,341] uses the organic solvent dimethylsulfoxide in the experiments. This solvent mixes completely with water. Ostendorf [353] applies a medium viscosity mixture of benzene and palatinol as well as the high viscosity mixture of glycerin and water. The refractive index of the mixtures matches the refractive index of the glass vessel and the glass stirrer, in order to avoid the deflection of the rays traversing the measurement volume. Turbines with six blades as well as simple cylindrical disks were used as stirrers. In the experiments, a small volume of the liquid is injected into the vessel which contains the same liquid at different temperature. The mixing of the two liquid volumes, the temperature equalization process and the thermal dissipation on the stirrer were measured. In Fig. 15.24, the interferometric images recorded for the same viewing angle at different time instants after the injection of the warm liquid are shown.

Apart from the mixing of liquids at different temperatures, Ostendorf [353] and Friedrich [305] investigate the energy dissipation during the mixing process. The resulting temperature field in the vicinity of the stirrer shows the dissipation of the energy introduced through the stirrer. The same optical and experimental set–up as presented in Fig. 15.23 is used in these investigations.

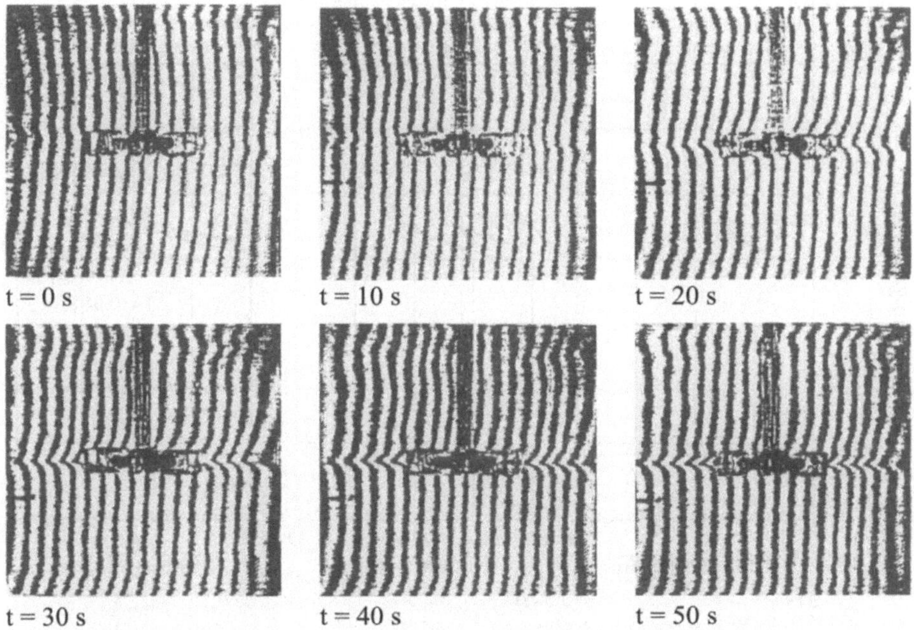

Fig. 15.25. Interferometric images recorded in the investigation of dissipation in the mixing process [353]

In Fig. 15.25, the interferometric images recorded at different instants of time in the investigation of dissipation during the unsteady mixing process are presented. A disk with six blades serves as a stirrer for the highly viscous glycerin solution. The deflection of the interference fringes in the vicinity of the stirrer increases with increased stirring time. The more distant interference fringes are not influenced by the stirring process. From the deflection of the interference fringes, the temperature fields in the vessel can be reconstructed at different time instants. The reconstructed temperature fields are presented in Fig. 15.26. After 12 s of stirring time, the temperature of the liquid in the vicinity of the stirrer has increased, as indicated in Fig. 15.26a. After 21 s (Fig. 15.26b) and 30 s, the volume of the liquid warmed up in the stirring process increases. The increased heat transfer in the radial direction can be recognized in Fig. 15.26. The spacing between the isotherms is a measure for the temperature gradient. It should be noted here that this technique allows the measurement of low temperature differences only. Significant change of the refractive index would cause a deflection of the light rays apart from the change of the optical pathlength. In this situation the ray path across the measurement volume is no longer linear thus the temperature field cannot be reconstructed from the interferometric images.

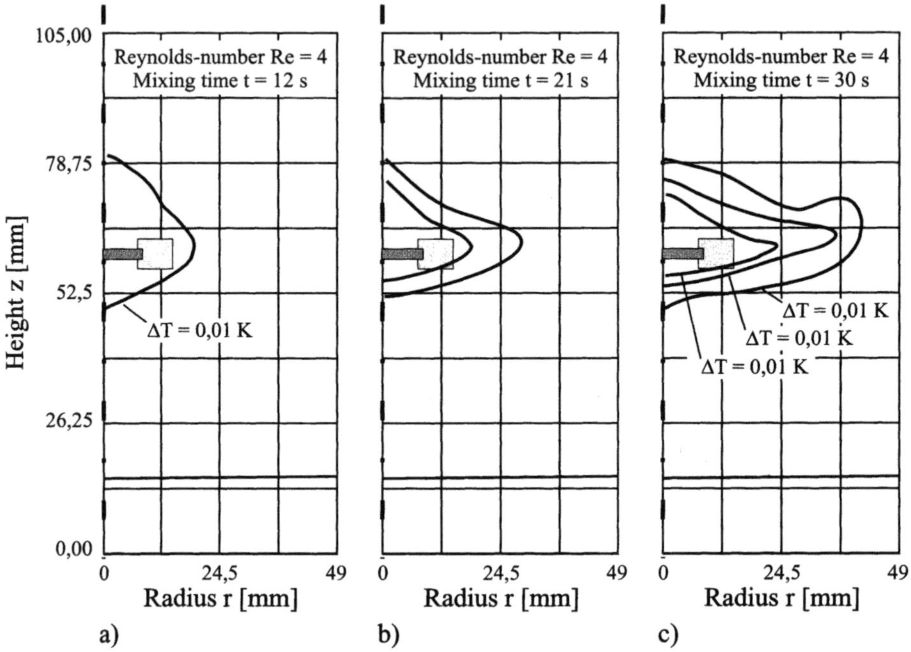

Fig. 15.26. Reconstructed temperature fields at different time instants during the unsteady stirring process. **a** after 12 *s*. **b** after 21 *s*. **c** after 30 *s*

The temperature fields were reconstructed using the sample method described in the preceding chapter. As the number of projections (four) available for the reconstruction is low, the sample method was combined with the iterative ART algorithm. The fidelity of reconstruction was investigated by Lübbe [338] using different test functions. The mean deviation is less than 5%, and the corresponding maximal deviation is below 20%.

15.4.2 Measurement of micro– and macromixing with the tomographical dualwavelenght tomography

The tomographic light absorption method is used by Renz [368] and Haarde [316] to measure the distribution of an injected inhibitior in storage tanks. This method is developed further by Buchmann and Mewes [284,285] to the tomographic dual wavelength photometry, which enables the simultaneous measurement of two dye concentration at multiple points inside stirred vessels. This method is used to separate the effects of micro– and macromixing by injecting mixture of an inert and a reacting dye into the vessel. The inert dye serves as tracer for the macromixing, whereas the vanishing of the reacting dyes show the micromixing.

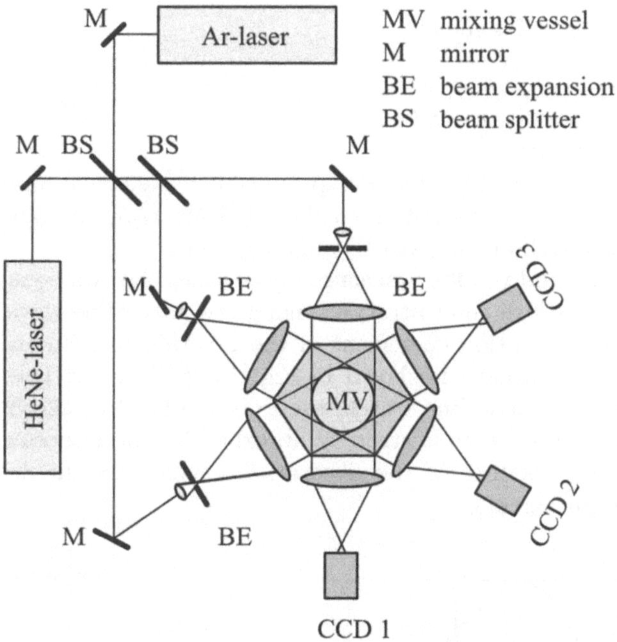

Fig. 15.27. Optical set–up for the tomographical dual wavelength photometry [284]

The optical set–up of tomographic dual wavelength photometry is shown in Fig. 15.27. Two laser beams of wavelength 514 nm and 632 nm are superimposed in a beamsplitter and then divided into three superimposed beams. The three beams are expanded to a diameter of 150 mm and are used to transilluminate the mixing vessel. The transparent mixing vessel itself is placed inside a hexagonal view bow, which is filled with refractive index matching liquid. This set–up enables the beams to transilluminate the cylindrical vessel without distortion and it also allows a temperature control of the vessel. The beams leave the vessel and give projections of the dye concentrations on RGB–CCD cameras. The spread of the dye in the vessel is so fast that all projections for the tomographic reconstruction must be photographed at the same time. This requirement limits the number of projections. Only three cameras are used. Their pictures are stored simultaneously in a computer. The cameras deliver pictures at fixed rate of 25 Hz, but the storage system can be adjusted so that for slower mixing processes the storage frequency is reduced.

The projections are used to determine the integrated dye concentrations along the paths of the beams. For this determination, a reference picture of the vessel without the dye is taken by every camera and is then compared with pictures of the dye projections. The camera signals for a given pixel and the two dye concentrations along the affiliated path are correlated by Lambert–Beer's law:

$$\ln\left(\frac{I_{514}}{I_{514}^{0}}\right) = -(\epsilon_{P,514}c_P + (\epsilon_{I,514}c_I)b \tag{15.30}$$

$$\ln\left(\frac{I_{632}}{I_{632}^{0}}\right) = -(\epsilon_{P,632}c_P + (\epsilon_{I,632}c_I)b \tag{15.31}$$

with I is the light intensity at the pixel for the appropriate wavelength, I_0 is the reference light intensity, ϵ the absorption coefficient, c the concentration along the path and b the length of the path through the vessel.

With the known ϵ and b and the four measured intensities the two equations yield the two integrated dye concentrations along the path. The wavelength dependence of the absorption coefficient of the reacting dye (iodine starch) and the inert dye (Ponceau) is shown in Fig. 15.28. The red laser beam is strongly absorbed by the iodine starch; the Ponceau hardly absorbs this beam. Conversly, the green beam is strongly absorbed by the Ponceau. This inverse wavelength dependence of the absorption coefficients enables precise concentration measurements.

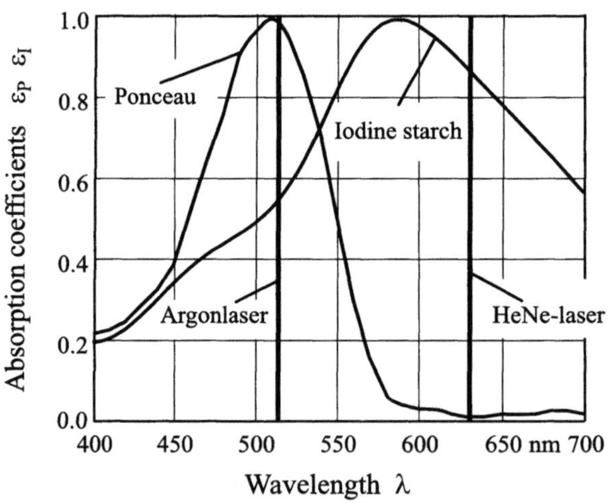

Fig. 15.28. Extinction of the dyes and laser wavelength [284]

The ART algorithm is used to reconstruct the three dimensional concentration fields from the measured concentrations along the paths. For the reconstruction the vessel is divided into 30 slices each divided again into 1220 adapted triangular elements (Fig.15.13) yielding a total of 36600 elements in the whole vessel. The concentration in each of these 36600 elements is calculated.

15.4.3 Tomographic measurements of flames with the Schlieren effect

Schwarz [374] uses tomography and the Schlieren effect to measure temperature distributions in flames. The Schlieren–method is a well known technique to measure variations of the index of refraction in transparent fluids. The measured index of refraction is used to calculate the density and temperature of the fluid. In the absence of tomography, only integrated measurements along the path of the light are possible. The use of tomography enables the determination of the three–dimensional temperature field in the flames.

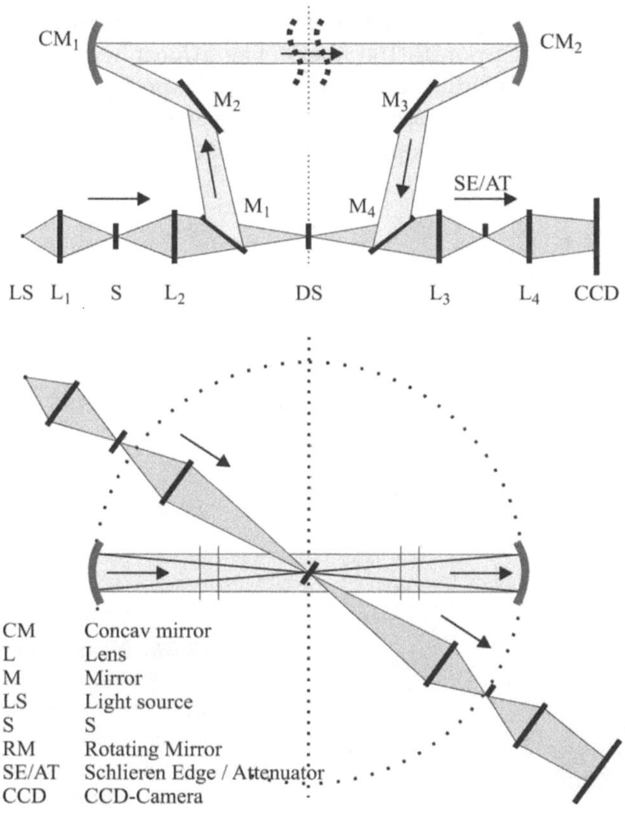

Fig. 15.29. Tomographical Schlieren set–up to measure temperature fields in flames [374]

The optical set–up shown in Fig. 15.29 is used to measure 20 Schlieren projections of the flame for the tomographical reconstruction. The projections are measured one after another by the rotating part of the set–up around the flame. Therefore, only one lightsource and one CCD–camera is needed. It must be noted that this set–up is limited to steady processes because the projections are obtained at different instants of time. The tomographical reconstruction of the temperature field out of the Schlieren projections is different from the scalar reconstruction described in the preceding sections. In this case the intensity of the light is constant, but the direction of the light changes due to the variation of the index of refraction in the flame. More details of this vectorial tomography are given in [289].

15.4.4 Chemical species tomography by near infra–red absorption

Carey et al. [291] use the near infra–red tomography to measure the distribution of chemical species in an internal combustion engine. The optical set–up around the combustion chamber is shown in Fig. 15.30 and consists of 28 fibers for launching and receiving the infra–red signals. In the future the set–up should be extended to incorporate 190 fibers. The atmosphere in the combustion chamber consists of different infra–red absorbing species (carbon dioxide, methane, water, hydrocarbons etc.) which makes it difficult to detect a specific species. The authors are mainly interested in the distribution

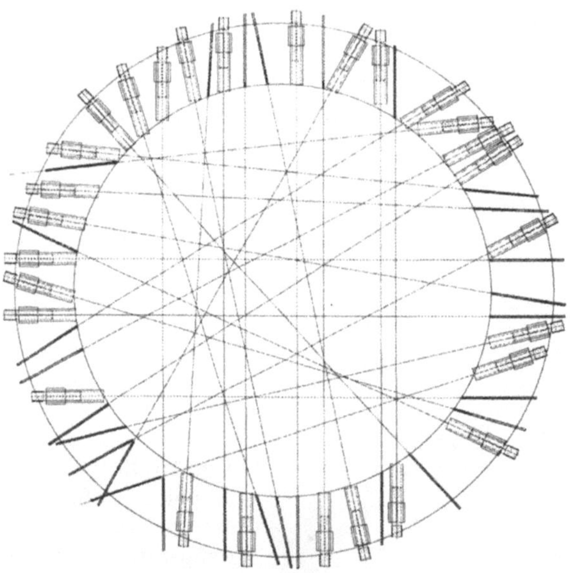

Fig. 15.30. Near infra–red tomographical set–up with 28 channels around the combustion chamber of an internal combustion engine [291]

of the unburned hydrocarbons and thus use the peak absorption wavelength of iso–Octane at 1700 *nm*. To correct for the distortion of the measured signals by blackbody radiation and scattering due to fuel droplets the reference wavelength of 1500 *nm* is also launched through each fiber. This wavelength is hardly absorbed by any of the chemical species in the chamber, so that change in the intensity of this wavelength is due to the effects mentioned above. To avoid the difficult optical separation of the two wavelengths at the receiver, the authors use an electronic separation. To obtain this the intensity of the two different wavelength is modulated at the sender with two different frequencies. These modulation frequencies are used to distinguish the signals which are detected by the same photodiodes. The coding/decoding of the signals is done by frequency division multiplexing (FDM), which enables a lock–in detection at the receiver.

16 Particle Image Velocimetry

Wolfgang Merzkirch

16.1 Introduction

Particle Image Velocimetry (PIV) belongs to the class of optical "whole–field" measuring techniques. The name of the method is self–explaining: The velocity distribution in a whole field of a fluid flow is determined by measuring the displacements Δs that the images of tracer particles experience during a time interval Δt. Local, instantaneous velocity values (and directions) $w(x, y) = \Delta s / \Delta t$ are measured simultaneously at many positions (x, y) in the field of view. The idea of determining flow velocities by measuring the displacement of tracer particles is not quite new and was in practical use long before the name *Particle Image Velocimetry* appeared. The new name arose from the possibility of registering the particle images in digital form and efficiently handling the large amount of planar, quantitative data with the techniques of digital image processing. The state of the art for the period prior to the PIV era is well documented in the article by Emrich [414].

The first attempts to perform PIV–like measurements were derived from optical speckle photography [400,412,424,445]. The name "speckle velocimetry" was used in these references, because, for the purpose of evaluation, the method of Young's fringes, as developed in the field of speckle metrology, was applied to the photographic double exposures of the particle images. These exposures are interrogated, point by point, with a thin laser beam, thus producing on a viewing screen a system of equidistant, parallel interference fringes. The spacing and direction of each parallel interference fringe must subsequently be measured for each interrogation point, in order to determine two components of the local velocity [416]. This procedure is an optical Fourier transformation of the measured signals now called particle image pairs. A disadvantage of speckle velocometry is the need of an additional experimental set–up. Further development of the PIV technique replaced speckle velocimetry with numerically executed Fourier transforms.The name speckle velocimetry was then abandoned, and in the mid eighties Particle Image Velocimetry became an established term.

The principle of taking PIV recordings is illustrated in Fig. 16.1. A laser beam is expanded in one plane by means of a cylinder lens, thus forming a light sheet which is directed in the flow field, where it illuminates the tracer particles with which the fluid is seeded. A camera whose viewing direction is normal to the light sheet records the motion of those particles moving in the plane of the light sheet. For this purpose, two or more recordings of the particle images are taken at different instants of time, separated by a time interval Δt. These exposures are recorded either on one or on several, consecutive frames. The content of these frames, once available in digital form, can be subjected to evaluation procedures for which different types of software exist. In the following, approaches to this kind of software as well as the hardware depicted in Fig. 16.1 will be discussed in some detail. Not included in the present review are the techniques of Particle Tracking Velocimetry (PTV) [407,426,444,452,457,476].

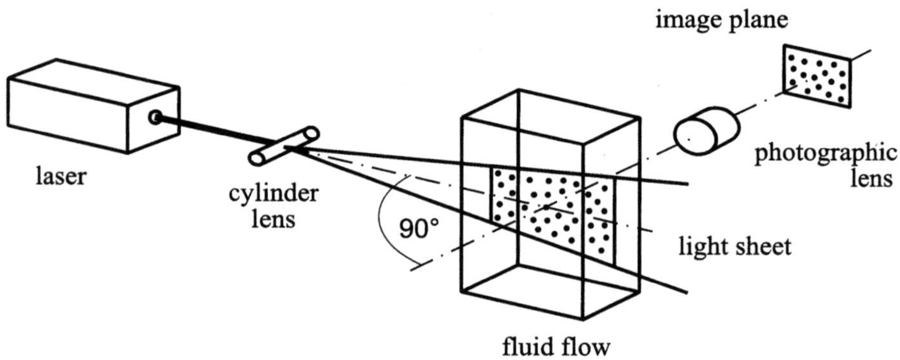

Fig. 16.1. Principle of PIV set–up

A number of reviews of the technique of Particle Image Velocimetry (PIV) has been published recently [399,404,423,436,465,477]. The method is still in a state of rapid development.

16.2 Hardware for the experimental set–up

The optical components for an experimental set–up are discussed here for the most frequently applied version of PIV, the production of pairs of particle images on the recording medium of the camera. By means of the hardware particle image recordings of a quality sufficient for the succeeding evaluation have to be secured. The following parameters may serve for defining such a quality: intensity and size of the particle images, minimum and maximum distance of particle images forming a pair (displacement Δs of the particle images), number of particle image pairs in the measurement domain, ratio of signal intensity to background noise (S/N ratio). These parameters are controlled by the following hardware components: light source, light sheet optics, tracer particles, imaging optics, recording system of the camera. The

S/N ratio may also be affected by components of the flow facility, e.g., optical viewing windows. Some characteristic features of these components will be briefly reviewed in the following.

In order to produce the desired pairs of particle images an illumination of the object according to the sketch shown in Fig. 16.2 is preferable.

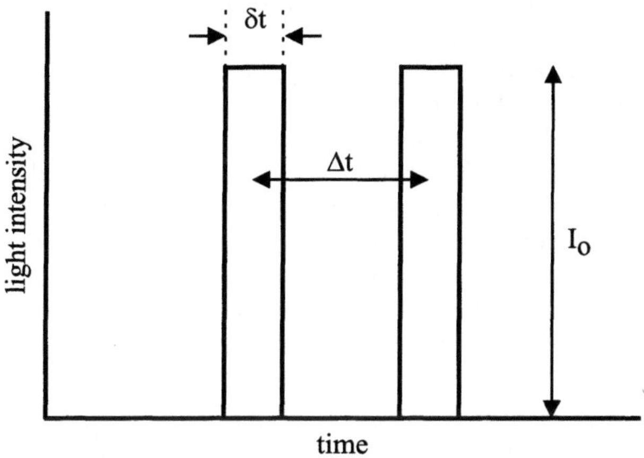

Fig. 16.2. Illumination provided by two laser pulses of width δt and separated by the interval Δt

Two short light pulses of duration δt and separated by a time interval Δt are generated with an intensity I_0 high enough to allow the particle images to be recorded. The particle velocity is approximated by $\Delta \boldsymbol{s}/\Delta t$ with $\Delta \boldsymbol{s}$ being the 2D vector of the particle displacement during the time interval Δt. The values necessary or allowable for δt and Δt depend, of course, on the velocity to be measured. The most comfortable way of controlling the values of δt and Δt is the use of a laser that can produce double pulses. Many PIV systems are equipped with a Nd:YAG laser that emits light at a wave length of $\lambda = 532\ nm$ in the frequency–doubled mode. This blue/green radiation is advantageous for two reasons. First, photographic film that can be used as a recording medium is very sensitive in this frequency range (in contrast to red emitted be HeNe–and ruby lasers). Secondly this radiation can stimulate fluorescence which is of interest when fluorescing tracer particles are used or when PIV measurements are combined with laser induced fluorescence (LIF) experiments. Nd:YAG lasers can produce double pulses with a selected time interval between pulses as low as $\Delta t \sim 1\ \mu s$ and at repetition rates between 10 and 20. The pulse duration δt is in the order of a few nanoseconds (up to 50 ns), and the intensity of the single pulse is for normal instruments between 70 and 300 mJ.

Continuous–wave (cw)–lasers may be applicable at low to moderate flow velocities and not too small tracer particles which might be the case in water flows. Then the necessary values for δt and Δt must be realized and controlled by either mechanical or electro–optical shutters, e.g. a Bragg cell.

The laser beam is expanded to form a light sheet that defines the measurement plane in the flow. This can be done by a cylinder lens or a simple glass rod as indicated in Fig. 16.1. The thickness of the sheet is then approximately equal to the diameter of the beam emitted by the laser, i.e. approximately 1 *mm*. Of course, the laser beam exhibits a Gaussian intensity profile, and in many references where number values for the light sheet thickness are given this thickness is not well defined. A thin light sheet might be desirable for improving the spatial resolution of the velocity measurement. This requires light sheet optics more refined than a simple glass rod. Prenel et al. [460] describe an optical system combining the telescopic arrangement of two spherical lenses with a cylinder lens. The area of the light sheet to be used effectively for the measurement and the sheet thickness can be controlled easily with one lens of the telescope. A light sheet thickness of 100 μm to 200 μm can be realized in this way. Making the light sheet parallel by means of a large diameter lens or a spherical mirror provides a uniform illumination intensity in the field of view. Prenel et al. [460] also report on a device that can rapidly move the sheet in a direction normal to its plane, so that the volume of the flow field can be scanned within a very short period of time. Brücker and Althaus [402] have adapted this principle for the use in a PIV system such that, when the recordings are taken with a digital high–speed camera [403], the flow volume can be scanned rapidly.

A flow with a strong velocity component normal to the plane of the light sheet can require the use of a sheet whose thickness is not very small, in order to secure that a sufficient number of tracer particles remains within the sheet during the time interval Δt. The particles may then experience a considerable displacement perpendicular to the sheet, i.e. in the viewing direction of the camera. From the images of the particles recorded at two different positions along the optical axis one may erroneously deduce an in–plane displacement that must be corrected for with simple rules of geometrical optics [485].

The application of evaluation algorithms as they are discussed in the following section requires the PIV recordings to be available in digital form. Therefore, the recording of the particle images by means of a digital (electronic) camera is most reasonable, unless a very high spatial resolution in terms of number of pixels is desired. This number is limited for each digital camera and, at this time, much smaller than the equivalent number characterizing the spatial resolution of photographic film. Here, a typical value is 200 *lines/mm*, whereby a "line" is comparable to a pixel. Ordinary CCD ("charge coupled device") cameras have a format of 512×512 pixels. Commercial PIV systems are most often equipped with a 1024×1024 pixels camera ("1k camera"). A few experimentalists could afford to perform PIV measurements with a 2k camera, and it is reported that 4k cameras will be available in the future, though at a high price. This 4k camera would still have half the number of pixels, in equivalent form, available on a 24×36 mm^2 film exposure. Yet, photographic film, which was exclusively used in the beginning of PIV, has two major drawbacks. First, the developing of the exposed film and the subsequent digitization by a scanning device interrupt the experimental process and make an on–line measurement impossible. Secondly, the two particle images forming a pair must be recorded on the same frame; thus a recording on two separate frames, which affords a has a number of advantages for the evaluation, is not possible.

The term "digital particle image velocimetry" (DPIV) has been introduced and is frequently applied for the direct digital redording of PIV images [481]. Kompenhans et al. [462] give a survey of digital PIV cameras. It is preferable not to use a camera operated in the interlaced mode. When combined with a double–pulsed laser the camera must be synchronized with the laser pulses. For the often preferred recording of the two exposures on separate frames, the synchronization must be such that the two laser pulses illuminate two successive frames taken by the camera (Fig. 16.3). The information recorded by the camera is transferred to a computer via a frame grabber. The possibility of performing time–resolved planar measurements is limited by this data transfer, as well as by the repetition rates available for double–pulsed lasers. An alternative approach for collecting data on the time–history of the flow is the use of a cw–laser, if applicable, in combination with a digital high–speed camera. This may however be at the expense of spatial resolution (number of pixels) which, for such type of cameras, decreases with increasing framing rate. For low–speed flows an inexpensive solution is the storing of the data on ordinary video tape [481]. Here, a resolution in time is achieved in terms of the available video frequency, and the total time during which data can be collected is nearly unlimited [455].

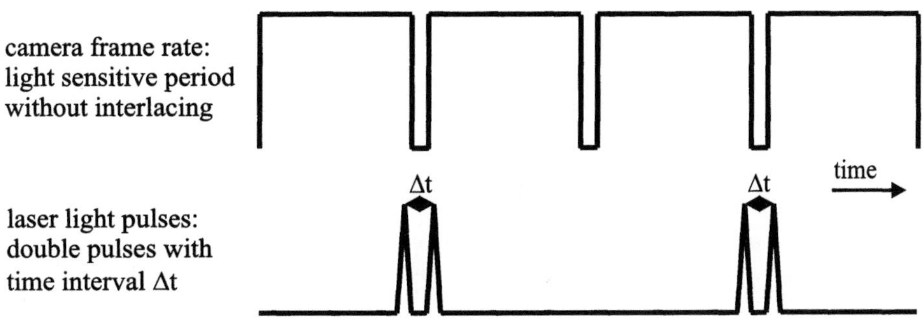

Fig. 16.3. Synchronization of camera frame rate with laser double pulses

A critical point for performing PIV measurements is the optical access to the flow, and this can be difficult for flow fields bounded by solid walls. Xiong and Merzkirch [485] report on the use of an endoscope for transferring the optical signal to the recording camera.

The problems met with when choosing tracer particles is in principle the same as discussed frequently in conjunction with LDA systems. For water flows, satisfactory experience has been made with spherical, almost neutrally–buoyant plastic particles of approximately 10 μm in diameter. The most common tracers for air flows are 1 $\mu m \times 2$ μm diameter liquid droplets which can be produced with a Laskin nozzle particle generator [465]. While olive oil droplets were very popular for a long time, it is now believed that DEHS ("di–2–ethyl–hexyl–sebacate", $C_{26}H_{50}O_4$) is a substance that better meets the requirement of non–toxicity. Problems of tracer particles and seeding have been discussed extensively by Melling [454].

The size of the particle images is larger or, for the very small particles, much larger than the size of the particles themselves, due to optical diffraction. For reasons of an accurate evaluation it is desirable to produce particle images of a size of 2 to 3 pixels. Another parameter that needs to be controlled with the particle generator is the number of tracer particles per volume in the flow. As will be discussed in the following section there is an optimum number regarding the evaluation accuracy.

16.3 Evaluation software

Once available in digital form, the PIV recording can be subjected to an evaluation by means of appropriate digital image processing methods. Prior to the evaluation the S/N ratio of the recording can be improved, if so desired or necessary, by means of standard digital filtering techniques [440]. Low–wave number background noise, such as those caused by light reflections, can be suppressed by high–pass filtering. In this way, the background noise is identified and can be subtracted from the image.

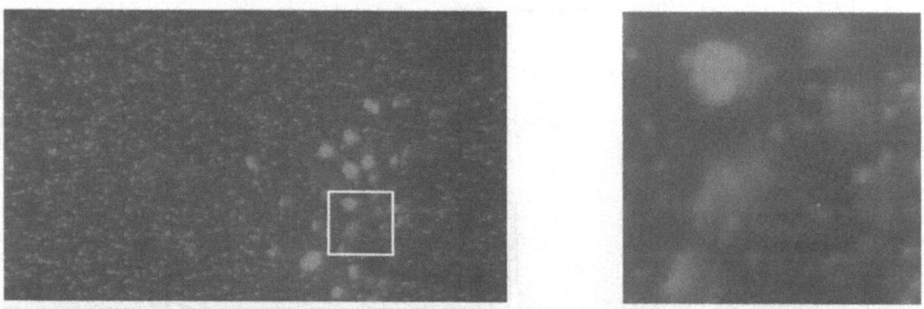

Fig. 16.4. PIV recording showing the images of tracer particles and big dispersed particles (left). The content of the indicated square window is shown in magnified form on the right [435].

Three additional filtering techniques for increasing the particle image contrast with respect to the background are demonstrated with the following example. Figure 16.4 is a PIV recording taken in a two–phase flow. The figure displays images of large solid particles dispersed in the continuous phase and

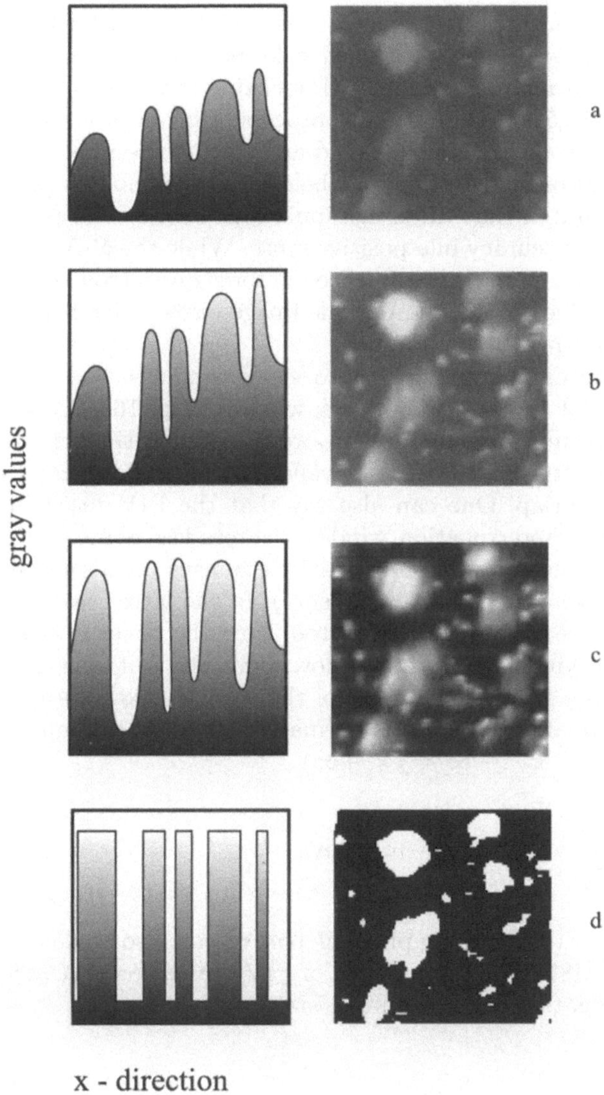

Fig. 16.5. a: Content of the window from Fig. 16.4 (right) and gray value distribution along a line in x–direction (left); b: spreading of gray values, c: removal of skewed gray value distribution; d: binarization [435]

small tracer particles with which the continuous phase is seeded. The filtering techniques are demonstrated with application to the quadratic window whose content is shown on the right side in magnified form. The first filtering step serves for spreading the range of gray values to the full range available in a digital image, i.e. from zero to 255. The original image does not cover this full range of gray values (Fig. 16.5 a); the result of this spreading is given in Fig. 16.5 b, where the distribution of the gray values along a horizontal axis through the image is shown on the left side. Minimum and maximum now reach the possible extreme values, but it can be seen that the distribution is still oblique. This skewed quality can be removed such that the difference between adjacent maxima and minima is increased, meaning that the contrast is improved locally (Fig. 16.5 c). The ultimate processing along this line is a binarization. This requires defining a threshold and setting the gray value of every pixel above the threshold to 255 and below the threshold to zero (Fig. 16.5 d). It must be stated that these manipulations of the contrast do not all affect the evaluation accuracy in a positive sense. While the S/N ratio is increased, the image of each particle is influenced in some uncontrolled way, and this might change the location of the particle image center with respect to which the displacement must be determined.

The digital PIV recording is subdivided into small windows, and a velocity value is determined at the position of each window (Fig. 16.6). These windows must have a certain minimum size. In order to refine the grid for whose nodal positions velocity results become available, the windows can be designed with a certain overlap. One can also say that the PIV recording is evaluated by moving an "interrogation window" across the plane of the recording. The window includes a number of particle images (two separate exposures) or particle image pairs (double exposure). In the latter case, the velocity is determined by means of a statistical tool from the particle image displacements (image pairs) included in the window. Here, the statistical tool is the autocorrelation of the gray values $g(i,j)$ of the $N \times M$ pixels forming the window. The autocorrelation function is normally defined in the infinite digitized plane $(-\infty < i < +\infty, \quad -\infty < j < +\infty)$

$$\Phi_A(m,n) = \sum_{i=-\infty}^{i=\infty} \sum_{j=-\infty}^{j=\infty} g(i,j) \cdot g(i+m, j+n), \tag{16.1}$$

with i, j, m, n being integers (designating pixel). $\Phi_A(m,n)$ can also be defined in some normalized form [481]. For the purpose of PIV evaluation $\Phi_A(m,n)$ is restricted to the interrogation window offinite size:

$$\Phi_A(m,n) = \sum_{i=0}^{M-1} \sum_{j=0}^{N-1} g(i,j) \cdot g(i+m, j+n), \tag{16.2}$$

and this requires the gray value function be repeated periodically outside the window:

$$g(i,j) = g(i+kM, j+lN), \tag{16.3}$$

with k, l being integers. The direct computation of $\Phi_A(m,n)$ is time consuming. The computation can be accelerated by applying the Fast Fourier Transformation (FFT), if the side lengths of the interrogation window, M and N, are powers of 2. This requirement might restrict considerably the

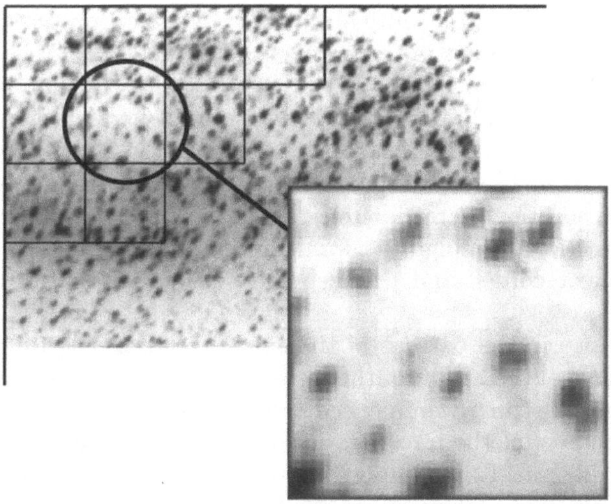

Fig. 16.6. Subdivision of a PIV double exposure into interrogation windows

choice of the interrogation window size. Gui and Merzkirch [429] have shown how this constraint can be circumvented.

The autocorrelation function has a central maximum for $m = n = 0$ that corresponds to the correlation of each pixel value with itself. Two second order maxima appear symmetrically on each side of the central maximum (Fig. 16.7). The positions of these second order maxima, (m^*, n^*) and $(-m^*, -n^*)$, designate the displacement Δs to be determined. The location of these maxima can be determined with sub–pixel accuracy. For this purpose it is assumed that the shape $\Phi(m,n)$ near the maximum can be approximated by a Gaussian profile whose peak is regarded as describing the correct position of the respective maximum [480,481].

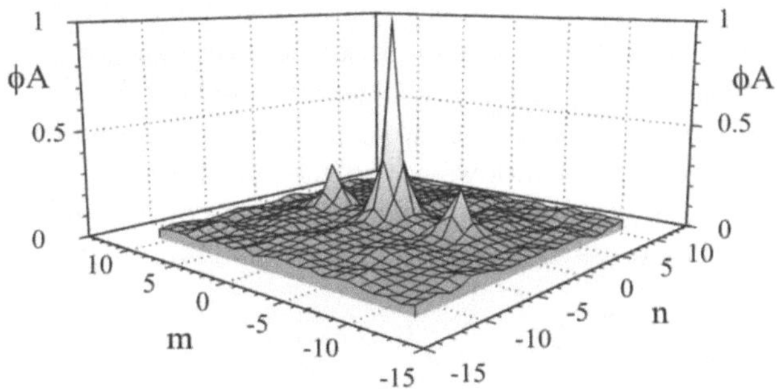

Fig. 16.7. Auto–correlation function $\Phi_A(m,n)$

The occurence of two second order maxima of the autocorrelation function is due to the fact that it is not known in the PIV double–exposure which particle image of a pair stems from the first, and which from the second exposure. A number of mechanical and optical devices have been proposed for removing this directional ambiguity in the evaluation [398,409,421]. These approaches either mark the particle images according to their sequence in forming the pair, or superimpose an artificial constant velocity which must be higher than the highest velocity in the real flow, similar to the Bragg cell technology for LDA. In any case, the disadvantages are the necessary additional equipment and additional error sources [461,486] and the problem of accounting for the directional ambiguity is much easier to handle when two separate exposures are recorded and the respective evaluation software is applied (see below).

The signal–to–noise ratio in the pattern of the autocorrelation function, for the present case the ratio of the height of the second order maxima to that of the next order maxima, is strongly influenced by the number of particle image pairs present in the area of the interrogation window. An optimum is just one pair but this is difficult, if not impossible, to realize by controlling the seeding rate. The worst case is the occurance of two pairs, because this causes an ambiguity in form of four second order maxima of equal height, i.e. a S/N ratio of one. Keane and Adrian [441] have shown that the S/N ratio increases again when the number of particle image pairs increases beyond 2, reaching a value comparable to the case with one pair for that number being 10 to 12. The number of particle images in the window can be influenced with certain limitations by the choice of the window size. Though the number of particle images can be increased by choosing a larger window, this increase occurs at the expense of the spatial resolution of the measurement. The interrogation window is equivalent to the measurement volume in LDA experiments in which a small size is desirable, particularly in flows exhibiting strong velocity gradients. If one considers dimensions of interrogation windows as typically used in PIV measurements and the thickness of the light sheet, then the resulting volume is indeed comparable to those dimensions reported for many LDA experiments, i.e. in the order of 1 mm^3. The measured velocity value (or vector) is assigned to the center of the interrogation window. It has been observed that, for decreasing interrogation window size, the PIV results exhibit a bias towards lower velocity values [429,464].

When recording the PIV image patterns on two separate frames, one subdivides both frames into interrogation windows as indicated in Fig. 16.6. The data fields are evaluated by cross–correlating the contents of corresponding windows of the two frames. The cross–correlation function is [442,481]:

$$\Phi_C(m,n) = \sum_{i=0}^{M-1} \sum_{j=0}^{N-1} g_1(i,j) \cdot g_2(i+m, j+n), \qquad (16.4)$$

with g_1, g_2 being the gray values of the pixels in the interrogation windows applied to the first and second frame, respectively (s. Eqs. 16.1 and 16.2). Φ has no central maximum like the auto–correlation function, unless the displacement or velocity is zero. This is of favorable influence on the S/N

ratio. The displacement to be measured is found from the position (m^*, n^*) of the highest maximum of Φ_c (Fig. 16.8). The measurement of the displacement direction is not ambiguous, because it is known which particle image pattern was first g_1, and which second g_2. It follows that the recording of two separate exposures and the subsequent evaluation by cross–correlation has definite advantages over the recording of a double–exposure and evaluation by auto–correlation.

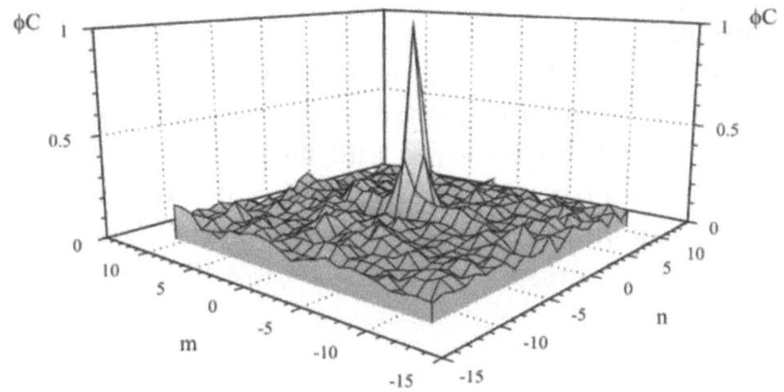

Fig. 16.8. Cross–correlation function $\Phi_C(m, n)$

There are upper and lower limits for the displacement or velocity value measurable with a given size of the interrogation window. The lower limit depends on the shape and size of the particle images and can be estimated to be of the order of half the diameter of a particle image. The upper limit follows from the Nyquist theorem and should be one half of the window diameter or less; Willert and Gharib [481] give an estimate of 1/3 of this diameter. This follows from the fact that, for increasing displacement or velocity, the portion of the window content that can be used for the correlation decreases; but this portion should not fall below a certain limit in order to maintain the desired degree of accuracy. This is, of course, a severe constraint for measuring high–speed flows. The solution to this problem is the window–offset as discussed in [480] and [482]. It requires moving the window in the second exposure in the direction and by an amount as estimated for the displacement to be measured. A refinement of this technique is described in [469]. The window offset can also serve for the purpose of increasing the accuracy for a sub–pixel evaluation of the particle image displacement. In this case the window is shifted by the integer number of pixels of the displacement to be measured, and in a second step of the evaluation the sub–pixel displacement is measured on the basis of a correlation. Since the measurement error depends on the magnitude of the displacement, the error is decreased considerably in this way [480].

The PIV results are affected by numerous error sources, ranging from imperfections of the set–up hardware to systematical errors caused by the evaluation software. Many of these error sources have been investigated systematically, yet it appears difficult, if not impossible, to add up all these possible errors and arrive at a precise value of the total error or uncertainty of a resulting velocity value. Investigations of the error sources comprise three–dimensional effects [471], the influence of the particle image shape and size [458], the S/N ratio of the PIV recordings [425], or systematical errors caused by the statistical approaches used for the evaluation [439]. Since a direct error analysis is difficult to perform, the accuracy achieved with an evaluation algorithm is often examined by applying the algorithm to an artificially constructed data set for which the (true) displacements (or velocities) are known.

A convenient way of presenting the resulting planar velocity distribution is a vector plot. The example given in Fig. 16.9 includes a number of obviously erroneous vectors that are caused by one or several of the various error sources. These erroneous vectors can be removed and then replaced by an average value of the neighboring vectors and/or by making use of the continuity equation. Statistical checks and digital filters serve for the detection, removal, and replacement of such "spurious" vectors [418,432,473,478].

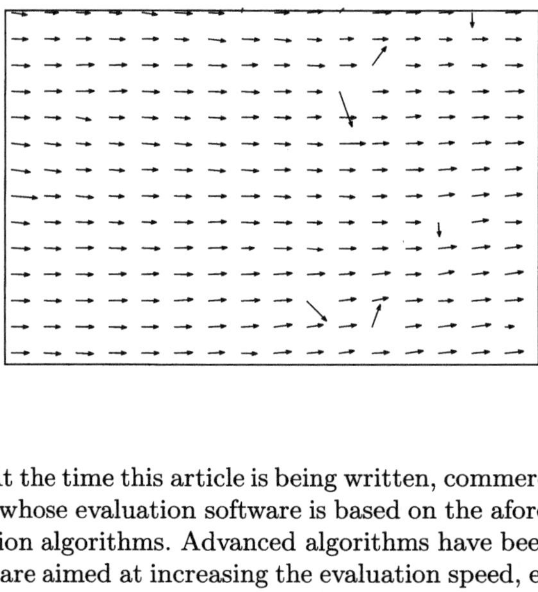

Fig. 16.9. PIV vector plot of a homogeneous parallel flow with spurious vectors

At the time this article is being written, commercial PIV systems are available whose evaluation software is based on the aforementioned "normal" correlation algorithms. Advanced algorithms have been developed and reported that are aimed at increasing the evaluation speed, e.g. for allowing an on–line measurement, improving the evaluation accuracy, or expanding the dynamic range. Rösgen and Totaro [466] evaluate the PIV images in several steps of different digital resolution and achieve a considerable acceleration of the evaluation procedure. Such an acceleration is also achieved with Hart's "sparse array image correlation" [431]. Advanced correlation schemes, i.e. modifications of the correlation function given in Eqs. 16.2 and 16.4, are described in [415,437,438,443,456,457]. These schemes also serve partly as search criteria for particle tracking. The "minimum quadratic difference" (MQD) method of [427] replaces the "normal" correlation by a least square algorithm for the purpose of image pattern identification and subsequent tracking of these patterns. Artificial neural networks have been used by [405,408,423,434] for the image pattern recognition and identification.

By means of artificially generated data sets of which the displacements of synthetic particle image patterns are known, accuracy and spatial resolution achievable with several of the listed algorithms have been investigated and compared. Figure 16.10 shows that the RMS error obtained with three different schemes is a function of the particle image displacement, while the dependence of the RMS error on the evaluation window size is given in Fig. 16.11 [430].

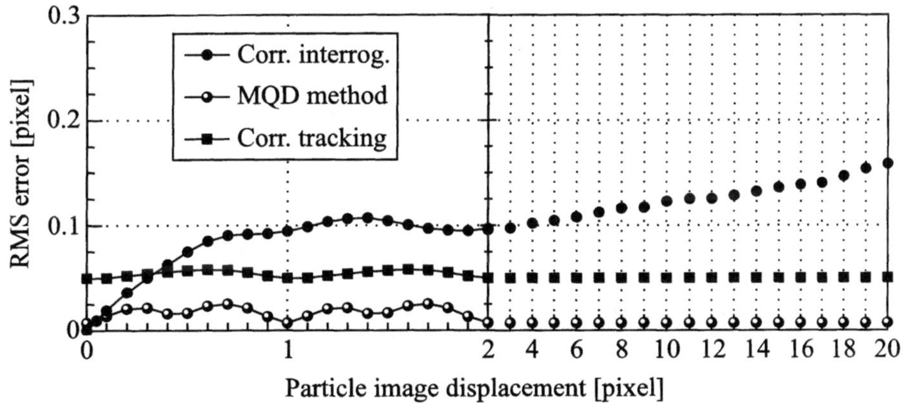

Fig. 16.10. Dependence of the RMS error on the particle image displacement for a 64 × 64 pixel interrogation window [430]

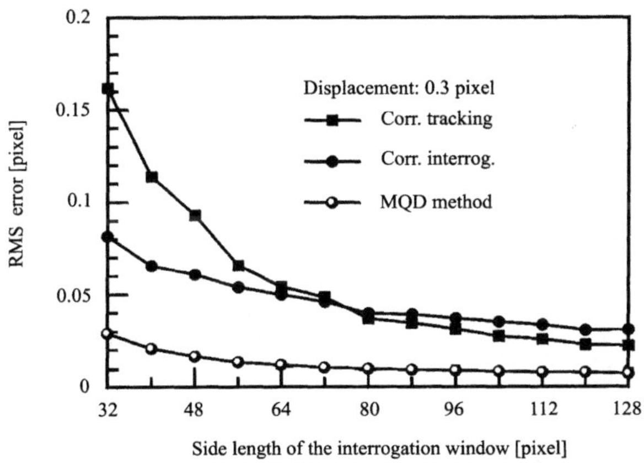

Fig. 16.11. Dependence of the RMS error on the interrogation window size [430]

16.4 Three–dimensional flow

An important extension of PIV systems is the performance of three–dimensional measurements. It is necessary to distinguish here between simultaneous measurements of the three components of the velocity (3C) at a position of the light sheet plane, and the simultaneous measurement of the velocity at different spatial positions of a three–dimensional measurement volume (3D). The latter is possible with holographic PIV [401,410,467] and advanced techniques of particle tracking [451,452], but only under the conditions of a low seeding rate.

In the first 3C measurements an attempt was made to use two or three parallel, differently colored or shaded light sheets and determine the third velocity component, normal to the light sheet plane, from the different signals resulting from the presence of the tracer particle in one of these sheets [406,462]. But the accuracy in measuring the third component remains low. Better results regarding the 3C resolution are obtained with stereoscopic systems operated with two CCD cameras that are viewing the flow scene (light sheet) under different angular directions. The abundance of knowledge in optics on stereoscopic viewing and recording has been adapted to 3C–PIV [419,447,459,483]. Two basic stereoscopic imaging conditions are known: the lens translation method with the camera lenses being parallel to the light sheet (object) plane, and the Scheimpflug configuration with the object, lens and image planes being at angles (Fig. 16.12). The stereoscopic arrangements are prone to different error sources whose effects are still the object of intensive research [446,472].

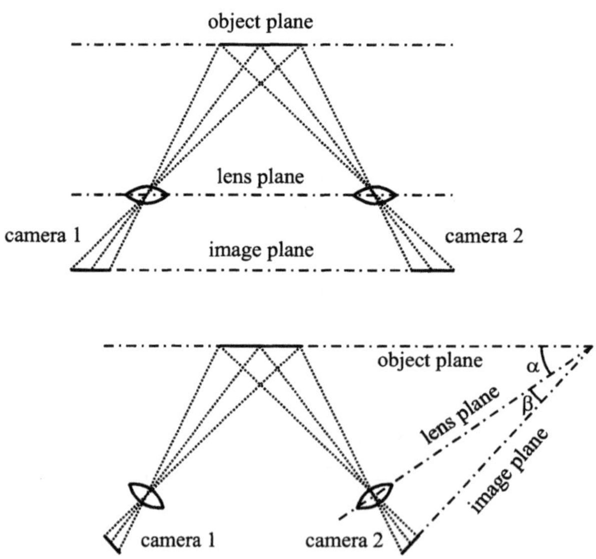

Fig. 16.12. Two principles for sterescopic image recording. a: lens translation method, b: Scheimpflug configuration (according to [483]).

16.5 Applications

Numerous applications of Particle Image Velocimetry have been reported, ranging from pure fluid mechanical investigations to studies of industrial flow problems. Here are just a few recent examples of extremely different fields of application:

- jet flow [420]
- turbine blade cooling [464]
- thermal plumes [475]
- thermocapillary convection [484]
- drag reduction by polymers [468]
- granular flows [449]
- microchannel flow [453]
- wind tunnel studies [463]

An application to a problem of two–phase flow is shown in Fig. 16.13. A system of two bubbles (dispersed phase) rises in water (continuous phase) where it induces a flow. The water is seeded with tracer particles. The velocities of the two phases are measured simultaneously by evaluating the PIV data with a digital mask separating the signals from the two phases [428]. The mask distinguishes between these signals if the particles characterizing the two phases have a significant difference in size. The bubbles are bigger than the thickness of the light sheet, and it is indicated in the figure where the light sheet intersects with the bubbles. Another means for separating the signals resulting from the two phases is the use of fluorescent tracer particles that emit light of a wavelength different from that scattered by the dispersed particles; the different radiation can then be separated by appropriate optical filters [433,474].

The most striking, novel results in applying PIV have been achieved regarding turbulent flow structures. Turbulence has since long been considered to be a time–dependent phenomenon. PIV has made it possible to get access to the spatial structure of turbulent fields. Vorticity, whose z–component is defined as:

$$\bar{\omega}_z = \frac{1}{2}\left(\frac{\delta u}{\delta y} - \frac{\delta v}{\delta x}\right) \tag{16.5}$$

with u, v being the velocity components in the x, y–plane, is considered to be a quantity of great interest for the modelling of turbulent shear flows. This quantity can be determined with PIV since the distribution of the two components u, v is measured in the x, y–plane (light sheet). Due to the applied evaluation grid, the values of u and v are only available at discrete positions of the plane. This might cause considerable errors in determining due to the necessary differentiation of the discrete data values. These error sources have been investigated intensively [397,417,450].

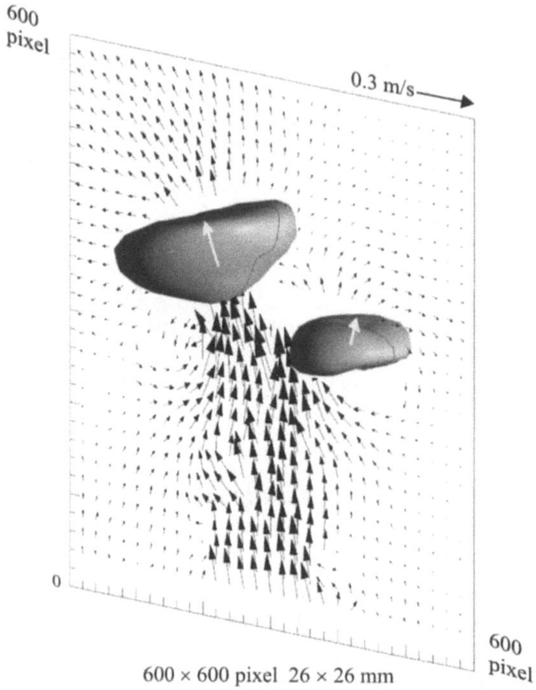

Fig. 16.13. Vector plot of the flow induced by two bubbles rising in stagnant water. The intersection of the light sheet with the large bubbles is shown, thus indicating the 3D position of the bubbles. The bubble velocities are also given. Experiments are described in [448].

Turbulent vertical structures appear in the field of the velocity fluctuations. In order to make them visible one must subtract a mean or constant velocity from an instantaneous velocity field. This way, the spatial structure of turbulent pipe and channel flow has been visualized quantitatively [413,479]. It has been demonstrated that other characteristic measures of turbulence can be determined from PIV data and that they are in good agreement with results derived by means of probe experiments or, where it is possible, with numerical simulations. The result of a PIV experiment in turbulent pipe flow at a pipe Reynolds number of 10^5 is presented in Fig. 16.14. The light sheet was through and along the axis of the straight pipe. Only a section covering the range $0 \leq x/R \leq 0.4$, $-0.1 \leq r/R \leq 0.3$ is shown. Here x is the axial, r the radial coordinate, and $R = 5\ cm$ the pipe radius. The axis is at $r = 0\ cm$, $r/R = 1$ the wall. The vector plot with the instantaneous velocity distribution on the left side (a) is dominated by the axial velocity component which is an order of magnitude higher than the fluctuations, particularly those in r–direction. These fluctuations and vertical structures become apparent if one subtracts the (theoretical) time–averaged velocity profile (b). The respective vorticity distribution (c) indicates high positive or negative vorticity values not only at the position of the vortical structures, but also for those locations exhibiting a strong shear rate.

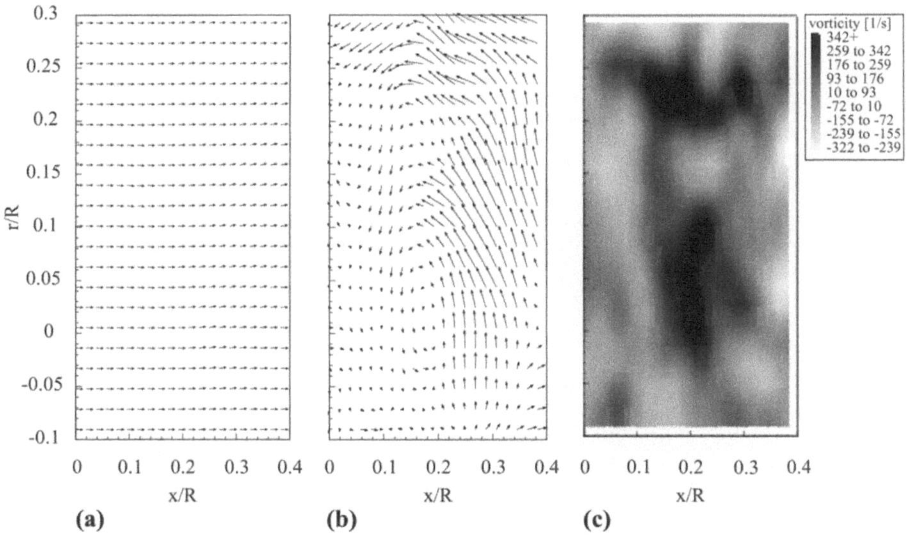

Fig. 16.14. Turbulent flow in a straight circular pipe of radius $R, r = 0$: axis, c: axial coordinate. a) PIV vector plot; b) Reynolds decomposition (subtraction of mean velocity); c) distribution of vorticity ω [470].

It is clear that such vortical structures also exist in turbulent convective heat transfer where they can cause a locally inhomogeneous heat transfer rate [411]. The merits of PIV in this area are that the spatial structures of turbulence can be elucidated, and it must be expected that the physical picture of turbulence, that so far was emphasized by the time dependence, will change considerably.

Acknowledgement

The material presented in this article is partly the result of doctoral thesis work performed at Lehrstuhl für Strömungslehre in Essen, and I am grateful for discussions with and contributions from Ruiqian Fei, Lichuan Gui, Ralph Lindken, Franka Schneider and Wei Xiong.

Nomenclature

Nomenclature to Chapter 2:

Symbol	Comment
a	distance between optical axis and edge of the schlieren–diaphragm
a'	deviation of light at the schlieren–diaphragm
c_v	velocity of light in the vacuum
c	velocity of light in the medium
f	focal length
n	refractive index of the fluid
p	pressure
r	radius
R	specific gas constant
t	time
T	temperature
x, y	coordinates
ε	angle between undeflected and deflected light
φ	angle between coordinate y and radius r
λ	wavelength
ρ	density

Nomenclature to Chapter 3:

Symbol	Comment
b	fringe distance
c_a, c_b	concentrations of gases
h	heat transfer coefficient
k	termal conductivity of the gas
l	length of the test section
M	molar mass
n	refractive index

n_0	refractive index averaged over a region
N	molar refractivity
N_m	molar refractivity for the mixture consisting of two different gases
p	pressure
$\dot{q}$	heat flux
R	specific gas constant
Re	Reynolds number
S	multiple of a wavelength
T	temperature
w	flow velocity
x, y	coordinates
λ	wavelength
λ_j	wavelength of He–Ne laser
λ_k	wavelength of Argon laser
ρ	density

Nomenclature to Chapter 4:

Symbol	Comment
a_k	coefficient
c_k	function
f_r	carrier frequency of the microscopic interference pattern
h	local heat transfer coefficient
k_{Fl}	thermal conductivity of the fluid
l	length of test section
M	molar mass
n, n_c, n_∞	refractive index (comparison beam, ambient)
N	molecular refractivity
p	pressure
r, R_0	radius
R	specific gas constant
Re	Reynolds number
S	multiple of a wavelength
T, T_∞	temperature, ambient temperature
x, y, z	coordinates
β_o, β_r	angle of the object beam (reference beam) measured from the perpendicular of the hologram plate
ϵ, ϵ_o	angle between comparison (object) beam and the test section axis
γ	angle between the gradient of n and the tangent of the ray
λ	wavelength
ρ	density

Nomenclature to Chapter 5:

Symbol	Comment
A_R, A_0	reference and object wave intensity
E	energy density
p_v	pressure
S	darkening
$\mathbf{T}_a$	amplitude transmittance
$\mathbf{T}_a^*$	conjugated amplitude transmittance
T_i	intensity transmittance
λ	wavelength
ρ	intensity relation

Nomenclature to Chapter 6:

Symbol	Comment
B	matrix of an image
F_{max}	maximum of normalized frequency distribution
g	grey value
i	horizontal coordinate
j	vertical coordinate
L_Z	break–up length
p	pixel
α	spray angle
β	angle towards the vertical

Nomenclature to Chapter 7:

Symbol	Comment
c	speed of light
E	photon energy

I_r	Intensity of Rayleigh scattered light
I_0	intensity of incident light
L	length of measuring volume
N	number density of molecules
x_i	part of component i
Ω	solid angle
λ	wavelength
λ_0	wavelength of incident light
ν	frequency
$\tilde{\nu}$	wave–number
σ	cross section

Nomenclature to Chapter 8:

Symbol	**Comment**
d_x, d_y, d_z	dimension of the measurement volume
d_f	fringe separation
d_l	diameter of the laser beam
f, F	focal length
f	frequency
k	turbulent kinetic energy
k	wavenumber
l_s	Kolmogorov micro scale
L	integral length scale
N_f	number of fringes
r	correlation distance
Re	Reynolds number
t	time
Tu	turbulence level
u'_{rms}	Root mean square turbulence intensity
U, u	flow velocity in main flow direction
v	particle velocity
V, v, W, w	flow velocity components perpendicular to main flow direction
ε	dissipation rate
λ	Wavelength
λ	Micro scale
λ_T	Taylor micro scale
ϕ	power spectral function
ν	kinematic viscosity
τ	time scale of the micro eddies

τ	time shift
$\mathcal{R}$	correlation coefficient
$\mathcal{T}$	integral time scale

Abbreviations

ACF	Auto correlation function
CFD	Computational fluid dynamics
DNS	Direct numerical simulation
FFT	Fast fourier transform
LDV	Laser–Doppler Velocimetry
PIV	Particle Image Velocimetry
PM	Photo–multiplier
PSD	Power Spectral Density
rms	Root–mean–square
SNR	Signal–to–noise ratio

Nomenclature to Chapter 9:

Symbol	Comment
$d_{meas.}$	measured droplet diameter
d_{real}	real droplet diameter
D1,D2,D3	photo detectors; photomultipliers
D_{10}	arithmetic mean diameter
m	relative refractive index; $m = n_p/n_m$
n_m	refractive index of the medium
n_p	refractive index of the particle
PDA	phase doppler anemometry
SNR	signal–to–noise–ratio
V_{10}	arithmetic mean velocity
Φ	phase difference
Φ_{12}	phase difference of signals detected by D1 and D2
Φ_{13}	phase difference of signals detected by D1 and D3
Φ	phase difference
λ	wave length of the light
Θ	half crossing angle of the incident laser beams
Ψ	elevation angle
φ	scattering angle

Nomenclature to Chapter 10:

Symbol	Comment
a	thermal diffusivity
B	magnetic displacement field
c_p	specific isobaric heat capacity
c_v	specific isochoric heat capacity
C_S	speed of sound
d	diameter
D	diffusion coefficient for mixtures
D	electric displacement field
e	sum of internal and kinetic energies
E	electric field
E_S	electromagnetic energy
$E_0(r,t)$	complex amplitude
E_S	resultant scattered field
$E_S^*(t)$	complex conjugated value of the field
E_{LO}	local oscillator
$g(\tau)$	normalized form of correlation function
$G(\tau)$	correlation function
H	magnetic field
i	momentary value of the signal
I	light intensity
I_S	scattering intensity
I_{LO}	background therm
j	running parameter
k	wave vector
n	refractive index of the fluid
P	dipole moment
q	scattering vector
R, r	radius
$S(\omega)$	frequency spectrum of the scattered light
t_1	constant time
Δt	fundamental sample time of the correlator
T	temperature
u	velocity vector
V	scattering volume
Ω	solid angle
ω	angular frequency of the wave
ω_0	frequency
α	molecular polarizability
β	sound absorption coefficient
Γ	linewidth

ϵ	dielectric constant
ϵ_0	mean macroscopic dielectric constant
λ	wavelength
λ_L	wavelength of light in vacuum
λ	thermal conductivity
ν	volume element
ρ	density
ρ_0	constant macroscopic equilibrium density
$\Delta\rho(t)$	fluctuating component of density
σ	free current density
τ	variable time
τ	characteristic decay time of the exponential function
τ	stress tensor

Nomenclature to Chapter 11:

Symbol	**Comment**
A	factor for the transition propability
$b_{J+\Delta J,J}$	Placzek coefficient
c	speed of light
C_1	proportionality factor
C_4	proportionality constant
C_5	factor, considering varying detection efficiency at two wavelengths
E	energy
h	Planck's constant
I_0	irradiance of the laser beam
I_R	intensity of Raman signal
J	rotational quantum number
k	Boltzmann's constant
l	length of the measuring volume
N	number density of molecules
$N_{v,J}$	number of molecules
Q_v, Q_J	rotational and vibrational partition functions
T	absolute temperature
T_{rot}	temperature of rotational conditions
v	vibrational quantum number
Ω_E	solid angle
α	molecular polarizability
α_e, β_e	rotation and vibration coupling

γ_v	factor for the vibrational transition
λ_0	wavelength
λ_R	wavelength at which the Raman signal appears
ν_0	wave number
ν_R	wave number of the scattered radiation
$\Delta\nu_R$	wave number shift
$\frac{d\sigma}{d\Omega}$	differential cross section of the molecule
$\omega_e x_e, \omega_e y_e$	deviations from an ideal harmonic oscillator

Nomenclature to Chapter 12:

Symbol	Comment
A	upper energy level
A_{21}	fluorescence transition
A_{2f}	first Einstein coefficient
B	Einstein's coefficient
C	constant
$c_{\bar{v}}(T)$	mean velocity of the molecule
e	electronic level
E	energy
E_{tot}	total energy content
$E_1(v, J)$	molecular energy of the state
FR	rate of fluorescence
h	Planck's constant
I_L	laser irradiance
I_V	laser spectral intensity
I_{LIF}	fluorescence intensity
I_v^{sat}	saturation intensity of the laser light
J	rotational level
k	Boltzmann's constant
n_{LIF}	number of photons
$n(x, y)$	density fields
N_L	Loschmidt's number
N	number of molecules
N_{tot}	total number density of molecules
$N_j(t)$	population of a specific energy level j
Q_{21}	intermolecular collisional quenching rate
$Q_{rot,vib}$	sum of all rotational and vibrational states
$R_{j,i}$	integral rate coefficient
$S(x, y)$	fluorescence intensity fields

T	system temperature
v	vibrational level
V	volume
X	lower energy level
Ω	solid angle
γ_S	mole fraction
η_c	detection efficiency
ν	frequency of the emitted light
$\Delta\nu$	spectral bandwidth
σ	collisional cross–section
τ	time constant

Nomenclature to Chapter 13:

Symbol	**Comment**
A	Absorption
A_{21}	Einstein coefficient for spontaneous emission
A, B, C	rotational constants
c	speed of light
E_m	discrete energy levels
E_{rot}	energy level of rotational state
E_{vib}	energy level of vibrational state
f_{12}	oscillator strength
F_m	fractional population
g_n	degeneracy of the initial state
I	moment of inertia of the molecule
I_0	initial light intensity
$I(\nu)$	light intensity behind the absorbing medium
J	rotational quantum number
$k(\nu)$	linear absorption coefficient
$k(\nu_0)$	line center absorption coefficient
m	absorber mass [kg]
M	absorber mass [AMU]
N_1	absorber concentration
p	actual pressure
p_0	standard pressure
Q	partition function
$\|R_{21}\|^2$	dipole transition element
S	line strength factor
T	transmission

T	temperature
T	actual temperature
T_0	standard temperature
v	vibrational quantum number
$V(x, y)$	tabulated Voigt function
x	normalized frequency separation from the line center
y	Doppler content
z	thickness of the absorbing medium
$\phi(\nu - \nu_0)$	line shape function
γ	line width (HWHM)
γ_L^0	pressure broadening coefficient
γ_{0L}	line width (HWHM) at 1 atm
λ	transition wavelength
ν	transition frequency (in wavenumbers or [Hz])
ν_0	classical vibrational frequency
ν_{21}	transition frequency
$\sigma(\nu)$	spectral absorption cross section
τ_{21}	lifetime
Ψ	wave function

Nomenclature to Chapter 14:

Symbol	**Comment**
A	receiver area
D^*	detectivity
e	degree of extinction
E	irradiance
$E_{(\lambda)}$	spectral irradiance
Δf	electronic measuring bandwidth
L	radiance
$L_{(\lambda)}$	spectral radiance
M	radiance emittance
$M_{(\lambda)}$	spectral radiant emittance
NEI	noise equivalent irradiance
R_S	responsivity of radiators
SQ	surface quality
T	temperature
T_{amb}	surrounding temperature
Ω	solid angle
α	absorptivity

ε	emissivity
ε'_n	hemispherical emissivity
ε'	directional total emissivity
$\varepsilon_{(\lambda)}$	hemispherical spectral emissivity
$\varepsilon'_{(\lambda)}$	directional spectral emissivity
$\tilde{\varepsilon}$	emissivity coefficient
λ	wavelength
λ_{max}	wavelength of maximum energy radiation
ρ	reflected radiation fraction
τ	degree of transmission
ϕ	angle of surface normal

Nomenclature to Chapter 15:

Symbol	**Comment**
a_i	slope of ray i
b_i	intercept of ray i with the y axis
c	molar concentration
f	field function
$\bar{f}$	mean value of the field function
f_j	element of the field function
$f_j^{(1)}$	first approximation of the field function
F	matrix of field function elements
F	two–dimensional Fourier transform of the field function
i	ray number
I	intensity of radiation
j	number of the field element
k, m, n	projection directions
l	length
m, n	grid–points in x und y directions
M	total number of rays
N^2	total number of field elements
P	uni–dimensional Fourier transform of projection Φ
P	matrix of projection values
s	coordinate in the rotated coordinate system
t	time
T	temperature
w	velocity
w_{ij}	coefficient matrix of beam i

W	coefficient matrix
x, y, z	coordinates in the cartesian coordinate system
α	correction term in the ART algorithm
δ	spacing in the projection plane
γ	angle increment in the fan–shaped beam
μ	absorption coefficient
Θ	angle of irradiation
ϱ	spatial coordinate in the projection plane
τ	coordinate in the Fourier plane
Φ	measured projection value

Subscripts

Symbol	Comment
i	beam, direction
j	element of the field matrix
k	number of iterations
o	initial state
s, ϱ	direction of the ray oriented rotated coordinate system
x, y	coordinates in the cartesian coordinate system

References

References

To Chapter 2:

1. Toepler A. (1864) Beobachtungen nach einer neuen optischen Methode. Max Cohen und Sohn, Bonn
2. Toepler A. (1867) Optische Studien nach der Methode der Schlierenbeobachtung. Poggendorfer Annalen, Vol. CXXXI
3. Schardin H. (1934) Das Toeplersche Schlierenverfahren, VDI-Forschungsheft 367, VDI-Verlag GmbH, Berlin
4. Gabler W. (1996) Gemischbildung, Flammenstabilisierung und Verbrennung in einer gestuften Überschallbrennkammer. PhD-Thesis, Technische Universität München, Utz-Verlag Wissenschaft, ISBN: 3-89675-131-X, Munich, Germany
5. Gabler W., Mayinger F. (1995) Non-intrusive Laser Based Optical Measurement-Techniques for Reacting and Non-reacting Supersonic Flows. Proc. Jsass 9th International Sessions in 33rd Aircraft Symposium, Hiroshima, Japan, 369–374
6. Eisen S., Ofner B., Mayinger F. (2000) Investigation of Common-Rail Fuel Injection Technique in DI-Diesel Engines, Proc. of the 1st Int. Symposium on Energy Engineering in the 21st Century, Vol. 4, ISBN 1-56700-132-7, Hong Kong, 1366–1378
7. Prechtl P. (2000) Analyse und Optimierung der innermotorischen Prozesse in einem Wasserstoff Dieselmotor. PhD-Thesis, Technische Univerität München
8. Prechtl P., Dorer F., Ofner B., Eisen S., Mayinger F. (1998) Modern Optical Measurement Techniques Applied in a Rapid Compression Machine for the Investigation of Internal Combustion Engine Concepts. ERCOFTAC Bulletin 38
9. Ardey N. (1998) Struktur und Beschleunigung turbulenter Wasserstoff-Luft-Flammen in Räumen mit Hindernissen. PhD-Thesis, Technische Universität München
10. Jordan M. (1999) Zündung und Verbrennung in vorgemischten turbulenten Freistrahlen. PhD-Thesis, Technische Universität München
11. Eder A., Edlinger B., Jordan M., Mayinger F. (1998) Investigation of Transient Flame Development using a Combination of Advanced Optical Measurement Techniques. In: Proc. of the 8th International Symposium on Flow Visualization, Sorrento, Italy

12. Gerlach C., Eder A., Jordan M., Ardey N., Mayinger F. (1999) Advances in Understanding of Flame Acceleration for the Improving of Combustion Efficiency, Energy Conservation Through Heat Transfer Enhancement of Heat Exchangers. Nato Advanced Study Institute, Series E, Applied Sciences, Vol. 355, ISBN 0-7923-5637-3, Kluwer Academic Publishers, Dordrecht
13. Eder A., Gerlach C., Mayinger F. (1999) Determination of Quantitative Criteria for the Transition from Deflagration to Detonation (DDT) in H_2/H_2O/Air Mixtures. In: Proc. of the 22nd Int. Symposium on Shock Waves, London
14. Kleine H. (1993) Verbesserung optischer Methoden für die Gasdynamik. PhD-Thesis, RWTH-Aachen

To Chapter 3:

15. Adams J.A., McFadden P.W. (1966) Simultaneous Heat and Mass Transfer in Free Convection with Opposing Body Forces. In: Am. J. of Chem. Eng. 12 (4)
16. Caulfield H.J. and Sun Li (1970) The Applications of Holography. Wiley, New York
17. Collier C.B., Burckhardt and Sun Li (1971) Optical Holography. New York: Academic Press
18. El-Wakil M.M., Ross P.A. (1960) A Two Wavelength Interferometric Technique for the Study of Vaporization and Combustion on Fuels, Liquid Rockets and Propellants. In: Progress in Astronautics and Rocketry. Vol.2 Academic Press, New York
19. El-Wakil M.M., Myers G.E. and Schilling R.J. (1986) An Interferometric Study of Mass Transfer from a Vertical Plate at Low Reynolds Numbers. In: J. of Heat Transfer 88
20. Gabor D. (1948) A New Microscopial Principle. Nature 161 ; Microscopy by Reconstructed Wavefronts. Proc. Roy. Soc. A 197; Microscopy by Reconstructed Wavefronts II. Proc. Roy. Soc. A 197 (1951);
21. Hauf W., Grigull U. (1970) Optical Methods in Heat Transfer. Advances in Heat Transfer 6
22. Jahn M., Reineke H.H. Free Convection Heat Transfer with Internal Heat Sources - Calculations and Measurements. In: Proceedings of the 5th. Int. Heat Transfer Conference Tokio 1974
23. Kiemle H. and Röss D. (1969) Einführung in die Technik der Holographie. Akademische Verlagsgesellschaft, Frankfurt/Main
24. Lübbe D. (1982) Ein Meßverfahren für instationäre, dreidimensionale Verteilungen und seine Anwendung auf Mischvorgänge. Phd Thesis TU Hannover
25. Panknin W. (1977) Eine holographische Zweiwellenlängen-Interferometrie zur Messung überlagerter Temperatur- und Konzentrationsgrenzschichten.
26. Smith H.M. (1969) Principles of Holography

To Chapter 4:

27. Bennett F.D. et al. (1978) Interferometric analysis of airflow about projectiles in free flight. J Appl Physics 23:3750–754
28. Chen Y.M. (1985) Wärmeübergang an der Phasengrenze kondensierender Blasen. PhD-Thesis Technische Universitt München
29. Collier R. (1971) Optical Holography. Academic Press, New York
30. Eckert E.R.G., Goldstein R.J. (1976) Measurements in Heat Transfer. Hemisphere Publishing Corp., New York
31. Francon M. (1972) Holographie. Springer Verlag, Berlin
32. Gabor D. (1948) A New Microscopic Principle. Nature 161:777–778
33. Hariharan P. (1984) Optical Holography. Cambridge University Press, Cambridge
34. Hauf W., Grigull U. (1970) Optical Methods in Heat Transfer. Advances in Heat Transfer Vol.6, Academic Press, New York
35. Kiemle H., Röss D. (1969) Einführung in die Technik der Holographie. Akademische Verlagsgesellschaft, Frankfurt
36. Leith E.N., Upatnieks J. (1962) Reconstructed Wavefronts and Communication Theory. J Opt Soc Am 52:1123–1130
37. Miller M.V. (1978) Optische Holographie. Verlag Karl Thiemig, München
38. Nordmann D. (1980) Temperatur, Druck und Wärmetransport in der Umgebung kondensierender Blasen. PhD-Thesis Technische Universität Hannover
39. Ostrowsky Y.I. (1980) Interferometry by Holography. Springer Verlag, Berlin
40. Panknin W. (1976) Eine holographische Zweiwellenlängen–Interferometrie zur Messung überlagerter Temperatur- und Konzentrationsgrenzschichten. PhD-Thesis Technische Universität Hannover
41. Schuhmann W., Dubas M. (1977) Holographische Interferometrie. Springer Verlag, Berlin
42. Smith H.M. (1975) Principles of Holography. John Wiley, New York
43. Smith H.M. (1977) Holographic Recording Materials. Springer Verlag, Berlin
44. Tilton L.W., Taylor J.K. (1938) Refractive index and dispersion of distilled water for visible radiation temperature $0° - 60°\ C$. N B S. Research Paper RP 1085, N B S J of Res 20:419–477
45. Vest C.M. (1979) Holographic Interferometry. John Wiley, New York
46. Wernicke G. (1982) Holographische Interferometrie: Grundlagen, Methoden und ihre Anwendung. Physik Verlag, Weinheim

To Chapter 5:

47. Agfa-Gevaert N.V. (1988) NDT/Holografie. DIS/NDT Technische Information (Fotografische Materialien für die Holografie). Mortsel, Belgien
48. Benton S.A. (1969) Hologram reconstruction with extended incoherent sources. J Opt Soc Am 59:1545–1546
49. Bergmann-Schaefer (1987) Lehrbuch der Experimentalphysik. In: Band III Optik H. Gobrecht (Herausgeber). de Gruyter Verlag
50. Bexn R. et al. (1976) Automatic assessment of aerosol holograms. J Aerosol Science 7:397–407

51. Brooks R. E., Heflinger L.O., Wuerker R.F. (1965) Interferometry with a holographically reconstructed comparison beam. Applied Physics Letters 7:248–249
52. BS 4803 (1982) Radiation Safety of Laser Products and Systems (A guide for protection of personal against hazards from laser radiation). British Standards Institution
53. Burch J.M. (1965) The application of lasers in production engineering. The Production Engineer 44:431–442
54. Chávez A. (1991) Holografische Untersuchungen an Einspritzstrahlen, Fluiddynamik und Wärmeübergang durch Kondensation, PhD-Thesis TU München, Germany
55. Chávez A. Mayinger F. (1988) Single - and double pulsed holography for the characterization of sprays of refrigerant R113 injected into its own saturated vapor. In: Shah R.K., Ganic E.N., Yang K.T. (Eds.) Experimental Heat Transfer Fluid Mechanics and Thermodynamics. Elsevier, 848–854
56. Chen Y.-M. (1985) Wärmeübergang an der Phasengrenze kondensierender Blasen. PhD-Thesis TU München
57. Collier R.J., Doherty E.T., Pennington K.S. (1965) Application of moire techniques to holography. Applied Physics Letters 7:223–225
58. De Belter M. (Agfa-Gevaert, Anwerpen), Nassenstein H., Mets H.J. (Agfa-Gevaert A.G., Leverkusen), Smits J.J.R. (Agfa-Gevaert N.V., Mortsel) (1967) Photographische Materialien für die Holographie. Holographie-Arbeitstagung, Battelle Institut e.V. Frankfurt/Main
59. Gabor D. (May 15 1948) A new microscopic principle. Nature 161:777–778
60. Feldmann O., Mayinger F. (1998) Laser Pulsed Holography Applied to the Three-Dimensional Two-Phase-Flow in Aerated Stirred Vessels. In: C. M. Carlomagno et al. (Eds.) CD Rom Proc. of the 8th Intl. Symp. on Flow Visualization, Sorrento, Italy, Sept. 1-4, 1998. Edinburgh
61. Gebhard P. (1996) Zerfall und Verdampfung von Einspritzstrahlen aus lamellenbildenden Düsen. PhD-Thesis TU München, Germany
62. Hariharan P. (1984) Optical Holography. Cambridge University Press
63. Haussmann G. (1979) Digitale Bildverarbeitung an dreidimensionalen Hologrammrekonstruktionen. Ein Verfahren zur Auswertung von Partikelverteilungen. PhD-Thesis TU Göttingen
64. Holographie-Arbeitstagung (1967) Battelle Institut e.V. Frankfurt/Main
65. Lauterborn W., Ebeling K.J. (1976) High-speed holocinematography of cavitation bubbles. In: XII Int. Congr. on High Speed Photography, Toronto SPIE, Canada 97:96–103
66. Leith E.N., Upatnieks J. (1962) Reconstructed wavefronts and communication theory. J Opt Soc Am 52:1123–1130
67. Leith E.N., Upatnieks J. (1964) Wavefront reconstructed with diffused illumination and three-dimensional objects. J Opt Soc Am 54:1195–1301
68. Mallow A., Chabot L. (1978) Laser Safety Handbuch. Van Nostrand Reinhold
69. Mayinger F., Panknin W. (1974) Holography in heat and mass transfer. In: 5th Int. Heat Transfer Conference at Tokyo, Japan, 1974 Vol.6:28–43
70. Mayinger F. (1988) Advances in optical interferometry for combined heat and mass transfer. In: Prep. 2nd. Int. Sym. on Heat Transfer at Beijing, China, 1988

71. Mayinger F., Feldmann O., Gebhard P. (1998) Evaluation of pulsed laser holograms of flashing sprays by digital image processing and holographic particle image velocimetry. Nuclear Engineering and Design 184:239-252
72. Nishida K., Nurakami N., Hiroyasu H. (1987) Holographic measurement of evaporating diesel sprays at high pressure and temperature. JSME Int. Journal Vol.30 259:107–115
73. Nordmann D. (1980) Temperatur, Druck und Wärmetransport in der Umgebung kondensierender Blasen. PhD-Thesis TU Hannover
74. Peterson D.A., Tankin R.S., Bankoff S.G. (1983) Holographic measurements of bubble size and velocity in a three-phase system. In: J.M. Delhaye and G. Cognet (Eds.) Measuring Techniques in Gas-Liquid Two-Phase Flows, IUTAM Sym. at Nancy, France, 1983. Springer Verlag 1984
75. Panknin W. (1977) Eine holographische Zweiwellenlängeninterferometrie zur Messung überlagerter Temperatur- und Konzentrationsgrenzschichten. PhD-Thesis TU Hannover
76. Schäfer M., Umhauer H. (1987) Realization of a concept for the complete evaluation of double pulse holograms of particulate phases in flows. Particle Characterization 4:166–174
77. Smeets G., George A. (1972) Doppelbelichtungs-Interferometrie. Bericht 39/72 Deutsch-Französisches Institut, Saint Louis, France
78. Spornik N. (1989) Modern holographic methods of flow visualization. In: 5th. Int. Symp. on Flow Visualization, Prague, Checoslovakia
79. Thompson B.J. (1974) Holographic particle sizing techniques. J Physics E.: Scientific Instruments 7:781–788
80. Tımko J.J. (1984) The investigation of transport phenomena by applied holography. In: Int. Symp. on Fluid Visualization at Bochum, Germany, 429–434
81. Trollinger J.D. (1975) Flow visualization holography. Optical Engineering 14:383–392

To Chapter 6:

82. Gonzales R.C., Wirtz, P. (1977) Digital Image Processing, Addison-Wesley, Massachussetts
83. Pavlidis, T. (1982) Algorithms for Graphics and Image Processing, Springer-Verlag, Berlin Heidelberg
84. Haberäcker P. (1987) Digitale Bildverarbeitung, Grundlagen und Anwendungen, 2nd edn. Hanser, München
85. Ligthart G., Groen C.A. (1982) A Comparison of Different Autofocus Algorithms. In. IEEE Trans.: 597-604

To Chapter 7:

86. Alonso M., Finn E.J. (1988) Quantum Physics. Eddison-Wesley, New York
87. Bergmann L., Schäfer C. (1978) Lehrbuch der Experimentalphysik. Band III Optik, Springer, Berlin

88. Bohren C.F., Huffman D.R. (1983) Absorption and Scattering of Light by Small Paticles. John Wiley, New York
89. Copeland R.A., Crosley D.A., Smith G.P. (1984) Laser Induced Fluorescence Spectroscopy of NCO and NH_2 in Atmospheric Pressure Flames. 20th Symposium (International) on Combustion, Combustion Institute, 1195-1203
90. Demtröder W. (1991) Laserspektroskopie. Springer, Berlin
91. Eckbreth A.C., Bonczyk P.A., Verdieck J.F. (1979) Combustion Diagnostics by Laser Raman and Fluorescence Techniques. Progress in Energy and Combustion Science Vol.5, 253
92. Fabelinski I. (1968) Molecular Scattering of Light. , Plenum Press, New York
93. Goulard, R. (1976) Combustion Measurements. Hemisphere Publishing Corp
94. Kerker, M.: The Scattering of Light and other Electromagnetic Radiation. Academic Press, New York
95. Laurendeau N.M. (1988) Temperature Measurements by Light-Scattering Methods. Prog. Energy Combust. Sci. Vol. 14, 147-170
96. Ledermann S. (1980) The Use of Laser Raman Diagnostics in Flow Fields and Combustion. Prog. Energy Combust. Sci. Vol.3, 1-34
97. Long D.A. (1977) Raman Spectroscopy. McGraw-Hill, London
98. Möller K.D. (1988) Optics. University Science Books, Mill Valley
99. Penner S.S., Wang C.P., Bahadori M.Y. (1984) Laser Diagnostics applied to Combustion Systems. 20th Symposium (International) on Combustion, Combustion Institute, 1149-1176
100. Ruck B. (1990) Lasermethoden in der Strömungsmesstechnik. AT-Fachverlag, Stuttgart

To Chapter 8:

101. Jordan M. (1999) Zündung und Verbrennung in vorgemischten turbulenten Freistrahlen. PhD-Thesis, Technische Universität München.
102. Ruck B. (1990), Lasermethoden in der Strömungsmeßtechnik, At Fachverlag.
103. Durst F., Melling A., Whitelaw J.H. (1987), Theorie und Praxis. der Laser-Doppler-Anemometrie, G. Braun, Karlsruhe.
104. Kuo K.K. (1986), Principles of Combustion, John Wiley & Sons, New York.
105. Reynolds A.J. (1974) Turbulent Flows in Engineering, John Wiley & Sons, London.
106. Rotta J.C. (1972) Turbulente Strömungen, Leitfäden der angewandten Mathematik und Mechanik, Bd.15, Teubner-Verlag, Stuttgart.
107. Bradshaw P. (1971) Introduction to Turbulence and its Measurement, Pergamon Press Ltd., Oxford.
108. Taylor G.I. (1935) Statistical Theory of Turbulence Part I/II, Proceedings of the Royal Society of London, Vol.873, no. A, pp.421-464.
109. Hinze J.O. (1953) Turbulence, McGraw-Hill, New York.
110. Tennekes H., Lumley J.L. (1987) A First Course in Turbulence, The MIT Press, Cambridge.
111. Kolmogorov A.N. (1941) The Local Structure of Turbulence in Incompressible Viscous Fluid for very Large Reynolds Numbers. In Dokl. Akad. Nauk. SSSR, vol. 30-32, (English translation in: Proceedings of the Royal Society of London, A434, 1991).

112. Chomiak J. (1979) Basic Considerations in the Turbulent Flame Propagation in Premixed Gases, Progress in Energy Combustion Science, Vol. 5, Pergamon Press Ltd., pp.207-221.
113. Abdel-Gayed R.G., Bradley D., Lung F.K.-K. (1975) Turbulence and Turbulent Flame Propagation – A Critical Appraisal, Combustion and Flame, Vol. 24, pp. 285-304.
114. Dryden H.L. (1943) Quant. Applied Mathematics, Vol.1.
115. Durst B. (2000) PhD-Thesis, Technische Universität München.

To Chapter 9:

116. Aizu Y., Durst F., Grehan G., Onofri F., Xu T.-H., (1993) PDA-system without Gaussian beam defects, Proceedings of 3rd International Conference on Optical Particle Sizing (OPS), Yokohama, Japan
117. Adrian R.J., Earley W.L. (1976) Evaluation of LDV Performance using Mie Scattering Theory, Proceedings of Symposium on Laser Anemometry, University of Minnesota, USA
118. Albrecht H.E., Borys M., Wenzel M., Wriedt Th. (1994), Influence of the Measurement Volume on the Phase Error in Phase Doppler Anemometry, Part 1: Reflective Mode Operation, Particle and Particle Systems Characterization 11, pp. 339 – 344
119. Albrecht H.E., Borys M., Wenzel M. (1996), Influence of the Measurement Volume on the Phase Error in Phase Doppler Anemometry, Part 2: Analysis by Extension of Geometrical Optics to the Laser Beam Refractive Mode operation, pp. 18 – 26, Particle and Particle Systems Characterization 13
120. Albrecht H.E., Bech H., Damaschke N., Feleke M. (1995), Die Berechnung der Streuintensität eines beliebig im Laserstrahl positionierten Teilchens mit Hilfe der zweidimensionalen Fouriertransformation, Optik 100, pp. 118 – 124
121. Bachalo W.D., Houser M.J. (1984) Phase Doppler Spray Analyzer for Simultaneous Measurements of Drop Size and Velocity Distributions, Optical Engineering 23, pp. 583 – 590
122. Bauckhage K., Flögel H.-H. (1984) Simultaneous Measurement of Droplet Size and Velocity in Nozzle Sprays. Proceedings of 2nd Int. Symp. on Appl. Laser Anemometry to Fluid Mechanics, July 2 – July 4, Lisbon
123. Bauckhage K., Flögel H.-H., Fritsching U., Hiller R. (1984) The Phase Doppler Difference Method, a New Laser Doppler Technique for Simultaneous Size and Velocity Measurements, Part 2: Optical Particle Characteristics as a Base for the New Diagnostic Technique, Particle and Particle Systems Characterization 5, pp. 66 – 71
124. Bohren C.F., Huffman D. (1983) Absorption and Scattering of Light by Small Particles, Wiley, New York, USA
125. Borys M., Schelinsky B., Albrecht H.E., Krambeer H. (1998), Light Scattering Analysis with Methods of Geometrical Optics for a Particle Arbitrarily Positioned in a Laser Beam, Optik 108, pp. 137 – 150
126. Damaschke N., Gouesbet G., Gréhan G., Mignon H., Tropea C. (1998), Response of PDA Systems to Non-Spherical Droplets, pp. 1752 –1760, Applied Optics 37

127. Durst F., Zaré M. (1975) Laser-Doppler Measurements in Two-Phase Flows, Proceedings of LDA-Symp., pp. 403-429, Technical University of Denmark, Copenhagen, P.O. Box 70, 2740 Skovlunde, Denmark
128. Gebhard P. (1996) Zerfall und Verdampfung von Einspritzstrahlen aus lamellenbildenden Düsen, Dissertation, TU-München
129. Gouesbet G., Gréhan G., Maheu B. (1991) Generalized Lorenz-Mie Theory and applications to optical sizing, Norman Chigier (Ed.) Combustion Measurements, pp. 339 – 384, Hemisphere Publishing
130. Gréhan G., Gouesbet G., Naqwi A., Durst F. (1991) Evaluation of the Phase Doppler System using Generalized Lorenz-Mie Theory, Int. Conf. on Multiphase Flows, Tsukuba, Japan
131. Hirleman E.D. (1996) History of Development of the Phase-Doppler Particle Sizing Velocimeter, Particle and Particle Systems Characterization 13, pp. 59 – 67
132. Lefebvre A. (1989) Atomization and Sprays, Combustion: An International Series, Hemisphere Publishing Corporation
133. Mie G. (1908) Beiträge zur Optik trüber Medien, speziell kolloidaler Metalllösungen, Ann. d. Physik 25, pp. 377 – 445
134. Naqwi A., Durst F. (1991) Light Scattering Applied to LDA and PDA Measurements, Part 1: Theory and Numerical Treatments, pp. 245 – 258, Particle and Particle Systems Characterization 8
135. Negus C.R., Drain L.E. (1990) Calculations of the Scattered Light from a Spherical Particle Traversing a Fringe Pattern Produced by Two Intersectin Laser Beams, J. Physics D: Appl. Phys., Vol. 15, pp. 375 – 402, Corrigendum Vol. 15, p. 2099
136. Raszillier H., Durst F. (1992) The Electromagnetic Scattering Problem of Laser-Doppler Anemometry, Part 1: Fluid Investigation, Particle and Particle Systems Characterization 9, pp. 105 – 115
137. Raszillier H., Durst F. (1992) The Electromagnetic Scattering Problem of Laser-Doppler Anemometry, Part 2: Particle Investigation, Particle and Particle Systems Characterization 9, pp. 186 – 194
138. Ofner B., Eisen S., Mayinger F. (1999) Performance of Common Rail Fuel Injection Systems in DI-Diesel Engines, Proceedings of 1st Int. Conf. on Engineering Thermophysics (ICET99), Bejing, China
139. Payri F., Desantes J.M., Arregle J. (1996) Characterization of D.I. Diesel Sprays in High Density Conditions, SAE-Paper 960774, CMT-Universidad Politecnica de Valencia, Spain
140. Saffmann M., Buchhave P., Tanger H. (1984) Simultaneous Measurement of Size, Concentration and Velocity of Spherical Particles by a Laser-Doppler Method, Proceedings of 2nd. Int. Symp. on Appl. of Laser Anemometry to Fluid Mechanics, July 2 – July 4, Lisbon
141. Sankar S.V., Weber B.J., Bachalo W.D. (1990) Sizing Fine Particles with Phase Doppler Interferometric Technique, Proceedings of 2nd. Int. Congress on Optical Particle Sizing, pp. 277 – 287, Arizona, USA
142. Tropea C., Xu T.H., Onofri F., Grehan G., Haugen P., Stiegelmeier M. (1996), Dual Mode Phase Doppler Anemometer, pp. 165 – 170, Particle and Particle Systems Characterization 13
143. Tropea C. (1999), The Development of the Phase Doppler Technique, pp. 279 – 292, Applied Optical Measurements, Springer Verlag, Germany

144. Van de Hulst H.C.(1981) Light Scattering by Small Particles, Dover Publications, Inc., New York

To Chapter 10:

145. Berne B.J., Pecora R. (1976) Dynamic Light Scattering. John Wiley, New York
146. Cheung H.M. (1988) Improvement Optics for Laser Light Scattering. In: NASA Laser Light Scattering Advanced Technology Development Workshop at Cleveland Ohio, Sept. 1988, 119–134
147. Chu, B. (1974) Laser Light Scattering. Academic Press, New York
148. Cummins H.Z., Pike E.R. (1974) Photon Correlation and Light Beating Spectroscopy. Plenum Press, New York
149. Cummins H.Z., Pike E.R. (1977) Photon Correlation Spectroscopy and Velocimetry. Plenum Press, New York
150. Einstein A. (1910) Theorie der Opaleszenz von homogenen Flüssigkeiten und Flüssigkeitsgemischen in der Nähe des kritischen Zustandes. Ann. Phys. IV, Vol. 33:1275–1298
151. Fixmann M. (1955) Molecular Theory of Light Scattering. J of Chem. Phys. 23:2074–2079
152. Kruppa B., Straub J. (1993) Dynamic Light Scattering: an Efficient Method to Determine Thermal Diffusivity of Transparent Fluids. Experimental Thermal and Fluid Science 6:28–38
153. Lightstone A.W. et al. (1988) Photon Counting Modules using RCA Silicon Avalanche Photodiodes. In: NASA Laser Light Scattering Advanced Technology Development Workshop at Cleveland Ohio, Sept. 1988, 65-80
154. Mountain R.D. (1965) Spectral structure of critical opalescence. Binary mixture. J Res Bur Stand 69A:523–525
155. Onsager L. (1931) Reciprocal Relations in irreversible processes I+II. Phys. Rev. 37:405–426
156. Pecora R. (1985) Dynamic Light Scattering: Applications of Photon Correlation Spectroscopy. Plenum Press, New York
157. Lord Rayleigh, Strutt J.W. (1871) On the light from the sky, its polarisation and colour. Phil. Mag. (4) 41:107–120, 274–279
158. Lord Rayleigh, Strutt J.W. (1818) On the Electromagnetic theory of light. Phil. Mag. (5) 12:81–101
159. Smoluchowski M. (1908) Molekular-kinetische Theorie der Opaleszenz von Gasen im kritischen Zustande sowie einiger verwandter Erscheinungen. Ann. d. Phys. 25:205–226
160. Tyndall J. (1869) On the blue colour of the sky and the polarisation of light. Phil. Mag. (4) 37:226–305
161. Zwanzig (1964) On the Validity of the Einstein - Smoluchowski Theory of Light Scattering. J Am Chem Soc 86:3489–3493

To Chapter 11:

162. Alonso M., Finn E.J. (1988) Quantum Physics. Eddison–Wesley, New York
163. Brackmann U. (1986) Lambdachrome Laser Dyes.Lambda Physik, Göttingen
164. Everall N., Jackson R.W., Howard J., Hutchinson K. (1986) Fluorescence Rejection in Raman Spectroscopy Using a Gated Intensified Diode–Array–Detector. Journal of Raman Spectroscopy Vol. 17, p. 415–423
165. Goulard R. (1976) Combustion Measurements. Hemisphere Publishing Corp.
166. Herzberg G. (1966) Molecular Spectra and Molecular Structure: I. Spectra of Diatomic Molecules. van Nostrand, New York
167. Lapp M., Hartley D.L. (1976) Raman Scattering Studies of Combustion. Combustion Science and Technology Vol. 13, p. 199–210
168. Laurendeau N.M. (1988) Temperature Measurements by Light–Scattering Methods. Prog. Energy Combust. Sci. Vol. 1, p. 147–170
169. Ledermann S. (1980) The Use of Laser Raman Diagnostics in Flow Fields and Combustion. Prog. Energy Combust. Sci. Vol.3 p. 1–34
170. Leipertz A. (1981) Laser-Raman-Spektroskopie in der Wärme und Strömungstechnik. Physik in unserer Zeit Vol.4, p. 107-115
171. Long D.A. (1977) Raman Spectroscopy. McGraw–Hill, London
172. Neti S., Anastaia C., Smith W., Chen J.C. (1984) Raman Scattering in Two–Phase Flows with Application to Temperature Measurements in Measuring Techniques in Gas–Liquid Two–Phase Flows. Springer, Berlin

To Chapter 12:

173. Bormann F. C., Nielsen T., Burrows M., Andresen P. (1997) Appl. Opt. 36:6129
174. Daily J. W. (1997) Prog. Energy Combust. Sci. 23:133
175. Daily J. W., Rothe E. W. (1988) Appl. Phys. B 67:1–10
176. Eckbreth A.C., Bonczyk P.A., Verdieck J.F. (1979) Combustion Diagnostics by Laser Raman and Fluorescence Techniques. Prog. Energy Comb. Sci. 5
177. Hanson R.K. (1986) Combustion Diagnostics: Planar Imaging Techniques. In: 18th Symp. (Intl.) on Combustion at the Combustion Institute Philadelphia, PA
178. Hanson R.K., Seitzmann J.M., Paul P.H. (1990) Planar Fluorescence Imaging of Gases. Applied Physics Vol.B50
179. Huber K.P., Herzberg G. (1979) Constants of Diatomic Molecules. Van Nostrand, New York
180. Knapp M., Luczak L., Beushausen V., Hentschel W., Manz P., Andresen P. (1996) In: 26th Symposium on Combustion, The Combustion Institute, 2589–2596
181. Koch A., Chryssostomon A., Andresen P., Bornscheuer W. (1992) Appl. Phys. B 31:7684
182. Kohse-Höinghaus K. (1994) Prog. Energy Combust. Sci. 20:203
183. Laurendeau N.M. (1988) Temperature Measurements by Light Scattering Methods. Prog. Energy Comb. Sci. 14
184. Nielsen T., Bormann F. C., Burrows M., Andresen P. (1997) Appl. Opt. 36:7960

185. Rothe E. W., Andresen P. (1997) Appl. Opt. 36:3971–4033
186. Rothe E. W., Gu Y., Chryssostomou A., Andresen P., Bormann F. (1998) Appl. Phys. B 66:251–258
187. Reckers W., Hüwel L., Grünefeld G., Andresen P. (1993) Appl. Opt. 32:907
188. Zhoa H., Ladommatos N. (1998) Prog. Energy Combust. Sci. 24:297–336 (1998)

To Chapter 13:

189. Abramovitz M., Stegun I.A. (1970) Handbook of Mathematical functions. Dover Publications, New York
190. Adler-Golden S., Lee J., Goldstein N. (1992) Diode Laser Measurements of Temperature-Dependent Line Parameters for Water Vapor Near 820 nm. J Quant Spectrosc Radiat Transfer 48: 527–535
191. Arroyo M.P., Hanson R.K. (1993) Absorption Measurements of Water Vapor Concentration, Temperature and Line Shape Parameters Using a Tunable InGaAsP Diode Laser. J Appl. Optics 32:6104–6116
192. Arndt R. (1965) Analytical line shapes for Lorentzian signals broadened by modulation. J Appl Phys 36:2522–2524
193. Cassidy D.T., Reid J. (1982) High sensitivity detection of trace gases using sweep integration and tunable diode lasers. J Appl Optics 21:2527–2530
194. Cassidy D.T. (1988) Trace gas detection using 1.3 μm InGaAsP diode laser transmitter modules. J Appl Optics 27:610–614
195. Cova S., Longini A. (1979) An introduction to signals, noise, and measurements. In: Analytical Laser Spectroscopy, Wiley Interscience, New York, 412–488
196. Davis P.B., Evenson K.M. (1975) Laser magnetic resonance spectroscopy of gaseous free radicals. In: Haroche S., Pebay-Peyroula J.C., Hänsch T.W., Harris S.E. (Eds.) Laser Spectroscopy II, Springer Lecture Notes in Physics, Vol.43
197. Dicke R.H. (1953) The effect of collisions upon the Doppler width of spectral lines. J Phys Rev 89:472–473
198. Ebert V., Sick V., Wolfrum J. (1996) Einsatz der Lasermeßtechnik zur Diagnostik und Optimierung von technischen Verbrennungsprozessen. Technisches Messen 63:268–277
199. Ebert V., Fitzer J., Gerstenberg I. et al. (1997) Fast In-situ-Monitoring of O2 in a Full-scale Waste Incinerator with NIR-Diode-Lasers. VDI-Berichte 1313:549–554
200. Ebert V., Hess P., Mashni M., Miklos A., Pitz H., Pleban K.U., Schäfer S., Sneider J. (1998a) Sensitive Detection of Methane with a 1.65μm Diode Laser by Photoacoustic and Absorption spectroscopy. J Appl Phys B 66:511–516
201. Ebert V., Fitzer J., Gerstenberg I., Pleban K.U., Pitz H., Wolfrum J., Jochem M., Martin J. (1998b) Online monitoring of water vapour with a fiber coupled NIR-diode laser spectrometer. In: 5th Intern. Symposium on Gas Analysis by Tunable Diode Lasers, VDI-Berichte 1366:145–154
202. Ebert V., Pleban K.U. and Wolfrum J. (1998c) In-situ Oxygen-Monitoring using Near-Infrared Diode Lasers and Wavelength Modulation Spectroscopy.

In: Laser Applications to Chemical and Environmental Analysis, Technical Digest (J Opt Soc Am, Washington DC), 206–209

203. Ebert V., Fitzer J., Gerstenberg I., Pleban K.U., Pitz H., Wolfrum J., Jochem M., Martin J. (1998d) Simultaneous Laser-based In-Situ-Detection of Oxygen and Water in a Waste Incinerator for Active Combustion Control Purposes. In: 27th Symposium (Int.) on Combustion, The Combus-tion Institute, Pittsburgh, 1301–1308
204. Ebert V., Schulz C., Volpp H.R., Wolfrum J., Monkhouse P. (1999) Laser Diagnostics of Combustion Processes. From Chemical Dynamics to Technical Devices, Israel Journal of Chemistry 39:1–24
205. Etzkorn T., Muris S., Fitzer J., Wolfrum J. (1993) Determination of absolute methyl and hydroxyl-radical concentrations in a low-pressure methane-oxygen flame. Chem Phys Lett 3/4 208:307–310
206. Etzkorn T. (1992) Laserspektroskopische Bestimmung absoluter CH3-und OH- Radikalkonzen-trationen in vorgemischten laminaren CH4/O2/NO-Niederdruckflamen. Diplomarbeit, Universit"at Heidelberg
207. Galatry L. (1961) Simultaneous effect of Doppler and foreign gas broadening on spectral lines. Phys Rev 122:1218–1223
208. Gough T.E., Scoles G. (1981) Optothermal infrared spectroscopy. In: McKelar A.R.W., Oka T., Stoichef B.P. Laser Spectrocopy V, Springer Series in Optical Sciences, Vol. 30
209. Grisar R., Riedel W.J., Ball D.R. (1987) A hydrogen fluoride analyzer using pulsed integrative spectroscopy. In: Monitoring of Gaseous Pollutants by Tunable Diode Lasers, Reidel, 153–158
210. Gudeman C.S., Saykally R.J. (1984) Velocity modulation infrared laser spectroscopy of molecular ions. Ann Rev Phys Chem 35:387–418
211. Hack W., Wagner H.G. (1983) Methoden zur Bestimmung von Radikalzuständen und -konzentrationen in der Gasphase. Z Anal Chem 316:124–134
212. Harris T.D. (1983) Laser intracavity enhanced spectroscopy. In: Klinger D.S. (Ed.) Ultrasensitive Laserspectroscopy, Academic Press, New York
213. Herbert F. (1974) Spectral line profile: A generalized Voigt function including collisional narrowing. J Quant Spectrosc Radiat Transfer 14:943–951
214. Hess P., Pelzl J. (Eds.) (1988) Photoacoustic and Photothermal Phenomena. Springer Series in Optical Sciences, Vol.58
215. Humlicek J. (1979) An efficient method for evaluation of the complex probability function: The Voigt function and its derivatives. JQRST 21:309–313
216. Hurst G.S., Payne M.G. (1988) Principles and Applications of Resonance Ionization Spectroscopy. Hilger, Bristol
217. Husson N. et al. (1992) Management and study of spectroscopic information: The GEISA program. JQSRT, 5/6, 48:509–518
218. Jacquinet - Husson N., Ari E., Ballard J. et al. (1998) The 1997 spectroscopic GEISA databank. JQSRT MAY 1999, 62 (2):205–254 see also: *http : //www.ara.polytechnique.fr/alexei¡ndex.html#overview* and *http : //ara01.polytechnique.fr/registration*
219. Langlois S., Birbeck T.P., Hanson R.K. (1994) Diode Laser Measurements of H2O Line Inten-sities and Self-Broadening Coefficients in the 1,4-μm Region. J Mol Spec 163:27–42
220. Letokhov V.S., Chebotayev V.P. (1977) Nonlinear Laser Spectroscopy. Springer Series Optical Sciences, Vol. 4, Berlin, Springer

221. Letokhov V.S. (1987) Laser Photoionisation Spectroscopy. Academic Press, Orlando
222. Levenson M.D. (1986) Introduction to Nonlinear Laser Spectroscopy. 2nd ed., Academic Press, New York
223. Lorentz H.A. (1908) Proc Amst Akad Sci 8:591ff
224. Meienburg W., Neckel H., Wolfrum J. (1990) In-situ measurement of ammonia with a 13CO2-waveguide laser system. Appl Phys B51:94–98
225. Möller W., Mozhukhin E. and Wagner H.G. (1986) High temperature reactions of CH3: 1. The reaction CH3+H2 -¿ CH4+H. Ber Bunsenges Phys Chem 90:854–861
226. Neckel H., Wolfrum J. (1989) IR Diode laser measurements of the NH3(n2) band at different temperatures. Appl Phys B49:85–89
227. Olivero J.J., Longbothum R.L. (1977) Empirical fits to the Voigt line width: A brief review. JQRST 17:233–236
228. Ouyang X., Varghese P.L. (1989) Reliable and efficient program for fitting Galatry and Voigt profiles to spectral data on multiple lines. Appl Optics 28:1538–1545
229. Pickett H.M. (1980) Effects of velocity averaging on the shapes of absorption lines. J Chem Phys 73: 6090–6094
230. Pickett H.M., Poynter R.L., Cohen E.A. et al. (1998) Submillimeter, Millimeter, and Microwave Spectral Line Catalog. JQSRT 60:883–890 and http://spec.jpl.nasa.gov/
231. Pierluissi J.H., Vanderwood P.C., Gomez R.B. (1977) Fast calculational algorithm for the Voigt profile. JQSRT 18:555–558
232. Pine A.S. (1980) Collisional narrowing of HF fundamental band spectral lines by Neon and Argon. J Mol Spectr 82:435–448
233. Poynter R.L. (1981) Submillimeter, millimeter and microwave spectral line catalogue. JPL Publication Rev.1, Jet Propulsion Laboratory, California Institute of Technology, Pasadena, 23–80
234. Radcig A.A. (1985) Reference Data on Atoms, Molecules and Ions. Springer, New York, Heidelberg
235. Reid J., Labrie D. (1981) Second harmonic detection with Tunable Diode Lasers - Comparison of experiment and theory. Appl Phys B26:203–210
236. Riedel W.J. (1976) Infrared gas spectroscopy system using pulsed lead chalcogenide diode lasers. SPIE Proc 99:17–21
237. Ritter K.J., Wilkerson, T.D. (1987) High-Resolution resolution Spectroscopy of the Oxygen A-Band. J of Molecular Spectroscopy 121:1–19
238. Rothman L.S. et al. (1992) The HITRAN molecular data-base, editions of 1991 and 1992. JQSRT 48: 5/6: 469–508
239. Rothman L.S., Rinsland C.P., Goldman A. et al. (1998) The HITRAN molecular spectroscopic database and HAWKS (HITRAN Atmospheric Workstation). JQSRT 60: 5: 665–710 and www.hitran.com and links therein
240. Schuler F., Rampp F., Martin J., Wolfrum J. (1994) TACCOS - A Thermography-Assisted Combustion Control System for Waste Incinerators. Comb and Flame 99:431–439
241. Sick V. (1992) Einsatz hochauflösender Laserspektroskopie zur quantitativen Analyse von Ge-genstrom-Diffusionsflammen. PhD-Thesis, Universität Heidelberg

242. Silver J.A. (1992) Frequency modulation spectroscopy for trace species detection: Theory and comparison among experimental methods. Appl Optics 31:707–717, (1992) + errata in Appl. Optics, 31:4927
243. Sobel'man I.I., Vainshtein L.A., Yukov E.A. (1995) Excitation of Atoms and Broadening of Spectral Lines. 2nd edition, Springer Series on Atoms and Plasmas Vol.15, Springer-Verlag, Heidelberg
244. Toth R.A. (1994) Measurements of $H_2^{16}O$ Line Positions and Strengths: 11610 to 12861 cm^{-1}. J Mol Spec 166:176–183
245. Travis J.C. (1985) Analytical optogalvanic spectroscopy in flames. In: Martellucci S., Chester A.N. (Eds.) Analytical Laser Spectroscopy, Plenum Press, NewYork
246. Ueda Y., Shimoda K. (1975) Infrared Laser Stark Spectroscopy. In: Haroche S., Pebay-Peyroula J.C., Hänsch T.W., Harris S.E. (Eds.) Laser Spectroscopy II, Springer Lecture Notes Physics, Vol.43, Springer
247. VanVleck J.H., Weisskopf V.F. (1945) On the shape of collision-broadened lines. Rev Mod Phys 17: 227–235
248. Varghese P.L., Hanson R.K. (1984) Collisional narrowing effects on spectral line shapes meas-ured at high resolution. Appl Optics 14:2376-2385
249. Wahlquvist H. (1961) Modulation broadening of unsaturated Lorentzian lines. J Chem Phys 35:1708–1710
250. Weber W.H., Tanaka K., Kanaka T. (feature eds.) (1987) Stark and Zeeman techniques in laser spectroscopy. J Opt Soc B4:1141–1226
251. Werle P. (1996a) Spectroscopic trace gas analysis using semiconductor diode lasers. Spectro-chimica Acta Part A52:805–822
252. Werle P. (1996b) Tunable Diode laser absorption spectroscopy: recent findings and novel approaches. Infrared Physics and Technology 37:59–66
253. Wolfrum J. (1992) Laser in der Reaktiontechnik - Analytik und Manipulation. Chem.-Ing.-Tech. 64 3:242–252
254. Zink L.R., Jennings D.A., Evenson K.M., Sasso A., Inguscio M. (1987) Stark spectroscopy using far-infrared radiation. J Opt Soc Am B 4:1173–1176

General Reading:

255. Atkins P.W. (1996) Physikalische Chemie. 2. Aufl., Wiley-VCH, Weinheim
256. Graybeal J.D. (1988) Molecular Spectroscopy. McGraw-Hill
257. Haken H., Wolf H.C. (1992) "Molekülphysik und Quantenchemie" and "Atomphysik", Springer
258. Kuhn H.G. (1969) Atomic Spectra. Longman
259. Thorne A.P. (1988) Spectrophysics. 2nd edn. Chapman and Hall, London, New York
260. Shimoda K. (1976) Line broadening and narrowing effects. In: Shimoda K. (Ed.) High Resolution Laser Spectros-copy, Topics in Applied Physics Vol. 13, Springer, Berlin
261. Walther H. (1976) Laser Spectroscopy of Atoms and Molecules. Topics in Applied Physics Vol.2, Springer

To Chapter 14:

262. Barr E.S. (1960) Historical Survey of the Early Development of the Infrared Spectral Region. Am. J. Phys. vol. 28:42-54

263. Bleicher M. (1986) Halbleiter Optoelektronik. Dr. Alfred Hüthig Verlag, Heidelberg
264. Haberäcker P. (1991) Digitale Bildverarbeitung. Fachbuchverlag, Leipzig
265. Haberäcker P. (1995) Praxis der digitalen Bildverarbeitung und Mustererkennung. Fachbuchverlag, Leipzig
266. Herrmann K., Walter, L. (1990) Wissensspeicher Infrarottechnik. Fachbuchverlag, Leipzig
267. Hudson D.R. Jr. (1969) Infrared System Engineering. John Wiley & Sons, New York
268. Glückert U. (1992) Erfassung und Messung von Wärmestrahlung. Franzis Verlag, München
269. Kruse P.W. (1962) Elements of Infrared Technology. John Wiley & Sons, New York
270. Lineweg F. (Herausgeber - Mitautor) (1976) Handbuch der technischen Temperaturmessung. Vieweg & Sohn, Braunschweig
271. Pavlidis T. (1982) Algorithms for Graphics and Image Processing. Springer, Berlin
272. Siegel R., Howell J.R. (1981) Thermal Radiation Heat Transfer.McGraw-Hill Book Company, New York
273. Stahl K., Miosga G. (1986) Infrarottechnik. Dr. Alfred Hüthig Verlag, Heidelberg
274. Touloukian Y.S., et al. (1972) Thermal Radiative Properties, Vol. 7 (Metallic Elements and Alloys), Vol. 8 (Nonmetallic Solids) Vol. 9 (Coatings). IFI / Plenum, Washington
275. Vanzetti R. (1972) Practical Applications of Infrared Techniques. John Wiley & Sohns, New York
276. Wu D. (Datong) (1996) Lockin - Thermographie für die zerstörungsfreie Werkstoffprüfung und Werkstoffcharakterisierung. Dissertation, Universität Stuttgart, Stuttgart

To Chapter 15:

277. Abdullah M.Z., Dyakowski T., Dickin F.J. and Williams R.A. (1993) Observation of Hydrocyclone Seperator Dynamics using Electrical Impedance Tomography. Proc. ECAPT '93, Karlsruhe, Germany, Process Tomograhpy – A Strategy for Industrial Exploitation – 1993, Edts. Beck, M.S. et al 93 – 96.
278. Achilles, D.: Die Fourier – Transformation in der Signalverarbeitung. Berlin: Springer Verlag 1978
279. Adams, M.F.; Anderson, A.P. Synthetic Aperture tomographic imaging for microwave diagnostics, Proc. Inst. Elec. Eng. Vol. 129(1982), 83 – 88.
280. Aitmehdi, R.; Anderson, A.P.; Ferrando, M.: The determination of dielectric loss tangent by microware phase tomography, Inrose Problems, Vol. 4(5)1988, 333 – 7345.
281. Anderson, A.H.; Kak, A.C.: Simultaneous algebraic reconstruction technique (SART): A superior implementation of the ART algorithm. Ultrason. Imaging 6 (1984) 81 – 94

282. Bidin, A.R.; Green, R.G.; Shackleton, M.E.; Stott, A.L.; Taylor, R.W.: Process tomography using electrodynamic sensors; Proc. ECAPT '93, Karlsruhe, Germany, Process Tomography – A Strategy for Industrial Exploitation – 1993, Edts. Beck, M.S. et al.
283. Boloney, J.C.; Hawley, M.S.: Noninvasive Control of Hyperthermia, in Methods of Hyperthermia Control, Editor: M. Gautherie, Springer – Verlag, Berlin, 1990, 35 – 111.
284. Buchmann, M.; Mewes, D.: Measurement of the local intensities of segregation with the tomographical dualwavelength photometry. Can J Chem Eng 76,3 (1998) 626 – 630
285. Buchmann, M.; Mewes, D.: The tomographical dualwavelength photometry – A new tool to distinguish micro – and macromixing. In: Lehner, M.; Mewes, D. (Eds.): Applied Optical Measurements, Springer, Berlin. 1999. 249 – 262
286. Bücherl, T.; Lierse, C.; Nuding, W.; Sauerwein, C.; Wiacker, H.: Bildgebende Radiometer an radioaktiven Abfallgebinden, DGZfP 47 (1995)1, 329 – 336.
287. Buonocore, M.H.; Brody, W.R.; Macovski, A.: A natural pixel decomposition for two – dimensional image reconstruction. IEEE Trans. Biomed. Eng. 28/2 (1981) 69 – 78
288. Bracewell, R.: The Fourier Transform and its Applications. New York: McGraw – Hill Book Company 1965
289. Braun, H.; Wernsdörfer A.: Tomographische Strömungsmessung. Technisches Messen 59 (1992) 408 – 417
290. Burns, D.H.: Optical Tomography for Three – Dimensional Spectroskopy, Appl. Spectroscopy, 1994, 12A – 19A
291. Carey, S. J.; McCann H.; Hindle F. P.; Ozanyan K. B.; Winterbone D. E.; Clough E.: Chemical Species Tomography by Near Infra – Red Absorption. First World Congress on Industrial Process Tomography. Buxton England 1999
292. Caspers, F.; Conoway, J.: (1982) Measurement of power density in a lossy material by means of electromagnetically acoustic signals for non – invasive determination of spatial thermal absorption in connection with pulsed hyperthermia; 12th – European Microwave Conference Helsinki, 13 – 17 Sept, 1982, Microwave Exhibitions and Publishers Ldt, pp 565 – 568.
293. Censor, Y.: Finite series – expansion reconstruction methods. Proceedings of the IDEE 71/3 (1983) 409 – 419
294. Cha, D.G.; Cha, S.S.: AIAA 21st Fluid Dynamics, Plasma Dynamics and Laser Conference. AIAA Paper No.90 Seattle (1990) 1517
295. Cha, S.S.; Sun, H.: Interferometric tomography of continuous fields with incomplete projections. Optics Letters 14/6 (1989) 299 – 301
296. Charalampopoulos, T.T.; Chang, R.; 1991, Agglomerate Parameters and Fractal Dimension of Soot using Light Scattering: Effects on surface growth, Comb and Flame, Vol. 89 – 99
297. Copley, D.C.; Eberhard, J.W.; Mohr, G.A.: Computed Tomography Part I : Introduction and Industrial Applications, JOM 46(1994)1, 14 – 26.
298. Cross, J.: Electrostatics; Adam Hilger, 1987.
299. Daries, G.; Spyron, N.M.; Hutchinson, I.G.; Huddleston, J. ???, Nucl. Instr. and Meth. A242 (1986), 615 – ???
300. Duwe, R.; Jansen, P.: Computer – Tomographie an Fässern mit radioaktivem Inhalt. Kernforschungsanlage Jülich GmbH (1989) Jül – Spez 498

301. Emmerman, P.J.; Goulard, R.; Santoro, R.J.: Multiangular absorption diagnostics of a turbulent argon – methane jet. J. Energy 4/2 (1980) 70 – 77
302. Faris, G.W.; Byer, R.L.: Beam – deflection optical tomography. Optics Letters 12/2 (1987) 72 – 74
303. Faris, G.W.; Byer, R.L.: Beam – deflection optical tomography of a flame. Optics Letters 12,3 (1987) 155 – 157
304. Fordham, E.J.; Hall, L.D.; Ramakrishnan, T.S.; Sharpe, M.R. and Hll, C.; 1993, Saturation gradients in drainage of porous media : NMR imaging measurements, AIChE J.; 39(9), 1431 – 1443.
305. Friederich, M.: Dissipation in gerührten nicht – newtonschen Flüssigkeiten. Dissertation Universität Hannover 1990
306. Garnero, C.; Hugorin, J.; Beaucondrey, N.: Limited – angle tomographic imaging using a constrained natural – pixel decomposition. Optica Acta 33 (1985) 659 – 671
307. Gilbert, P.F.C.: Iterative methods for the three – dimensional reconstruction of an object from projections. J. Theor. Biol. 36/1 (1972) 105 – 117
308. Gladden, L.F.; 1994, NMR in chemical engineering. Chem. Eng. Sci.; 49(20), 3339 – 3408.
309. Gladden, L.F.; 1995, Industrial Applications of NMR Imaging. Eng. J.; 56, 149 – 158.
310. Goodman, J.W.: Introduction to Fourier Optics New York: McGraw – Hill 1968
311. Gordon, R.: A tutorial on ART. IEEE Trans. Nucl. Science 71 (1974) 78 – 93
312. Gordon, R.; Bender, R.; Herman, G.T.: Algebraic reconstruction techniques (ART) for three – dimensional electron microscopy and X – ray photography. J. Theor. Biol. 29 (1970) 471 – 481
313. Gordon, R.; Herman, G.T.: Three – dimensional reconstruction from projections: A review of algorithms. Int. Rev. Cytol. 38 (1974) 111 – 151
314. Grimm, R.: Industrielle CT – Anlagen : 2.Sem. on CT, 1991, München, DGZfP, 64 – 73.
315. Guo, T.C.; GUO, W.W.: Physics of image formation by microwave scatering, Med. Imag.; SPIE Proc.; 767(1987) 1 – 10.
316. Haarde, W.: Das Vermischen mit Hilfe von Flüssigkeitsstrahlen. Dissertation, Universität Hannover 1989
317. Harding, G.; Kosanetzky, J.: X – ray scatter image in non – destructive testing : 10P Short meetings series 19(1988) 1 – 15.
318. Hauck, A.: Ultrasonic Tomography for the Non – Intrusive Measurement of Flow – Velocity – Fields. VDI Berichte Nr. 768 (1989)
319. Herman, G.; Rowland, S.: Three methods for reconstructing objects from X – rays: a comparative study. Comput. Graphics Image Processing 2 (1973) 151 – 178
320. Herman, G.T.: ART : Mathematics and applications. J. Theor. Biol. 42 (1973) 1 – 32
321. Herman, G.T.; Lent, A.: Iterative reconstruction algorithms. Comput. Biol. Med. 6 (1976) 273 – 294
322. Herman, G.T.: Image Reconstruction from Projections – Implementation and Applications. Berlin: Springer Verlag 1979

323. Herman, G.T.: Image Reconstruction from Projections – The Fundamentals of Computerized Tomography. New York: Academic Press 1980
324. Hesselink, L.: Optical tomography. Handbook of flow visualisation Chapter 20. New York: W. – J. Yang Hemispere Publishing Corporation 1989
325. Hiller, D.; Ernest, H.: Tomography using transmission and reflection mod application to medical diagnostics, Ac. Imag. Vol.12, Proc. 12th Int. Symp.; New York Plem Press, 1982, 557 – 563.
326. Hori K.; Fujimoto T.; Kawaniski K.; Nishikawa H. : Advanced high speed x – ray CT scanner for measurements and visualization of multiphase flow; OECD/ CSNI Specialist Meeting, Santa Barbara, 1997
327. Hunter, J.C.; Collins, M.W.: Three – dimensional refractive index field reconstruction from holographic interferograms. Int. J. Optoelectronics 4/2 (1989) 95 – 132
328. Inouye, T.: Image reconstruction with limited angle projection data. IEEE Trans. Nucl. Sci. 26/2 (1979) 2666 – 2669
329. Isaksen, *.; Nordtvedt, J.E.; 1992, Capacitance Tomography : Reconstruction Based on Optimization Theory. Proc. ECAPT '92, Manchester, United Kingdom, Process Tomography – A Strategy for Industrial Exploitation – 1992, Edts. Beck, M.S. et al, 178 – 189.
330. Jacobi, J.H.; Larsen, L.E.; Hast, C.T.: (1979) Water – immersed microwave antennas and their application to microwave interrogation of biological targets, IEEE Trans Microwave; Theory Tech. 27(1), 70 – 78.
331. Junginger, H. – G.; v. Haeringen, W.: Calculation of three – dimensional refrac – tive index field using phase integrals. Optics Communications 5/1 (1972) 1 – 4
332. Kak, A.C.; Slaney, M.: Principles of Computerized Tomographic Imaging. New York: IEEE Press 1988
333. Kawaji, M.; Ahmad, W.; DeJesus, J.M.; Sutharsham, B.; Lorencez, C.; Oiha, M.: Flow visualization of two phase, Eng. Des.; 141(1993), 343 – 355.
334. Klug, F.; 1993, Ein Meßverfahren zur impedanzgestützten Rekonstruktion der Gemischzusammensetzung in Mehrphasenströmungen, Fortschr. – Ber VDI 3(327), VDI – Verlag Düsseldorf
335. Lewitt, R.M.: Reconstruction algorithms: Transform methods. Proceedings of the IEEE 71/3 (1983) 390 – 408
336. Link, R.; Nuding, W.; Wiacker, H.: Industrielle Computer Tomographie, Atomwirtschaft 10(1992), 495 – 497.
337. Liu, T.C.; Merzkirch, W.; Obeste – Lehn, K.: Optical tomography applied to speckle photographic measurement of asymmetric flows with variable density. Experiments in fluids 7 (1989) 157 – 163
338. Lübbe, D.: Ein Meßverfahren für instationäre, dreidimensionale Verteilungen und seine Anwendung auf Mischvorgänge. Dissertation Universität Hannover 1982
339. Maleki, M.H.; Devaney, A.J.: Phase retrieval and intensity – only reconstruction algorithms for optical diffraction tomography, J. Opt. Soc. Am. A10, 1993, 1086 – 1092
340. Mayinger, F.; Panknin, W.: Holography in heat and mass transfer. Proc. 5th Int. Heat Transfer Conf. Tokyo (1974)
341. Mayinger, F.; Lübbe, D.: 1984 Ein tomographisches Meßverfahren und seine Anwendung auf Mischvorgängen und Stoffaustausch, Wärme und Stoffübergang, Vol.18 49 – 59.

342. Mengü, M.P.; Dutta, P.: Scattering Tomography and its Application to sooting diffusion flames, Trans. ASME J. Heat Transfer, 1994, 144 – 151
343. Mewes, D.; Ostendorf, W.; Friederich, M.; Haarde, W.: Tomographic measurement techniques for process engineering studies. Ed. N.P. Cheremisinoff: Handbook of Heat and Mass Transfer Houston: Gulf Publishing Company 1989
344. Mewes, D.; Ostendorf, W.: Einsatz tomografischer Meßverfahren für verfahrenstechnische Untersuchungen. Chem. Ing. Tech. 55/11 (1983) 856 – 864
345. Michael, Y.C.; Yang, K.T.: Recent developments in axial tomography for heat transfer and fluid flow studies. J. of Experimental Thermal and Fluid Science 3 (1991) 5
346. Morriss, S.C.; Hill, A.D.: Ultrasonic Imaging and relocimetry in two – phase flow, Trans ASME, J.: Energ. Res. Tech.; 6(115), 1993, 108 – 116.
347. Moshfeghi, M.: Ultrasound Reflection – Mode Tomography using Fan – Shaped – Bean Insonification, IEEE Trans. on Ultrasonics, Ferroelectrics and Frequency Control, Vol. UFFC – 33,3(5)1986, 299 – 314.
348. Neumann, H.; Lorenz, S.; Leirer, W.: Infrarot – Thermography durch Fenster bei niedriegen Temperaturen , Wärme – und Stoffübertragung, 29(1994), 219 – 225.
349. Norton, S.S.; Linzer, M.: Ultrasonic Reflectivity Tomography : Reconstruction with circular transducer array , Ultrasonic imaging, Vol.1, 1979, 154 – 184.
350. Olsen, R.G.; Lin, J.C.: Microwave puls induced resonance in spherical head models; IEEE Trans. Microwave Theory Tech. 29(1981) 10, 1114 – 1117.
351. Onoe, M.; Tsao, J.W.; Yamada, H.; Kogure, H.; Kawamura, H.; Yoshimatsu, M.: Computed tomography for measuring annual rings of a live tree. Proceedings of the IEEE 71/7 (1983) 907 – 908
352. Ostendorf, W.; Mewes, D.; Mayinger, F.: A Tomographical method using holographic interferometry for the registration of three – dimensional unsteady temperature profiles in laminar and turbulent flow , Heat – Transfer 1986, Proc. 8th Int. Heat Transfer Conference, edts.: C.L. Tiens, Hemisphere Publ. Corp.; N.Y.; Vol. 2 (1986), 519 – 524.
353. Ostendorf, W.: Einsatz der optischen Tomographie zum Messen von Temperaturfeldern in Rührgefäßen. Dissertation Universität Hannover 1987
354. Panknin, W.: Einige Techniken und Anwendungen der holografischen Durchlicht – Interferometrie. Chemie Technik SD 3 (1974) 219 – 225
355. Parker, D.J.; Hawkesworth, M.R.; Beynon, T.D.: Process Applications of Emission Tomography, Proc. ECAPT Portugal, Oporto, 1994, Process Tomography – A Strategy for Industrial Exploitation, Edts. Beck, M.S. etal , 1994, 478 – 489.
356. Parker, D.S.; Hawkesworth, M.R.; Broadbent, C.J.; Fowles, P.; Fryer, T.D.; Mc Neal, P.A.: Industrial positron – based imaging : principles and applications, Nucl. Instr. and Meth. A349 (1994) 583 – 592.
357. Persson, S.; Östman, E.: Use of computed tomography in nondestructive testing of polymeric materials. Applied Optics 24/23 (1985) 4095 – 4104
358. Phillipp, H.; Winkelhofer, E.; Fraidl, G.K.: Verbrennungsdiagnostic mit Computertomographie, MTZ 56(3)1995, 148 – 157.
359. Pichot, C.; Jofre, L.; Peronnot, G,; Boloney, J.C.: Active microwave imaging of inhomogeneous Godies, IEEE Trans. Ant. Propojat.; AP – 33, 1985, 416 – 425.

360. Proceedings of Physics and Engineering in Computerized Tomography. Newport Beach California. IEEE Transactions Nuclear Science 26/2 (1979) 2661 – 3636
361. Proceedings of Computerized Tomography. IEEE 71 (1983) 291 – 435
362. Proceedings of Topical Meeting on Industrial Applications of Computed Tomography and NMR Imaging. Canada: Hecla Island 1984
363. Radon, J.: Über die Bestimmung von Funktionen durch ihre Integral – werte längs gewisser Mannigfaltigkeiten. Berichte Sächsische Akademie der Wissenschaften Leipzig Math – Phys. Kl. 69 (1917) 262 – 267
364. Rangayyan, R.; Dhawan, A.P.; Gordon, R.: Algorithms for limited – view computed tomography: An annotated bibliography and a challenge. Applied Optics 24/23 (1985) 3950 – 3957
365. Reimers, P.; Heidt, H.; Stade, J.; Weise, H. – P.: Beispiele für die Anwendung der Computer – Tomographie (CAT) in der zerstörungsfreien Materialprüfung. Materialprüf. 22/5 (1980) 214 – 217
366. Reinecke N.; Buchmann M.; Petritsch G.; Mewes D.: Enhancement of resolution for reconstruction algorithms for limited views absorption spectoscopy Ed. ImechE London: Optical Methods and Data Processing in Heat and Fluid Flow. C516 (1996) 048 383 – 391
367. Reinecke, N.: Petritsch G.; Schmitz D.; Mewes D.: Tomographic Measurement Techniques – Visualisation of Multiphase Flows Chem. Eng. Technol. 21,1 (1998) 7 – 18
368. Renz, R.: Das diskontinuierliche Vermischen von Flüssigkeitsstrahlen in zylindrischen Behältern . Dissertation, Universität Hannover 1992
369. Riesemeier, H.; Goebbels, J.; Illerhaus, B.; Onel,Y.; Reimers, P.: 3 – D Microcomputertomography für die Werkstoffentwicklung und Bauteilprüfung, DGZfP, 37(1993) 280 – 287.
370. Rowley, P.D.: Quantitative interpretation of three – dimensional weakly refractive phase objects using holographic interferometry. J. Opt. Soc. Am. 59/11 (1969) 1496 – 1498
371. Santoro, R.J.; Semerjian, H.G.; Emmerman, P.J.: Optical tomography for flow fields diagnostics. Int. J. Heat Mass Transfer 24/7 (1981) 1139 – 1150
372. Sato, T.; Saski, K.; Nakamura, Y.; Linzer, M.; Norton, S.J.: Tomographic image reconstruction from limited projections using iterative revisions in image and transform spaces. Appl. Optics 10/3 (1981) 395 – 399
373. Sato, T.; Saski, K.; Nakamura, Y.; Linzer, M.; Norton, S.J.: Tomographic image reconstruction from limited projections using coherent optical feedback. Appl. Optics 20/17 (1981) 3073 – 3076
374. Schwarz, A.: Multitomographische Temperaturmessung in Flammen mit einem Schlierenmeßaufbau. Fortsch.Ber. VDI Reihe 8 Nr. 444. VDI – Verlag, Düsseldorf. 1995
375. Seiraffi, M.A.: Ultraschall – Computertomography zur Messung von Zweiphasenströmungen, Fortschr. – Ber. VDI, 8(327), VDI – Verlag 1993.
376. Smith, K.T.; Kleinert, F.: Mathematical foundations of computed tomography. Applied Optics 24/23 (1985) 4000 – 4012
377. Snyder, R.; Hesselink, L.: Optical tomography for flow visualisation of the density field around a revolving helicopter rotor blade. Applied Optics 23/20 (1984) 3650 – 3657
378. Snyder, R.; Hesselink, L.: High speed optical tomography for flow visualisation. Applied Optics 24/23 (1985) 4046 – 4051

379. Stokes, J.A.; Alvar, K.R.; Corey, R.L.; Corstello, D.G.; John, J.; Kocimski, S.; Lurie, N.A.; Thayer, D.D.; Trippe, A.P.; Young, J.C.: Some new applications of collimated photon scattering for non – destructive examination, Nucl. Instr. and Methods 193(1982) 261 – 267.
380. Subbarao, P.M.V.; Munshi, P.; Muralidhar, K.: Performance evaluation of iterative tomographic algorithms applied to reconstruction of a three – dimensional temperature field. Numerical Heat Transfer, Part B 31(1997) 347 – 372
381. Sweeney, D.W.: Interferometric measurement of three – dimensional temperature fields. Dissertation University of Michigan 1972
382. Takenaka, N.; Fujii, T.; Ono, A.; Motomura,Y.; Toruno,A.: Visualization and Measurement of Multi – phase and Liquid Metal Flow by Real – time Neutron Radiography, Proc. analyze – Japaneese Symp. on Multi – Phase Flow, 23. – 25. 8. 1994, Karlsruhe, Germany, 147 – 161.
383. Tuffs, A.: Neue Techniken für den Blick in den Körper. AGF – Jahresheft (1990)
384. VDI: Große Schmiedteile mit Echo – Tomographie untersucht. VDI – Nachrichten 32/ 12 (1989)
385. Vogg, H.: Strahlendiagnostik – in der Medizin ja, für die Verfahrenstechnik nein? Ein kritischer Vergleich. Chem. – Ing. – Tech. 55,6 (1983) 467 – 474
386. Wang, M.; Dickin, F.J.; Williams, R.A.; Waterfall, R.C.; Beck, M.S.; 1994, Electrical Resistance Tomography on Metal Walled Vessels. Proc. ECAPT '94, Oporto, Portugal, Process Tomography – A Strategy for Industrial Exploitation – 1994, Edts. Beck, M.S. et al, 183 – 170.
387. Wang, S.Y.; Huang, Y.B.; Pereira, V.; Gryte, C.G.: Application of computed tomography to oil recovery from porous media. Applied Optics 24/23 (1985) 4021 – 4027
388. Webb, A.G.; Jezzard, P.; Hall, L.D.; Ng, S.: Detection of inhomogenities in rubber samples using n.m.r. imaging. Polymer Communications 30/12 (1989) 363 – 366
389. Wernike, G.; Osten, W.: Holographische Interferometrie. Weinheim: Physik Verlag 1982
390. Wetberg, T.C.; Stamnes, J.J.: Recent results in quantitative optical diffraction tomography; Proc ECAPT '95, Bergen, Norway, 6. – 8.4.1994, Processs Tomography – 1995 Implementation for Industrial Proesses, Edts. M.S. Beck et al.; 118 – 126
391. Wolf, J.: Ein Messverfahren zur Untersuchung von Blasensäulenreaktoren. Die Ultraschalltomography, Forschr. – Ber. VDI, 8(166), VDI – Verlag 1988.
392. Xie, C.G.; Huang, S.M.; Hoyle, B.S.; Lenn, C.P.; Beck, M.S.; 1992, Transputer – Based Electrical Capacitance Tomography for Real – Tie Imaging of Oilfield Flow Pipelines. Proc. ECAPT '92, Manchester, United Kingdom, Process Tomography – A Strategy for Industrial Exploitation – 1992, Edts. Beck, M.S.et al, 281 – 294.
393. Yan, H.; Yuen, W.W.; Theofamous, T.G.: The use of Flourescence in the measurement of local liquid content in the multiphase flows, Nuc. Eng. Des.141(1993), 315 – 328.
394. Yu, Z.Z.; Peyton, A.J.; Beck, M.S.; Conway, W.F.; Xu, L.A.; 1993, Imaging systems based on elektromagnetic tomography (EMT). Electronic Letter, 29(7).

395. Zhang, Y.; Ruff, C.A.: Three – dimensional temperature measurements in enclosures by using multiview interferometric tomography, Meas. Sci. Technol. 5(1994) 495 – 502.
396. Zhu, P.; Peix, G.; Babot, D.; Muller, J.: X – ray imaging and characterization of materials: Compton backscatter techniques; Proc ECAPT '95, Bergen, Norway, 6. – 8.4.1995, Process Tomography – 1995 Implementation for Industrial Processes, Edts. M. S. Beck et al.; 331 – 340.

To Chapter 16:

397. Abrahamson S., Lonnes S. (1995) Uncertainty in calculating vorticity from 2D velocity fields using circulation and least-squares approaches. Exp Fluids 20:10–20
398. Adria R.J. (1986) Image shifting technique to resolve directional ambiguity in double-pulsed velocimetry. Appl Opt 25:3855–3858
399. Adrian R.J. (1991) Particle-imaging Techniques for experimental fluid mechanics. Annu Rev Fluid Mech 23:261–304
400. Barker D.B., Fourney M.E. (1977) Measuring fluid velocities with speckle patterns. Optics Letters 1:135–137
401. Barnhart D.H., Adrian R.J., Papen G.C. (1994) Phase-conjugate holographic system for high-resolution particle image velocimetry. Appl Opt 33:7159–7170
402. Brücker C., Althaus W. (1992) Study of vortex breakdown by particle tracking velocimetry (PTV). Part 1: Bubble-type vortex breakdown. Exp Fluids 13:339–349
403. Brücker C. (1995) Digital Particle Image Velocimetry (DPIV) in a scanning light sheet: 3D starting flow around a short cylinder. Exp Fluids 19:255–263
404. Buchhave P. (1994) Particle image velocimetry. In: Lading L., et al (Eds.) Optical Diagnostics for Flow Processes, Plenum Press, New York
405. Carasone F., Cenedese A., Querzoli G. (1995) Recognition of partially overlapped particle images using the Kohonen neural network. Exp Fluids 19:225–232
406. Cenedese A., Paglialunga A. (1989) A new technique for the determination of the third velocity component with PIV. Exp Fluids 8:228–230
407. Cenedese A., Paglialunga A. (1990) Digital direct analysis of a multiexposed photograph in PIV. Exp Fluids 8:273–280
408. Chen P.H., Yen J.Y., Chen J.L. (1998) An artificial neural network for double exposure PIV image analysis Exp Fluids 24:373–374
409. Coupland J.M., Pickering C.J.D., Halliwell N.A. (1987) Particle image velocimetry: theory of directional ambiguity removal using holographic image separation. Appl Opt 26:1576–1578
410. Coupland J.M., Halliwell N.A. (1992) Particle image velocimetry: Three-dimensional fluid velocity measurements using holographic recording and optical correlation. Appl Opt 31:1005–1007
411. Drobniak S., Elsner J.W., El-Kassem E.S.A. (1998) The relationship between coherent structures and heat transfer processes in the initial region of a round jet. Exp Fluids 24:225–237

412. Dudderar T.D., Simpkins P.G. (1977) Laser speckle photography in a fluid medium. Nature 270:45–47
413. Eggels J.G.M., Unger F., Weiss M.H., Westerweel J., Adrian R.J., Friedrich R., Nieuwstadt F.T.M. (1994) Fully developed turbulent pipe flow: a comparison between direct numerical simulation and experiment. J Fluid Mech 268:175–209
414. Emrich R.J. (1983) Flow field measurement by tracer photography. Exp Fluids 1:179–184
415. Fincham A.M., Spedding G.R. (1997) Low cost, high resolution DPIV for measurement of turbulent fluid flow. Exp Fluids 23:449–462
416. Fomin N.A. (1998) Speckle Photography for Fluid Mechanics Measurements. Springer, Berlin
417. Fouras A., Soria J. (1998) Accuracy of out-of-plane vorticity measurements derived from in-plane velocity field data. Exp Fluids 25:409–430
418. Fujita I., Kaizu T. (1995) Correction method of erroneous vectors in PIV. J Flow Visual Image Processing 2:173–185
419. Gaydon M., Raffel M., Willert C., Rosengarten M., Kompenhans J. (1997) Hybrid stereoscopic particle image velocimetry. Exp Fluids 23:331–334
420. Gogineni S., Visbal M., Shih C. (1999) Phase-resolved PIV measurements in a translational plane wall jet: a numerical comparison. Exp Fluids 27:126–136
421. Grant I., Liu A. (1990) Directional ambiguity resolution in particle image velocimetry by pulse tagging. Exp Fluids 10:71–76
422. Grant I., Pan X. (1995) An investigation of the performance of multi layer, neural networks applied to the analysis of PIV images. Exp Fluids 19:159–166
423. Grant I. (1997) Particle image velocimetry: a review. Proc Instn Mech Engrs 211:55–76
424. Grousson R., Mallick S. (1977): Study of flow pattern in a fluid by scattered laser light. Appl Opt 16:2334–2336.
425. Guezennec Y.G., Kiritsis N. (1990) Statistical investigation of errors in particle image velocimetry. Exp Fluids 10:138–146
426. Guezennec Y.G., Brodkey R.D., Trigui N., Kent J.C. (1994) Algorithms for fully automated three-dimensional particle tracking velocimetry. Exp Fluids 17:209–219
427. Gui L., Merzkirch W. (1996a) A method of tracking ensembles of particle images. Exp Fluids 21:465–468
428. Gui L., Merzkirch W. (1996b) Phase separation of PIV measurements in two-phase flow by applying a digital mask technique. ERCOFTAC Bull 30:45–48
429. Gui L., Merzkirch W. (1998) Generating arbitrarily sized interrogation windows for correlation-based analysis of particle image velocimetry recordings. Exp Fluids 24:66–69
430. Gui L., Merzkirch W. (1999) A comparative study of the MQD method and several correlation-based PIV evaluation algorithms. Exp Fluids 28
431. Hart DP (1996) Sparse array image correlation. In: Proc. 8th Intl Symp Appl Laser Techn Fluid Mech, Lisbon
432. Hartmann J., Köhler J., Stolz W., Flögel H.H. (1996) Evaluation of instationary flow fields using crosscorrelation in image sequences. Exp Fluids 20:210–217
433. Hassan Y.A., Blanchat T.K., Seeley C.H., Canaan R.E. (1992) Simultaneous velocity measurements of both components of a two-phase flow using particle image velocimetry. Int J Multiphase Flow 18:371–395

434. Hassan Y.A., Philip O.G. (1997) A new artificial neural network tracking technique for particle image velocimetry. Exp Fluids 23:145-154
435. Hilgers S. (1996) Darstellung der fluidmechanischen Prozesse in Blasensäulen-Reaktoren durch die Particle-Image-Velocimetry. Dissertation, Universität Essen
436. Hinsch KD (1993) Particle image velocimetry. In: Sirohi R.S. (Eds.) Specle Metrology, M Dekker, New York, 235–323
437. Huang H.T., Fiedler H.E., Wang J.J. (1993) Limitation and improvement of PIV. Part I: Limitation of conventional techniques due to deformation of particle image patterns. Exp Fluids 15:168–174
438. Huang H.T., Fiedler H.E., Wang J.J. (1993) Limitation and improvement of PIV. Part II: Particle image distortion, a novel technique. Exp Fluids 15:263–273
439. Huang H., Dabiri D., Gharib M. (1997) On errors of digital particle image velocimetry. Meas Sci Technol 8:1427–1440
440. Jähne B. (1989) Digitale Bildverarbeitung. Springer, Berlin
441. Keane R.D., Adrian R.J (1990) Optimization of particle image velocimeters. Part I. Double-pulsed systems. Meas Sci Technol 1:1202–1215
442. Keane R.D., Adrian R.J. (1992) Theory of cross-correlation analysis of PIV images. Appl Sci Res 49:191–215
443. Kemmerich T., Rath H.J. (1994) Multi-level convolution filtering technique for digital laser speckle velocimetry. Exp Fluids 17:315–322
444. Khalili B. (1991) Study of the intake tumble motion by flow visualization and particle tracking velocimetry. Exp Fluids 10:230–236
445. Lallement J.P., Desailly R., Froehly C. (1977) Mesure de vitesse dans un liquide par diffusion coherente. Acta Astronautica 4:343–356
446. Lawson N.J., Wu J. (1997) Three-dimensional particle image velocimetry: experimental error analysis of a digital angular stereoscopic system. Meas Sci Technol 8:1455–1464
447. Lecerf A., Renao B., Allano D., Boukhalfa A., Trinite M. (1999) Stereoscopic PIV: validation and application to isotropic turbulent flow. Exp Fluids 26:107–115
448. Lindken R., Gui L., Merzkirch W. (1999) Velocity measurements in multiphase flow by means of particle image velocimetry. Chem Eng Technol 22:202–206
449. Lueptow RM., Akonur A., Shinbrot T. (1999) PIV for granular flows. Exp Fluids 27
450. Lourenco L., Krothapalli A .(1995) On the accuracy of velocity and vorticity measurements with PIV. Exp Fluids 18:421–428
451. Maas H.G., Gruen A., Papantoniou D. (1993) Particle tracking velocimetry in three–dimensional flows. Part I: Photogrammetric determination of particle coordinates. Exp Fluids 15:133–146
452. Malik N.A., Dracos T., Papantoniou D.A. (1993) Particle tracking velocimetry in three-dimensional flows. Part II: Particle tracking. Exp Fluids 15:279–294
453. Meinhart CD., Wereley ST., Santiago J.G. (1999) PIV measurements of a microchannel flow. Exp Fluids 27
454. Melling A. (1997): Tracer particles and seeding for particle image velocimetry. Meas Sci Technol 8:1406–1416

455. Merzkirch W., Mrosewski T., Wintrich H. (1994) Digital particle image velocimetry applied to a natural convective flow. Acta Mechanica (Suppl.) 4:19–26
456. Nogueira J., Lecuona A., Rodriguez P.A. (1999) Local field correction PIV: on the increase of accuracy of digital PIV systems. Exp Fluids 27:107–116
457. Okamoto K., Hassan Y.A., Schmidl W.D. (1995) New tracking algorithm for particle image velocimetry. Exp Fluids 19:342–347
458. Prasad A.K., Adrian R.J., Landreth C.C., Offutt P.W. (1992) Effect of resolution on the speed and accuracy of particle image velocimetry interrogation. Exp Fluids 13:105–116
459. Prasad A.K., Adrian R.J. (1993) Stereoscopic particle image velocimetry applied to liquid flows. Exp Fluids 15:49–60
460. Prenel J.P., Porcar R.E.l., Rhassouli A. (1989) Three-dimensional flow analysis by means of sequential and volumic laser sheet illumination. Exp Fluids 7:133–137
461. Raffel M., Kompenhans J. (1995) Theoretical and experimental aspects of image shifting by means of a rotating mirror system for particle image velocimetry. Meas Sci Technol 6:795–808
462. Raffel M., Gharib M., Ronneberger O., Kompenhans J. (1995a) Feasibility study of three-dimensional PIV by correlating images of particles within parallel light sheet planes. Exp Fluids 19:69–77
463. Raffel M., Kompenhans J., Wernert P. (1995b) Investigation of the unsteady flow field above an airfoil pitching under deep dynamic stall conditions. Exp Fluids 19:103–111
464. Raffel M., Kost F. (1998) Investigation of aerodynamic effects of coolant ejection at the trailing edge of a turbine blade model by PIV and pressure measurements. Exp Fluids 24:447–461
465. Raffel M., Willert C., Kompenhans J. (1998): Particle Image Velocimetry. A Practical Guide. Springer, Berlin
466. Rösgen T., Totaro R. (1995) Two-dimensional on-line particle image velocimetry. Exp Fluids 19:188–193
467. Royer H. (1997) Holography and particle image velocimetry. Meas Sci Technol 8:1562–1572
468. Sadeh M., Strauss K., Schneider T. (1997) A combined PIV/LIF system for the measurement of heterogeneous drag reduction effects in a pipe flow. Exp Fluids 22:292–299
469. Scarano F., Riethmuller F.M. (1999) Iterative multigrid approach in PIV image processing with discrete window offset. Exp Fluids 26:513–523
470. Schneider F., Merzkirch W., Rettich T. (1999) PIV visualization of coherent structures for analyzing an ultrasonic flow metering approach. In: Proc 8th Intl Conf Laser Anemometry - Advances & Applications, EALA
471. Sinha S.K. (1988) Improving the accuracy and resolution of particle image or laser speckle velocimetry. Exp Fluids 6:67–68
472. Soloff S.M., Adrian R.J., Liu Z.C. (1997) Distortion compensation for generalized stereoscopic particle image velocimetry. Meas Sci Technol 8:1441–1454
473. Song X., Yamamoto F., Iguchi M., Murai Y. (1999) A new tracking algorithm of PIV and removal of spurious vectors using Delaunay tesselation. Exp Fluids 26:371–380
474. Sridhar G., Katz J. (1995) Drag and lift forces on microscopic bubbles entrained by a vortex. Phys Fluids 7:389–399.

475. Volino R.J., Smith G.B. (1999) Use of simultaneous IR temperature measurements and DPIV to investigate thermal plumes in a thick layer cooled from above. Exp Fluids 27:70–78
476. Wernet M.P., Pline A. (1993) Particle displacement tracking technique and Cramer-Rao lower bound error in centroid estimates from CCD imagery. Exp Fluids 15:295–307
477. Westerweel J. (1993) Digital Particle Image Velocimetry - Theory and Application. Delft University Press, Delft
478. Westerweel J. (1994) Efficient detection of spurious vectors in particle image velocimetry data. Exp Fluids 16:236–247
479. Westerweel J., Draad A.A., van der Hoeven J.G.T., van Oord J. (1996) Measurement of fully developed pipe flow with digital particle image velocimetry. Exp Fluids 20:165–177
480. Westerweel J., Dabiri D., Gharib M. (1997) The effect of a discrete window offset on the accuracy of cross-correlation analysis of digital PIV recordings. Exp Fluids 23:20–28
481. Willert C.E., Gharib M. (1991) Digital particle image velocimetry. Exp Fluids 10:181–193
482. Willert C. (1996) The fully digital evaluation of photographic PIV recordings. Appl Sci Res 56:79–102
483. Willert C. (1997) Stereoscopic digital particle image velocimetry for application in wind tunnel flows. Meas Sci Technol 8:1465–1479
484. Wozniak K., Wozniak G., Rösgen T. (1990) Particle image velocimetry applied to thermocapillary convection. Exp Fluids 10:12–16
485. Xiong W., Merzkirch W. (1999) PIV experiments using an endoscope for studying pipe flow. J Flow Visualization and Image Processing
486. Zhang Z., Eisele K. (1995) The two-dimensional velocity shift caused by the use of a rotating mirror in PIV flow field measurements. Exp Fluids 20:106–111
487. Nordmann, D. Mayinger, F. (1981) Temperatur, Druck und Wärmetransport in der Umgebung kondensierender Blasen, VDI Forschungsheft, 605, VDI Verlag Düsseldorf.
488. Chen, Y.M. (1985) Wärmeübergang an der Phasengrenze kondensierender Blasen, Diss. Techn. Universität München.
489. Chen, Y.M., Mayinger, F. (1992), Measurements of Heat Transfer at the Phase-Interface of Condensing Bubbles, Int. J. Multiphase Flow, No 18, No 6, pp. 877–890.

Index

MIX
Papier aus verantwortungsvollen Quellen
Paper from responsible sources
FSC® C105338

If you have any concerns about our products,
you can contact us on
ProductSafety@springernature.com

In case Publisher is established outside the EU,
the EU authorized representative is:
Springer Nature Customer Service Center GmbH
Europaplatz 3, 69115 Heidelberg, Germany

Printed by Libri Plureos GmbH
in Hamburg, Germany